Global Consistency of Tolerances

The 6th CIRP International Seminar on Computer-Aided Tolerancing

The University of Twente, Collegezalengebouw, Enschede, The Netherlands
March 22-24, 1999

Sponsored by
CIRP (International Institution for Production Engineering Research)
Stichting LPT (foundation Laboratory of Production and Design Engineering)
CIPV (Center for Integrated Manufacturing and Development)
University of Twente

International Program Committee

P. Bourdet (LURPA, ENS-Cachan, France)
A. Clément (Dassault Systèmes, France)
H. Elmaraghy (University of Windsor, Canada)
L.E. Farmer (University of New South Wales, Australia)
C. Fortin (Ecole Polytechnique de Montreal, Canada)
I. Ham (Pennsylvania State University, USA)
F. Kimura (University of Tokyo, Japan)
V. Srinivasan (IBM, USA)
R. Weill (Technion, Israel)

National Organizing Committee

F.J.A.M. van Houten (Chairman)
H.J.J. Kals (Co-chair)
O.W. Salomons
D. Lutters
I. Dos Santos

Global Consistency of Tolerances

Proceedings of the 6th CIRP International Seminar on
Computer-Aided Tolerancing,
University of Twente, Enschede, The Netherlands,
22–24 March, 1999

Edited by

Fred van Houten

and

Hubert Kals

Department of Mechanical Engineering,
University of Twente,
Enschede, The Netherlands

KLUWER ACADEMIC PUBLISHERS

DORDRECHT / BOSTON / LONDON

A C.I.P. Catalogue record for this book is available from the Library of Congress.

ISBN 0-7923-5654-3

Published by Kluwer Academic Publishers,
P.O. Box 17, 3300 AA Dordrecht, The Netherlands

Sold and distributed in North, Central and South America
by Kluwer Academic Publishers,
101 Philip Drive, Norwell, MA 02061, U.S.A.

In all other countries, sold and distributed
by Kluwer Academic Publishers,
P.O. Box 322, 3300 AH Dordrecht, The Netherlands

Printed on acid-free paper

Printed in the Netherlands

Contents

PREFACE

For a long time, tolerancing has been one of the most difficult and least understood activities in design. Every designer knows that the proper functioning of a mechanical device depends on the size of the clearances in the joints between the parts. However, it appears to be rather problematic to assign clearances in complex assemblies in such a way that all functional and assemblability requirements are guaranteed simultaneously. Clearances, which essentially are attributes belonging to pairs of parts, have to be converted into tolerances, which are attributes of single parts. This process has to be carried out with manufacturability, exchangeability and maintainability considerations in mind.

The relations between the tightness of tolerances and the manufacturing cost are usually non-linear and do show discontinuities when the limits of process capabilities are exceeded. Moreover, the relations between tightness of tolerances and the quality of functioning are not always clear. Because of this, it is quite impossible for a human being to define fully consistent interrelated sets of requirements for squareness, distance, concentricity, straightness etc. Usually, tolerances are specified for every functional requirement without much attention for the side effects. This leads easily to inconsistent tolerancing schemes, which contain excessively tight tolerance values. As a result, the parts become expensive and the proper functioning of the assembly is still not guaranteed.

Global Consistency of Tolerances is becoming a hot topic, now computer-based geometric modelling is getting mature. Presently it is possible to build virtual product models with thousands of parts, which can have very complex shapes and geometric relationships. However, the specification of allowable deviations from the nominal geometry is still a big problem. Until now, the CADCAM systems used in industry have provided inadequate support in the definition, analysis and synthesis of tolerances. Most systems are even not capable of representing them properly.

Tolerancing has now become an important issue for CADCAM vendors. The latest generation of CAD/CAM systems is using advanced geometric modelling and constraint satisfaction kernels. The technology used in these kernels can be applied to macro geometric aspects as well as micro geometric aspects.

The creation of virtual product models only for visualization and marketing purposes does not pay off. The real profit can be made if the data can be used for the subsequent downstream processes. For instance, the process planning function can be automated to a high degree, if consistent micro geometry specifications are available.

A number of research groups is working on tolerancing issues for several years now. They try to bridge the gap between the mathematical formulation of the problem and the practical aspects in terms of computer representation, automatic or computer assisted specification, consistency and completeness analysis and tolerance set optimization.

Some of these efforts have already resulted in commercially available tolerancing software packages, which are being used in industry. Most of them are focussing on tolerance analysis and optimization. Also efforts are being undertaken to bring the tolerancing standards up to the present requirements.

Because the existing tolerance standards are still based on 2D drawing and conventional machining, they do not hold any more in the world of 3D CAD, NC machining and CMM. Moreover, with the ongoing miniaturization, the limited possibilities for specifying shape tolerances starts to become a nuisance. Several ISO TC's and working groups are proposing additions and alterations.

Different types of industries deal in different ways with tolerancing issues. In mass pro-duction, tolerancing is mainly based on statistical techniques, focussing on part exchangeability versus manufacturing cost issues. In small batch manufacturing tolerancing is usually based on worst case scenarios for assemblability. In both cases however, there can also be an interest in functional degradation caused by wear and the resulting increase or decrease of clearances with time. Kinematic and dynamic simulation of the effect of changing clearances on the virtual product's behavior can be used to predict malfunctions due to deterioration. In this way it becomes possible to establish robust designs (a well functioning assembly, constituted of loosely tolerated parts). Simulation can also support design decisions about aspects of wear, like tribological conditions, material selection with respect to part re-use or recycling and maintainability of the product.

Many companies do not have a clear methodology for tolerance specification. Most designers use their intuition and experience. Sometimes they follow general guidelines. As a consequence, most product designs do not contain all functionally relevant tolerances while some tolerance values are too tight. A clear and reliable method for tolerance specification is required as part of an overall tolerance management strategy. Tolerance management includes all design, manufacturing and inspection activities, striving to control and optimize the effect of geometrical variation. In this way Computer-Aided Tolerancing tools should be used to increase geometric robustness during concept design and to assign tolerances on the basis of sensitivity, manufacturability and cost during the detailing phase.

Within the context of robust design and life cycle engineering, education about consistent tolerancing becomes a very important item. With a consistent tolerancing theory at hand it becomes possible to teach robust design as a science.

The contents of this book originates from a collection of selected papers presented at the 6th CIRP International Seminar on Computer-Aided Tolerancing (CAT), organized by CIPV (The Center for Integrated Manufacturing and Development) of the University of Twente. The Seminar was sponsored by Stichting LPT (The foundation Laboratory of

Production and Design Engineering) and the University of Twente (where the seminar took place at 22 - 24 March 1999).

Previous CIRP seminars on Computer-Aided Tolerancing have been held in the USA 1989, Israel 1991, France 1993, Japan 1995 and Canada 1997. The seminars are intended to consolidate and advance the understanding of tolerancing as distinct technology and an important aspect of the design and manufacturing of mechanical products. CIRP is the International Institution for Production Engineering Research

The 6th CIRP CAT seminar focused in particular on research and development and application of techniques, which support the global consistency of tolerances. The topic of global consistency was proposed by Professor André Clément, a pioneer of research on tolerancing. His excellent keynote paper, the first one in the book, sets the scene for the other contributions.

It is our intention that the book will serve as a proper introduction to the field of consistent Computer-Aided Tolerancing. We think that it is also interesting for users of CADCAM systems who want to extend their knowledge about advanced modelling of micro geometric issues. And we hope it will also provide a good starting point for future research work.

We want to express our sincere thanks to the authors of the keynote papers, all the other contributors, to the members of the international program committee and the organizing committee, in particular Dr.Ir. O.W. Salomons, Ir. D. Lutters and Mrs. I. Dos Santos for their substantial effort to make the publishing of this book possible.

Fred J.A.M. van Houten
Hubert J.J. Kals

Global Consistency of Dimensioning and Tolerancing

André CLEMENT
Dassault Systèmes
9, quai Marcel Dassault, 92156 Suresnes Cedex, FRANCE
Tel : 33 1 40 99 42 87 Fax : 33 1 40 99 42 13
e.mail : andre_clement@ds-fr.com

Alain RIVIERE and Philippe SERRE
Groupe de Recherche en Ingénierie Intégrée des Ensembles Mécatroniques
3, rue Fernand Hainaut, 93407 Saint-Ouen Cedex, France
Tel : 33 1 49 45 29 20 Fax : 33 1 49 45 29 29
e. mail : ariviere@ismcm-cesti.fr
e. mail : philippe.serre@ismcm-cesti.fr

Abstract : The general objective of this paper is to define the problem areas linked to the specification and overall consistency of a dimensioning diagram and to propose a solution principle to solve this problem.

Firstly, we will analyse and propose a solution to the question of the overall consistency of the dimensioning diagram of a mechanical part.

In particular, we will show the importance of a specification which is too often implicit - that of the dimension of space.

This will be shown by the existence of decline constraints which will be expressed according to two mathematical formalisms, one linked to volume or hypervolume calculation and the other to the utilisation of the metric tensor.

And, it is by satisfying these decline constraints that we will be able to guarantee the consistency of a dimensioning diagram.

Different vectorial, tensorial and affine expressions of dimensioning and geometric tolerancing will be examined in detail and compared.

Finally, we will consider the transposition of this consistency analysis to a tolerancing diagram; the problem posed, as well as some elements of solution, will be covered in detail.

Keywords : dimensioning, tolerancing, consistency, metric tensor, chirality.

1. FOREWORD

In CAD/CAM, the principal function of 3-D software is the definition of a physically producible mechanical part and not the representation on a screen of a simple optical illusion, such as the famous perpetually ascending staircase.

The only known industrial solution to this essential difficulty consists of using CSG modelling, i.e. Boolean operations on closed solids. Despite the inflexibility of CSG in relation to B.REP-type modelling, it is the only type used since it guarantees the feasibility of the resultant solid. The solid obtained is said to be "consistent".

There is less topological difficulty involved in 2D since, on one hand, a closed contour can be specified and, on the other hand, the operator can "see" the auto-intersections of the result and is able to verify the consistency of the object himself.

However, even in 2D, the difficulty reappears when we want to modify the object by varying one of its dimensional or angular specifications. The very crux of the problem of parametric or variational geometry lies in the fact that, even when we start off with a consistent object, it is extremely difficult to obtain a consistent object once a dimensional specification has been modified.

The aim of this study is to seek the actual causes of such a phenomenon and to give the explicit rules for consistency. We will demonstrate that it is always a question of a specification deficiency. The properties of the space in which the object is produced as regards topology, orientation capacity and dimensional nature are implicitly included in the analytic specifications of the Cartesian reference system. Consequently, to render these specifications explicit in the dimensional description of an object involves reconsidering the properties of the Euclidean space, then defining new types of specifications.

The globally consistent dimensioning of an object will be explicitly composed of:
* explicit specification of the dimension of the affine space: 1D, 2D, 3D and the associated vector space,
* dimensional and angular specifications: the vector space has Euclidean metrics and triangular inequality must be constantly verified,
* and topological specifications: the type of object must be included in the specification: sphere, torus, etc.

However, a new type of specification must be added to this, which is rarely recognised as such, the chirality, although all aircraft, automobiles and many commonly used objects have a left-hand side and a right-hand side.

2. SUMMARY OF THE IMPLICIT PROPERTIES OF THE CARTESIAN REFERENCE POINT

Analytic geometry is based on the Cartesian reference system. In 3-D, this reference system is composed of 4 points in the affine space E^3 (OABC) and an associated, direct normed space (i, j, k). On one hand, the association between these 2 spaces is written by

OA = i; OB = j; OC = k and, on the other hand, by applying an arbitrary external rule in order to choose a direct trihedron (for example, the rule of the 3 fingers of the right hand).

Following this choice, any vector development will be consistent, provided that the direction of the direct trihedron is never changed. However, should the association between the 2 spaces be made inadvertently with the fingers of the left hand, all analytical expressions will still be correct and consistent but a mirror image will be obtained. This phenomenon is known as chirality. In practical terms, the physical, chemical and biological world is chiral. What is absolutely remarkable and little known is that left or right-hand chirality is encountered independently of the spatial dimension. The phenomenon is well known in 3-D space, but it is also identical in both 2-D and n-D space. Mirror image (i.e. symmetry in relation to a hyperplane) is possible in any Euclidean space. To ascertain the orientation of this space involves indicating on which side of the mirror we are located. To do this, we have to give a reference point on either side of this plane.

The need to explain the various properties of the Cartesian reference system (dimension (1-D, 2-D, 3-D), metrics and orientation) arises in CAD/CAM when we no longer wish to define geometry by means of procedural expressions carefully programmed by geometers, based on "operator" choices consistent with the initial trihedron. We now require the operator alone to define this geometry in full by means of a series of dimensional specifications, involving lengths and angles which are consequently true, independently of the direction of the reference system and spatial dimension. A family of results are thus obtained, the majority of which are aberrant. The most striking example already observed by C. Hoffman is the combinatory explosion of the orientation of angles which are solutions to an apparently complete and consistent dimensioning system. A perfectly constrained geometry problem (the same number of non-linear equations as unknowns) may have an exponential number of discrete angular solutions according to the number of constraints (n). However, this is not due to chirality alone. We can demonstrate this with a well-known example (Boumard then Pierra and, finally, Hoffman).

EXAMPLE:

Is the contour below with 6 « situation points», (A, B,…, F) perfectly defined for a geometer, if we know 3 angles (α_1, α_2 and $\pi/2$), 2 lengths (d1, d2) and the connecting radius (r) ?

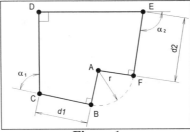

Figure 1

4

The positive response found in documentation is not sufficient. In fact, neither the orientation of the angles is specified, nor the type of continuity C1 of the connecting fillet, as shown in the two examples in Figure 2:

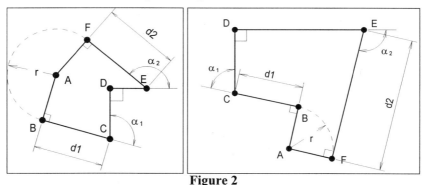

Figure 2

Finally, the global chirality of the object is not specified (cf Figure 3),

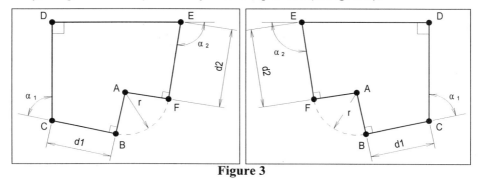

Figure 3

nor, moreover, is the dimension of the space in which the object is deployed (see, for example, the same object in 3-D in Figure 4, paragraph V):

The global specification of an object must, therefore, include specifications that enable all these possibilities to be differentiated and allow one of them to be chosen with precision. We can see that consistent dimensional specifications are only a small part of the overall problem.

Two location points form an affine space with a dimension of 1; 3 points form an affine space with a dimension of 2; 4 points form an affine space with a dimension of 3, and, in general, N points form an affine space with a dimension of N-1.

No constraint exists on the relative positions of these points while a spatial dimension other than that called for by the number of points considered is not imposed (This is the case, for example, for a robot arm with N segments, all of which can move freely). However, as soon as this assembly of points is immersed in a space with a smaller dimension, relations between the positions of the different points are created. For example, in Figure 1, the 6 location points are implicitly immersed in the 2D page of this text, in Figure 4-§V, the 6 points are implicitly immersed in a 3D space. (In the case of the robot, it is what happens when the robot grasps on object linked to its base: the angles of the different articulations are then interlinked).

We propose to enumerate and explain these relations, on one hand by an initial, rather intuitive analysis, covering the lengths in the affine space and, on the other hand, by a much more general analysis covering the angles in the associated normed vector space.

3.   MATHEMATICAL ANALYSIS OF GLOBAL CONSISTENCY IN THE AFFINE SPACE

The point concept and Euclidean distance concept between 2 points are given. Aided by these 2 concepts, we propose to define the accurate specifications required to adequately construct a basic topological object: the point, the segment, the triangle and the tetrahedron. These basic topological objects are known as "simplices"; mathematicians have demonstrated that these are sufficient to define any topology. Simplex method approximation is sufficient to define the position, orientation and topology of a real object.

Elementary examples

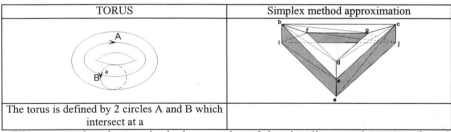

Closed lateral cylindrical surface	Simplex method approximation
Each triplet (a,b,c) and (d,e,f) defines a circle, the cylinder is defined by the 2 circles	

TORUS	Simplex method approximation
The torus is defined by 2 circles A and B which intersect at a	

We assume that the topological properties of the simplices are known and analyse here the other properties of the simplices: dimensions and orientations in an affine Euclidean space.

3.1 Point-Space (Dimension: zero)

A point-space is a set of points denoted P(A, B,...) such that the distance of any pair of points in this set is equal to zero.

A set of this kind can be immersed in a straight-line-space or a plane space, or in any space with a larger dimension.

3.2 Straigth-Line-Space (Degree of calibration: one; Dimension: one)

A straight-line-space is a set of points denoted D(A,B,C...) such that the 3 distances of any triplet of points in this set are linked by a relation.

In a 1-dimension Euclidean space, (numeric straight-line R) to define the relative position of 3 points A, B, C, we only need to know 2 lengths (AB and BC for example) and the 3rd length L = AC can be deduced since there is ALWAYS a relation (AC = AB +BC in the case of the figure or symmetrically in the other configurations: AB = AC + CB ; BC = BA + AC) which link these 3 lengths.

This fundamental relationship is the basis for numerous work projects on tolerancing consistency and single direction assembly. It is also widely used in object specification, but in such an intuitive manner that we not only forget its origin but also its extension to spaces with higher dimensions.

Let us determine the origin of this relation.

Two points A and B define a straight-line, a Euclidean space with a dimension of 1. Three points A, B, C define a plane, a Euclidean space with a dimension of 2. The 3 lengths a = AB ; b= BC ; and CA = c of the triangle ABC verify a certain number of inequalities due to the triangle inequality axiom of the Euclidean norm.

For example for « b », we have:

$$| a - c | \leq b \leq a + c$$

The maximum value (right-hand chirality) or the minimum value (left-hand chirality) of « b » will be obtained when the 3 points are aligned in a certain order, i.e. when the spatial dimension changes back from 2 to 1 with a certain orientation.

This decrease in spatial dimension, which we will now call the "folding back" of the space, results in the creation of an internal relation oriented between the dimensions of the figure thus constructed. This relation can be expressed by writing that the surface of triangle ABC is zero (which reflects the fact that the 3 points are aligned: 2D space is folded back into 1D space).

This is an absolutely general analysis: any folding back of space stems from an oriented dimensional relation.

We can say that there is 1 degree of calibration, since any pair of points can form a unit length. (This corresponds to the unit vector module in Cartesian geometry)

We can say that it is 1-dimensional: since on the 3 distances linked by a constraint, with one length being calibrated, only one arbitrary length remains, as the latter can be immediately deduced. We will see later on that the relative positions of the 3 points (chirality) can be stipulated further by taking into account the fact that distances are positive numbers.

3.3 Plane-Space (Calibration degree: three; Dimension: two)

A plane-space is a set of points denoted PLANE(A, B, C...) such that the 6 distances of any quadruplet of points in this set are linked by a relation.

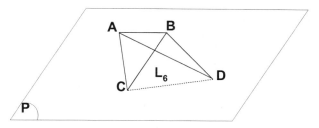

In a 2-dimensional space (the Euclidean plane) to define the relative position of 4 points A, B, C, D, we only need to know 5 lengths (a, b, c, d, e), and the 6th length (L_6) can be deduced since there is ALWAYS a relation which links these 6 lengths.

This relation will be expressed by writing that the volume of the tetrahedron ABCD is zero (**which reflects the fact that these 4 points are in the same plane: 3D space is folded back into 2D space**). This volume can be expressed in the form of the determinant:

$$V = k \cdot \begin{vmatrix} 0 & 1 & 1 & 1 & 1 \\ 1 & 0 & a^2 & b^2 & c^2 \\ 1 & a^2 & 0 & d^2 & e^2 \\ 1 & b^2 & d^2 & 0 & L^2 \\ 1 & c^2 & e^2 & L^2 & 0 \end{vmatrix}$$

We obtain the relation between the distances of the 4 coplanar points by writing that V = 0. This relation is much more complex than in the one-dimension case, and has to be processed by a computer.

Application of the method to very simple geometry:

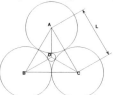

Given 3 circles with radius R and centres A, B and C, and a circle with a radius r and a centre D, all on the same plane. The following local tangency constraints are imposed:

AB = 2R ; BC = 2R ; CA = 2R ; CD = R + r ; BD = R + r ; AD = R + r.

As the points A, B, C and D are in the same plane, a global consistency constraint is imposed by writing: V = 0. The relation which links R and r can thus be deduced, i.e.

$$R = r.(3 + 2\sqrt{3}).$$

We can say that there are 3 degrees of calibration since a triplet of points taken from among the 4 points, can form a triangle with sides whose measurement is arbitrary. The other 3 triangles are then calibrated. (This corresponds to the 3 independent components of the metric tensor of 2D Cartesian geometry: the modules of 2 unit vectors and the cosine of their angle)

We can say that it is 2-dimensional: since, of the 6 distances linked by a constraint, with 3 lengths being taken as the standard measure, 2 lengths remain independent, and the last can be deduced. We will see later that we can stipulate further the relative positions of the 4 points by taking into account the fact that the distances are positive numbers.

A plane set of this kind can be immersed in any space of a larger dimension. Conversely, it can be said that a 3D space can be folded back into a plane-space when the previous constraint with 4 points in space is applied.

3.4 Euclidean-Space[3] (Calibration degree: 6 ;Dimension: three)

A EUCLIDEAN-space[3] is a set of points denoted E^3(A, B, C, D, E...) such that the 10 distances of any quintuplet of points of the set are linked by a relation.

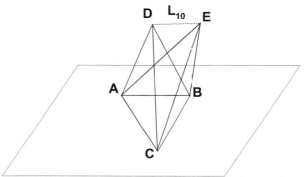

In a 3-dimensional Euclidean space, to define the relative positions of 5 points A, B, C, D, E, we only have to know 9 lengths (a, b, c, d, e, f, g, h, j, k), the 10th length L can be deduced since there is ALWAYS a relation which links these 10 lengths.

Expression of this relation will be obtained by writing that the hypervolume of the object ABCDE of the 4-dimensional space is zero (the 5 points are immersed in the 3-dimensional space: 4D space is folded back into 3D space).

This hypervolume was calculated in the last century in its most general form by de Tilly. It is presented in the form of the determinant below which will be declared as zero in a 3-dimensional space.

$$W = k \cdot \begin{vmatrix} 0 & 1 & 1 & 1 & 1 & 1 \\ 1 & 0 & a^2 & b^2 & c^2 & d^2 \\ 1 & a^2 & 0 & f^2 & g^2 & h^2 \\ 1 & b^2 & f^2 & 0 & j^2 & k^2 \\ 1 & c^2 & g^2 & j^2 & 0 & L^2 \\ 1 & d^2 & h^2 & k^2 & L^2 & 0 \end{vmatrix}$$

It is a second degree equation at L, for example, and, as previously, the max. or min. value of L allows the requisite chirality to be chosen.

We can say that there are 6 degrees of calibration, since any quadruplet of points can form a tetrahedron with any 6 sides. (This corresponds to the 6 independent components of the metric tensor of 3D Cartesian geometry).

We can say that it is 3-dimensional: since, of the 10 distances linked by a constraint, with 6 lengths being taken as the standard measure, 3 lengths remain independent and the 10th can be deduced. We will see later that we can stipulate further the relative positions of the 6 points by taking into account the fact that the distances are positive numbers.

A plane set of this kind can be immersed in any space of a larger dimension. Conversely, it can be said that a EUCLIDEAN-4D space can be folded back into a 3D space when the previous constraint with 5 points is applied..

4. MATHEMATICAL ANALYSIS OF THE GLOBAL CONSISTENCY IN THE VECTOR SPACE

In the Cartesian approach, the metric tensor is associated with a packet of vectors that are general orthogonal, in an equal number to the dimension of the metric space under consideration (for example 3 in 3D, 2 in 2D…).

Conversely, in the approach proposed here, the metric tensor is constructed on all the vectors under consideration and the dimension of the space is specified **explicitly** by imposing the rank adapted to this tensor.

By definition, specifying a valid metric tensor leads to constructing a geometrically consistent object.

4.1 Metric tensor of a packet of vectors - the general case

4.1.1 Calculation of the vector module using the components of the covariant metric tensor

Given a Cartesian reference point (O, i, j, k) and a point P with **contravariant** coordinates, x, y, z.

$$\overrightarrow{OP} = \begin{bmatrix} \vec{i} & \vec{j} & \vec{k} \end{bmatrix} \cdot \begin{bmatrix} x \\ y \\ z \end{bmatrix}$$

The vector module $\overrightarrow{OP}$ can be calculated easily, by the tensorial product: $\overrightarrow{OP}^{T} \otimes \overrightarrow{OP}$ which introduces the metric tensor covariant $\overline{\overline{G}}$

$$\overrightarrow{OP}^2 = \begin{bmatrix} x & y & z \end{bmatrix} \cdot \begin{bmatrix} \vec{i}\cdot\vec{i} & \vec{i}\cdot\vec{j} & \vec{i}\cdot\vec{k} \\ \vec{j}\cdot\vec{i} & \vec{j}\cdot\vec{j} & \vec{j}\cdot\vec{k} \\ \vec{k}\cdot\vec{i} & \vec{k}\cdot\vec{j} & \vec{k}\cdot\vec{k} \end{bmatrix} \cdot \begin{bmatrix} x \\ y \\ z \end{bmatrix}$$

We assume: $$\overrightarrow{OP}^2 = \vec{X}^T \cdot \overline{\overline{G}} \cdot \vec{X}$$

The corresponding position of $\overrightarrow{OP}$ is defined to within a rotation, a **symmetry** and a translation when we know $\overline{\overline{G}}$ and $\overrightarrow{OP}^2$. Stated differently: **the metric tensor is an invariant for the translations, rotations and symmetries of the affine space**. It therefore allows us to specify the relative positions of a set of points to within an isometry. **This is our aim!**

Invariance for translations and rotations is an advantage which justifies the use of the metric tensor; invariance in symmetry is a disadvantage that has to be taken into account (Cf. Figure 3 et 4). Designating one of the 2 terms of a symmetry is called « choosing the left or right-hand chirality ». The chirality of a specification is chosen in relation to an initial orientation of the affine space.

In Cartesian methodology, there is no other option. In the methodology proposed here, all the other bi-points, triangles or tetrahedrons are thus constructed by retaining or reversing this initial chirality according to operator requests. These relative chiralities can be inscribed into the structure of the metric tensor without any difficulty.

In 1D, we can see that the bi-point has the same (or opposite) direction as the initial bi-point.

In 2D, we can say that the triangle has the same (or opposite) direction path as the initial triangle.

In 3D, we can say that the tetrahedron has the same (or opposite) direction as the initial tetrahedron.

4.1.2 Calculation of the covariant metric tensor

In the case of a set of N points A, B…, using couples of bi-points oriented arbitrarily, (for example $\overrightarrow{AB} = \vec{a}$) we define the vectors $\vec{a}, \vec{b}, \vec{c}, \vec{d}, \ldots$ as any number but at least equal to $N - 1$ (forming either a sequence, circuit, or a network), and we choose unit/base vectors on these vectors, such as, for example, $\vec{a} = a.\vec{e_a}$.

NOTE. Base vectors « $\vec{e_a}$ »are called « covariants » as well as the corresponding metric tensor g_{ij} (« lower » indices). The preceding components« a » on these base vectors are called « affines » or « contravariants » (no index here, but, in general, « upper »). The orthogonal projections on the base vectors are the « covariant » components.

Example of a packet of 5 vectors of lengths: $\vec{a}, \vec{b}, \vec{c}, \vec{d}, \vec{e}$

$$\vec{a} = \begin{pmatrix} a & 0 & 0 & 0 & 0 \end{pmatrix} \cdot \vec{e_a} \text{ with } \vec{e_a} = \begin{pmatrix} 1 & 0 & 0 & 0 & 0 \end{pmatrix}$$

$$\vec{b} = \begin{pmatrix} 0 & b & 0 & 0 & 0 \end{pmatrix} \cdot \vec{e_b} \text{ with } \vec{e_b} = \begin{pmatrix} 0 & 1 & 0 & 0 & 0 \end{pmatrix}$$

$$\vec{c} = \begin{pmatrix} 0 & 0 & c & 0 & 0 \end{pmatrix} \cdot \vec{e_c} \text{ with } \vec{e_c} = \begin{pmatrix} 0 & 0 & 1 & 0 & 0 \end{pmatrix}$$

$$\vec{d} = \begin{pmatrix} 0 & 0 & 0 & d & 0 \end{pmatrix} \cdot \vec{e_d} \text{ with } \vec{e_d} = \begin{pmatrix} 0 & 0 & 0 & 1 & 0 \end{pmatrix}$$

$$\vec{e} = \begin{pmatrix} 0 & 0 & 0 & 0 & e \end{pmatrix} \cdot \vec{e_e} \text{ with } \vec{e_e} = \begin{pmatrix} 0 & 0 & 0 & 0 & 1 \end{pmatrix}$$

We can form the corresponding covariant metric tensor g_{ij}

$$\overline{\overline{G}} = \begin{pmatrix} \vec{e_a} & \vec{e_b} & \vec{e_c} & \vec{e_d} & \vec{e_e} \end{pmatrix}^T \otimes \begin{pmatrix} \vec{e_a} & \vec{e_b} & \vec{e_c} & \vec{e_d} & \vec{e_e} \end{pmatrix}$$

Any non-diagonal term of the tensor equals $\cos\left(\vec{e_c}, \vec{e_d}\right)$. It is an oriented angle of between $[0, +\pi]$; a reference orientation is then necessary.

The various **contravariant** components, a, b, are quantities with signs, and the sign depends only on the arbitrary orientation of the unit vectors in relation to the affine vectors.

Unlike what happens in a Cartesian reference, here the rank of this tensor does not generally equal its size since the base vectors are not independent. **The rank will be equal to the spatial dimension in which the vectors are immersed.**

4.1.3 Calculation of the contravariant components (also called affines)

All the vectors, linear combinations of these base vectors, will have **the coefficients of these linear combinations** as **contravariant** components.

Example: the vector module $\vec{a} + \vec{b} + \vec{c}$ is equal to:

$$\begin{pmatrix} a & b & c & 0 & 0 \end{pmatrix}^T \cdot \overline{\overline{G}} \cdot \begin{pmatrix} a & b & c & 0 & 0 \end{pmatrix}$$

Interpretation in the affine space

With the previous hypotheses, the contravariant coordinates of a vector are always presented in the form of a series of lengths and 0 according to whether or not the vectors pass through the corresponding points.

These coordinates then change **numerically** if the numeric value changes, and change **sequence** if the specification changes.

4.1.4 Calculation of covariant components

The v^i contravariant (or affine) components of $\vec{V}$ represent the oblique projections $v^i \cdot \vec{e_i}$ of $\vec{V}$ on the axes to within the norm of the base vectors. **Here all the vectors considered are taken as base vectors; consequently their non-zero contravariant component is always equal to their module**

The v_i covariant components of $\vec{V}$ represent the orthogonal projections $\vec{V} \cdot \vec{e_i}$ of $\vec{V}$ on the axes, to within the norm of the base vectors.

A vector $\vec{V}$ is entirely determined if one or other of its components are given. Here we always know its contravariant components: this is the declaration of the specification. If we are looking for covariant components: this is a reversion to the Cartesian reference point.

We know that there is the following relation between the covariant and contravariant components of a vector $\vec{V}$

$$v_i = g_{ij} \cdot v^j$$

and we know the covariant metric tensor $g_{ij} = \vec{e_i} \cdot \vec{e_j}$ when we know a specification, therefore we know the lengths and angles between these vectors.

4.2 System for the global dimensioning of N points

The points belong to the affine space, the metric tensor belongs to the associated vector space. With N-1 vectors defined by N-1 pairs of points, we can determine the associated vector space exactly, in the form of a metric tensor. Conversely, knowledge of the metric tensor allows all the distances and angles of this set of points to be specified to within a translation, rotation or symmetry of the set.

The metric tensor gives « geometric semantics » to a set of numeric values, independently of any Cartesian reference point. This could be an efficient means of exchange between CAD/CAM systems; we will demonstrate that it is an efficient way to verify the completeness of a specification. The 4 rules below should be perceived as the necessary, adequate conditions for the global consistency of a geometric object.

Firstly, the rank of the metric tensor has to be established in accordance with the affine spatial dimension with which it is associated. We must then ensure that the tensor is a metric tensor, that is to say that it corresponds to a defined positive quadratic form.

Hence the two metric tensor construction rules that stem from the properties of the vector space:
- **RANK RULE, (symbol $\Re$ = 0, 1, 2, 3),**
- **METRICS RULE (symbol m $\geq$ 0),**

Finally, certain linear combinations are known, in particular the relative chirality and the closed contours.

Hence the 2 additional rules that stem from the properties of the affine space.
- **ZERO VECTOR SUM RULE (symbol Σ = 0),.**
- **RELATIVE CHIRALITY RULE (symbol χ= G or D),.**

4.2.1 Rank Rule (symbol $\Re$),

All the vectors must be of a certain rank (1, 2 , 3). That is to say that all the minors greater than the rank $\Re$ chosen are nil. In particular, we will express the fact that the running vector is a linear combination of the 1, 2, or 3 preceding vectors in order to ensure the transitivity of the rank constraint:

All the angles used are vector angles, such that $[0 \leq \theta \leq \pi]$.

- $\Re = 1$ - In 1D, we can express the co-linearity and the chirality of N vectors $\overrightarrow{OA}, \overrightarrow{OB}, \dots, \overrightarrow{OM}$ by writing that the angle between any doublet of vectors equals 0 or π:

For example for the 4 aligned points O, A, B, M, we specified the following chirality (order OAM and OAB)

$$angle(\overrightarrow{OA},\overrightarrow{OM}) = 0; \qquad angle(\overrightarrow{OA},\overrightarrow{OB}) = 0;$$

- $\Re = 2$ - In 2D, we can express the co-planarity and chirality of N vectors $\overrightarrow{OA},\overrightarrow{OB},....,\overrightarrow{OM}$ by writing the Chasles relation on any triplet of their oriented angles:

$$angle(\overrightarrow{OA},\overrightarrow{OM}) = angle(\overrightarrow{OA},\overrightarrow{OB}) + angle(\overrightarrow{OB},\overrightarrow{OM})$$

- $\Re = 3$ - In 3D, we can express the co-spatiality of N vectors $\overrightarrow{OA},\overrightarrow{OB},....,\overrightarrow{OM}$ by writing that the 6 angles oriented between any quadruplet of vectors are subject to the following angular relation: The determinant of their metric tensor is nil :

$$\begin{vmatrix} 1 & \cos\alpha & \cos\beta & \cos\gamma \\ \cos\alpha & 1 & \cos\varphi & \cos\theta \\ \cos\beta & \cos\varphi & 1 & \cos\phi \\ \cos\gamma & \cos\theta & \cos\phi & 1 \end{vmatrix} = 0$$

4.2.2 Metric Rule (symbol m ≥ 0),

The rank rule allows a tensor of a certain rank $\Re$ to be constructed. However, this will only be a metric tensor if, from this rank up to rank 1 included, its determinant is positive, as well as all its minors.

As all the constructions possible will be in 1D, 2D or 3D space, we only have to study 3 cases

- **1D SPACE: $m_1 \geq 0$.**

In 1D, only non-zero minors are diagonal terms. The metric rule is simple, therefore: all the diagonal terms must be positive. (They represent the square of the module of each vector)

- **2D SPACE: $m_2 \geq 0$**

In 2D, (we already have $m_1 \geq 0$), and the non-zero minors are all 2 X 2 determinants extracted from the tensor band and form:

$$\begin{vmatrix} 1 & \cos\alpha \\ \cos\alpha & 1 \end{vmatrix}$$

This determinant will always be positive if the terms outside the diagonal are comprised between [-1, +1] in order to represent a cosine.

We will constantly use this property by putting all these terms in this form.

- **SPACE 3D: $m_3 \geq 0$**

In 3D, we already have $m_1 \geq 0$ and $m_2 \geq 0$ non-zero minors are all 3 X 3 determinants extracted from the tensor band and form:

$$\begin{vmatrix} 1 & \cos\alpha & \cos\beta \\ \cos\alpha & 1 & \cos\gamma \\ \cos\beta & \cos\gamma & 1 \end{vmatrix}$$

and this determinant will be positive if, and only if, the following strict inequalities (or any circular permutation) are verified:

$$|\alpha - \beta| < \gamma < \alpha + \beta$$

4.2.3 Zero Vector Sum Rule (symbol Σ),

We very often know an additional relation on the modules and angles of these vectors: their vector sum is the zero vector (for example, it is the vectorial constraint associated with the closing of an affine space contour). The covariant components of this zero vector are zero.

Example with 3 vectors.

$$\vec{a} + \vec{b} + \vec{c} = \vec{0}$$

By scalar multiplication by $\vec{a}$ we obtain: $a^2 + a \cdot b \cdot \cos(\alpha) + a \cdot c \cdot \cos(\beta) = 0$

where $|a^2 \quad a \cdot b \cdot \cos(\alpha) \quad a \cdot c \cdot \cos(\beta)|$

represents a line (or a column) of the metric tensor of the 3 non-unit vectors. By scalar multiplication by b or c we would obtain the same relation on the other 2 lines (or columns) of this metric tensor. By means of association, the relation is extended to any number of vectors, hence the rule.

The sum of certain elements of a line (or column) of the metric tensor is zero when the vector sum of the corresponding vectors is zero

4.2.4 Relative Chirality Rule (symbol χ = G or D),

In 1D, after choosing the first vector corresponding to a fixed bi-point in the affine space, chirality can be easily declared since the scalar product of 2 co-linear unit vectors with the same direction equals +1 and with opposite directions – 1.

In 2D (in 3D resp.), after choosing the first triangle (first tetrahedron resp.) corresponding to a fixed triangle (tetrahedron resp.) in the affine space, chirality is declared by the orientation of the angles relative to these references.

5. EXAMPLE OF VERIFICATION OF A SPECIFICATION IN A 2-D ORIENTED VECTOR SPACE

We propose to verify the consistency of the specification of the example in Figure 1. For each new point, the operator uses the reference point constituted by the last 3 points created. For each new point, the operator gave the length of the bi-point created and the oriented angle in relation to the last vector (or the relative «left - right-hand » chirality and the lengths of 2 of the last 3 bi-points). We took the orientation of the first angle arbitrarily to be $-\pi/2$. All other orientations are relative to this one. We obtain the primary values of the tensor band below:

$$\begin{Vmatrix} 1 & \cos(-\pi/2) & \cdot & \cdot & \cdot & \cdot \\ \cos(-\pi/2) & 1 & \cos(-\alpha 1) & \cdot & \cdot & \cdot \\ \cdot & \cos(-\alpha 1) & 1 & \cos(+\pi/2) & \cdot & \cdot \\ \cdot & \cdot & \cos(+\pi/2) & 1 & \cos(-\alpha 2) & \cdot \\ \cdot & \cdot & \cdot & \cos(-\alpha 2) & 1 & \cos(-\pi/2) \\ \cdot & \cdot & \cdot & \cdot & \cos(-\pi/2) & 1 \end{Vmatrix}$$

The secondary values, represented by the points in the above example, calculated in the examples below, must verify the 4 rules of global consistency.

Verification of the metric rule is immediate for all non-zero minors.

Verification of the rank rule is also immediate: the others minors are in fact zero.

Verification that the contour is closed will consist of checking that the vector sum $S = AB + BC + CD + DE + EF + FA$ is zero. Subsequently, we necessarily have the 6 scalar relations below which have to be zero, since they are the covariant components $s_i = g_{ij} \cdot s^j$ of a zero vector.

In the case in Figure 2-a, we obtain the following 6 rank 2 equations: (the matrix is symmetrical)

$$\begin{Vmatrix} 1 & \cos(-\pi/2) & \cos(-\pi/2-\alpha 1) & \cos(-\alpha 1) & \cos(-\alpha 1-\alpha 2) & \cos(-\alpha 1-\alpha 2-\pi/2) \\ \cdot & 1 & \cos(-\alpha 1) & \cos(-\alpha 1+\pi/2) & \cos(\pi/2-\alpha 1-\alpha 2) & \cos(-\alpha 1-\alpha 2) \\ \cdot & \cdot & 1 & \cos(+\pi/2) & \cos(\pi/2-\alpha 2) & \cos(-\alpha 2) \\ \cdot & \cdot & \cdot & 1 & \cos(-\alpha 2) & \cos(-\alpha 2-\pi/2) \\ \cdot & \cdot & \cdot & \cdot & 1 & \cos(-\pi/2) \\ \cdot & \cdot & \cdot & \cdot & \cdot & 1 \end{Vmatrix} \begin{Vmatrix} r \\ d1 \\ CD \\ DE \\ d2 \\ r \end{Vmatrix} = \begin{Vmatrix} 0 \\ 0 \\ 0 \\ 0 \\ 0 \\ 0 \end{Vmatrix}$$

In the case in Figure 2-b, we obtain the following 6 rank 2 equations: (the matrix is symmetrical)

$$\begin{Vmatrix} 1 & \cos(-\pi/2) & \cos(-\pi/2-\alpha 1) & \cos(-\alpha 1-\pi) & \cos(-\alpha 1-\alpha 2-\pi) & \cos(-\pi/2-\alpha 1-\alpha 2) \\ \cdot & 1 & \cos(-\alpha 1) & \cos(-\alpha 1-\pi/2) & \cos(-\pi/2-\alpha 1-\alpha 2) & \cos(-\alpha 1-\alpha 2) \\ \cdot & \cdot & 1 & \cos(-\pi/2) & \cos(-\pi/2-\alpha 2) & \cos(-\alpha 2) \\ \cdot & \cdot & \cdot & 1 & \cos(-\alpha 2) & \cos(-\alpha 2+\pi/2) \\ \cdot & \cdot & \cdot & \cdot & 1 & \cos(+\pi/2) \\ \cdot & \cdot & \cdot & \cdot & \cdot & 1 \end{Vmatrix} \begin{Vmatrix} r \\ d1 \\ CD \\ DE \\ d2 \\ r \end{Vmatrix} = \begin{Vmatrix} 0 \\ 0 \\ 0 \\ 0 \\ 0 \\ 0 \end{Vmatrix}$$

The case of the global chirality of Figure 3-a and Figure 3-b, is slightly different. We obtain the following 2 systems of 6 rank 2 equations: (the matrix is symmetrical)

$$\varepsilon \cdot \begin{Vmatrix} 1 & \cos(-\pi/2) & \cos(-\pi/2-\alpha1) & \cos(-\alpha1-\pi) & \cos(-\pi-\alpha1-\alpha2) & \cos(-3\pi/2-\alpha1-\alpha2) \\ . & 1 & \cos(-\alpha1) & \cos(-\alpha1-\pi/2) & \cos(-\pi/2-\alpha1-\alpha2) & \cos(-\pi-\alpha1-\alpha2) \\ . & . & 1 & \cos(-\pi/2) & \cos(-\pi/2-\alpha2) & \cos(-\pi-\alpha2) \\ . & . & . & 1 & \cos(-\alpha2) & \cos(-\alpha2-\pi/2) \\ . & . & . & . & 1 & \cos(-\pi/2) \\ . & . & . & . & . & 1 \end{Vmatrix} \cdot \begin{Vmatrix} r \\ d1 \\ CD \\ DE \\ d2 \\ r \end{Vmatrix} = \begin{Vmatrix} 0 \\ 0 \\ 0 \\ 0 \\ 0 \\ 0 \end{Vmatrix}$$

with $\varepsilon = 1$ or $\varepsilon = -1$. The 2 systems of equations, mathematically equivalent, **are** indicated by this sign, but it is impossible to specify one or other of them without external reference.

The case in Figure 4 is even more complex. We deployed the initial specification in a 3-D space. We obtain the $7^{\times}7$ tensor below into which an additional line and column have been inserted, corresponding to the creation of the new vector DA. We are, in fact, confronted by 2 sub-tensors corresponding to the 2 plane parts, with the vector DA in common with 2 new angles $\beta1 = (CD, DA)$ in the first sub-tensor and $\beta2 = (DA, DE)$ in the second. The tensor obtained must be specified as rank 3 by adequate values in the boxes left blank .

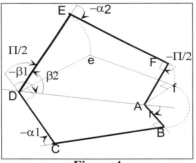

Figure 4

$$\begin{Vmatrix} 1 & \cos(-\pi/2) & \cos(-\pi/2-\alpha1) & \cos(-\pi/2-\alpha1-\beta1) & . & . & . \\ . & 1 & \cos(-\alpha1) & \cos(-\alpha1-\beta1) & . & . & . \\ . & . & 1 & \cos(-\beta1) & \cos(\pi/2) & . & . \\ . & . & . & 1 & \cos(\beta2) & \cos(\beta2-\alpha2) & \cos(-\alpha2+\beta2-\pi/2) \\ . & . & . & . & 1 & \cos(-\alpha2) & \cos(-\alpha2-\pi/2) \\ . & . & . & . & . & 1 & \cos(-\pi/2) \\ . & . & . & . & . & . & 1 \end{Vmatrix} \cdot \begin{Vmatrix} AB \\ BC \\ CD \\ DA \\ DE \\ EF \\ FA \end{Vmatrix} = \begin{Vmatrix} 0 \\ 0 \\ 0 \\ 0 \\ 0 \\ 0 \\ 0 \end{Vmatrix}$$

The preceding 6 relations can be further exploited, by noting that the zero status of the covariant coordinates must be accompanied by a tolerance, due to the inaccuracy of calculations with floating numbers on the computer: this is the micro-tolerancing of CAD/CAM objects. This system of equations is rank 2; consequently, we can assert that the 6 tolerances are not independent and that we can calculate 4 of them by using 2.

This simple comment explains current difficulties for modelling errors in CAD-CAM, since these intrinsic relations are not taken into account at present. Furthermore, unlike the current method, these tolerances are independent of the position of the object in the Cartesian reference point. It would be a considerable advantage for the improvement of STEP exchange standards.

6.   DIMENSIONAL AND TOLERANCING SPECIFICATIONS

The general objective of this section is to present a generalised expression of form, dimension and tolerancing specification for any kind of geometric entity.

After a brief review of the exhaustivity of dimensional specifications evidenced in the study of the TTRS model, some scalar expressions of these dimensional constraints will be given in detail, followed by a general structure for a geometric entity and, finally, the general expression of a geometric tolerancing specification will be given.

6.1 Review of the exhaustivity of dimensional specifications:

Since the first seminar arranged by the CIRP in Jerusalem in December 1989 up to the latest seminar in Toronto in April 1997, we have been presenting the TTRS model in detail, together with its applications to the dimensioning and tolerancing specification.

This is a review of the principal results evidenced by this study:

- the existence of 7 classes of surface (this result is now officially accepted by ISO),
- the existence of 28 cases of association between elementary surfaces or TTRS,
- the existence of 44 cases of reclassification for the new entity constituted by an association,
- and, finally, the result of prime interest to us here, an exhaustive list of 13 dimensional constraints:
    - C1: point $\leftrightarrow$ point, coincidence
    - C2: point $\leftrightarrow$ point, distance
    - C3: point $\leftrightarrow$ plane, distance
    - C4: point $\leftrightarrow$ line, coincidence
    - C5: point $\leftrightarrow$ line, distance
    - C6: plane $\leftrightarrow$ plane, parallelism, distance
    - C7: plane $\leftrightarrow$ plane, angle
    - C8: plane $\leftrightarrow$ line, perpendicularity
    - C9: plane $\leftrightarrow$ line, parallelism, distance
    - C10: plane $\leftrightarrow$ line, angle
    - C11: line $\leftrightarrow$ line, coincidence
    - C12: line $\leftrightarrow$ line, parallelism, distance
    - C13: line $\leftrightarrow$ line, angle and distance

We will now give details of the various types of scalar expression of these constraints, then use them to specify the form, dimensioning and tolerancing of any geometric entity.

6.2 Scalar expression of the 13 dimensional constraints:

6.2.1 Cartesian methodology

We have already stipulated the vectorial expression of these 13 constraints (cf).

In this methodology, in order to express a vectorial expression in scalar terms, we express each vector as a linear combination of base vectors, then, by using the scalar product, we project this vector onto each of the base vectors. When there are several reference systems, we reduce this to a unique system by translation and rotation.

The fundamental scalar expression is then the algebraic measurement a of a vector $\overrightarrow{X}$ in relation to a reference unit vector $\overset{.}{i}$:

$$\overrightarrow{X} = a \cdot \overset{.}{i} \Leftrightarrow \overrightarrow{X} \cdot \overset{.}{i} = a$$

We thus obtain scalar expressions where it is unfortunately very difficult to differentiate the intrinsic values of geometric objects from the values stemming from the choice of projection system.

We are then looking for another form of scalar expression that does not involve a projection reference point, but, on the contrary, is completely intrinsic to the geometry of the part under study.

6.2.2 Affine methodology

In this methodology, we use the module of a vector as a fundamental scalar expression:

$$\overrightarrow{X} \to \left\| \overrightarrow{X} \right\|$$

An immediate advantage is that the measurement of a vector module, in practical terms, is the measurement of the distance between 2 points. This is the basic metrological operation. However, the fundamental advantage of this approach is the intrinsic nature of this expression, since module X of a vector is independent of any reference point used to calculate it;

However, we will see in the following paragraph that the scalar expression of dimensional constraints by following this methodology, even if it is always possible, is sometimes not very simple to express.

6.2.3 Tensorial methodology

In this methodology, we use the metric tensor as the fundamental scalar expression of a set of vectors:

$$\left\{ \overrightarrow{X_i} \right\} \to \overline{\overline{G}}$$

We find the same properties as regards independence of the reference point used, with a very simple expression that can be clearly differentiated from the intrinsic characteristics of geometric objects.

We will see that, in particular, angle and distance characteristics are clearly separated.

As regards observance of immersion constraints in a space of one, two or three dimensions, it should be noted that:

- immersion is respected in the Cartesian approach, but in an implicit manner, by using the Cartesian co-ordinates of these points and vectors,
- immersion is respected in the affine approach by the explicit expression of the zero-value of the hypervolumes corresponding to larger dimensions of the work space,
- immersion is respected in the tensorial approach by the explicit specification of the rank of the metric tensor under consideration.

6.3 Examples of expression of dimensional constraints:

In this paragraph, we give the scalar expression of three-dimensional constraints from the thirteen previously mentioned. These three constraints are the constraints C5 (point ↔ line, distance), C9 (plane ↔ line, parallelism, distance) and C13 (line ↔ line, angle and distance).

We have chosen only to give the scalar expressions corresponding to the affine and tensorial methodologies since the expression of constraints according to Cartesian methodology are well known.

- Expression of the constraint C5 point ↔ line, distance

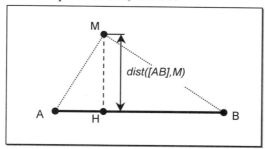

Figure 5 : Constraint C5

Given a straight-line [AB], a point M and H the orthogonal projection of the point M on the straight-line.

<u>Affine methodology</u>

The constraint C5 is expressed as follows $dist([AB], M) = 2 \cdot \dfrac{S_{ABM}}{L_{AB}}$

with $S_{ABM} = \dfrac{1}{4} \cdot \sqrt{(L_{AB} + L_{BM} + L_{MA}) \cdot (-L_{AB} + L_{BM} + L_{MA}) \cdot (L_{AB} - L_{BM} + L_{MA}) \cdot (L_{AB} + L_{BM} - L_{MA})}$

It should be noted that S_{ABM} represents the place surface encompassed by points A, B and M, and L_{AB} represents the length of the segment connecting the points A and B.

Tensorial methodology

In the base $\left(\overrightarrow{ab}, \overrightarrow{hm}\right)$ the tensor is expressed like this $\qquad G = \begin{pmatrix} \overrightarrow{ab} \cdot \overrightarrow{ab} & \overrightarrow{ab} \cdot \overrightarrow{hm} \\ \overrightarrow{ab} \cdot \overrightarrow{hm} & \overrightarrow{hm} \cdot \overrightarrow{hm} \end{pmatrix} = \begin{pmatrix} 1 & 0 \\ 0 & 1 \end{pmatrix}.$

The distance is then expressed simply as being the norm of the vector $\overrightarrow{HM}$, we thus obtain the following expression $\qquad dist([AB], M) = \left\|\overrightarrow{HM}\right\|$

To have the complete expression of this constraint, a constraint C4 (point ↔ line, coincidence) must be added between the point H and the straight-line [AB].

- Expression of the constraint C9 point ↔ line, parallelism, distance

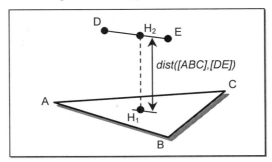

Figure 6 : Constraint C9

Given a plane [ABC], a straight-line [DE], H_1 a point belonging to the plane and H_2 a point belonging to the straight-line.

Affine methodology

The constraint C5 is expressed like this $\qquad dist([ABC], [DE]) = 3 \cdot \dfrac{V_{ABCD}}{S_{ABC}} = 3 \cdot \dfrac{V_{ABCE}}{S_{ABC}}$

with

$$V_{ABCD} = \frac{1}{3 \cdot L_{AB}} \cdot \sqrt{\left(\frac{1}{2} \cdot L_{AB} \cdot L_{CD} \cdot \sin\alpha + S_{ABC} + S_{ABD}\right) \cdot \left(-\frac{1}{2} \cdot L_{AB} \cdot L_{CD} \cdot \sin\alpha + S_{ABC} + S_{ABD}\right) \cdot \left(\frac{1}{2} \cdot L_{AB} \cdot L_{CD} \cdot \sin\alpha - S_{ABC} + S_{ABD}\right) \cdot \left(\frac{1}{2} \cdot L_{AB} \cdot L_{CD} \cdot \sin\alpha + S_{ABC} - S_{ABD}\right)}$$

It should be noted that V_{ABCD} represents the volume encompassed by the points A, B, C et D, V_{ABCE} represents the volume encompassed by the points A, B, C and E and S_{ABC} represents the plane surface encompassed by the points A, B and C.

Tensorial methodology

In order to express the distance, we use the two points H_1 and H_2, such that H_1 is a running point on the plane [ABC] and H_2 a running point on the straight-line [DE].

In the base $\left(\overrightarrow{ab}, \overrightarrow{ac}, \overrightarrow{de}, \overrightarrow{h_1 h_2}\right)$ the tensor is expressed like this

$$G = \begin{pmatrix} \vec{ab} \cdot \vec{ab} & \vec{ab} \cdot \vec{ac} & \vec{ab} \cdot \vec{de} & \vec{ab} \cdot \overrightarrow{h_1 h_2} \\ \vec{ac} \cdot \vec{ab} & \vec{ac} \cdot \vec{ac} & \vec{ac} \cdot \vec{de} & \vec{ac} \cdot \overrightarrow{h_1 h_2} \\ \vec{de} \cdot \vec{ab} & \vec{de} \cdot \vec{ac} & \vec{de} \cdot \vec{de} & \vec{de} \cdot \overrightarrow{h_1 h_2} \\ \overrightarrow{h_1 h_2} \cdot \vec{ab} & \overrightarrow{h_1 h_2} \cdot \vec{de} & \overrightarrow{h_1 h_2} \cdot \vec{de} & \overrightarrow{h_1 h_2} \cdot \overrightarrow{h_1 h_2} \end{pmatrix} = \begin{pmatrix} 1 & \cos(\alpha) & \cos(\alpha + \varepsilon \cdot \beta) & 0 \\ \cos(\alpha) & 1 & \cos(\varepsilon \cdot \beta) & 0 \\ \cos(\alpha + \varepsilon \cdot \beta) & \cos(\varepsilon \cdot \beta) & 1 & 0 \\ 0 & 0 & 0 & 1 \end{pmatrix}$$

The distance is then expressed simply as being the norm of the vector $\overrightarrow{H_1 H_2}$, we thus obtain the following expression $dist([ABC],[DE]) = \left\| \overrightarrow{H_1 H_2} \right\|$

In order to have the full expression of this constraint, a constraint of coincidence must be added (point ↔ plane, distance = 0) between H_1 and the plane [ABC] and a constraint C4 (point ↔ line, coincidence) between the point H_2 and the straight-line [DE].

- Expression of the constraint C13 point ↔ line, parallelism, distance

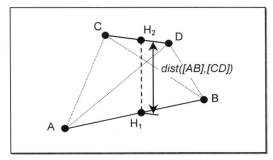

Figure 7 : Constraint C13

Given two straight-lines [AB] and [CD], H_1 a point belonging to the straight-line [AB] and H_2 a point belonging to the straight-line [CD].

<u>Affine methodology</u>

The constraint C13 is expressed like this $dist([AB],[CD]) = \dfrac{2 \cdot 3 \cdot V_{ABCD}}{L_{AB} \cdot L_{CD} \cdot \sin([AB],[CD])}$

with $\sin([AB],[CD]) = \dfrac{\sqrt{4 \cdot L_{AB}^2 \cdot L_{CD}^2 - ((L_{AC} + L_{BD}) - (L_{AD} + L_{BC}))^2}}{2 \cdot L_{AB} \cdot L_{CD}}$

it should be noted that V_{ABCD} represents the volume encompassed by the points A, B, C et D, L_{AB} represents the length of the segment linking the points A and B and L_{CD} represents the length of the segment linking the points C and D.

<u>Tensorial methodology</u>

In order to express the distance, we use the two points H_1 and H_2, such that H_1 is a running point on the straight-line [AB] and H_2 a running point on the straight-line [CD]. In the base $\left(\vec{ab}, \vec{cd}, \overrightarrow{h_1 h_2} \right)$ the tensor can be expressed like this

$$\begin{pmatrix} \overrightarrow{ab}\cdot\overrightarrow{ab} & \overrightarrow{ab}\cdot\overrightarrow{cd} & \overrightarrow{ab}\cdot\overrightarrow{h_1h_2} \\ \overrightarrow{cd}\cdot\overrightarrow{ab} & \overrightarrow{cd}\cdot\overrightarrow{cd} & \overrightarrow{cd}\cdot\overrightarrow{h_1h_2} \\ \overrightarrow{h_1h_2}\cdot\overrightarrow{ab} & \overrightarrow{h_1h_2}\cdot\overrightarrow{cd} & \overrightarrow{h_1h_2}\cdot\overrightarrow{h_1h_2} \end{pmatrix} = \begin{pmatrix} 1 & \cos(\alpha) & 0 \\ \cos(\alpha) & 1 & 0 \\ 0 & 0 & 1 \end{pmatrix}$$

The distance is then expressed simply as being the norm of the vector $\overrightarrow{H_1H_2}$, we thus obtain the following expression $dist([AB],[CD]) = \left\| \overrightarrow{H_1H_2} \right\|$

In order to have the complete expression of this constraint, the two constraints C4 (point ↔ line, coincidence) must be added between the point H_1 and the straight-line [AB] and between the point H_2 and the straight-line [CD].

6.4 Information model for a geometric entity:

The data structure which will serve as a support for our model is presented below. This structure defines what a geometric object is (GEOMETRIC ENTITY) as well as an association of several geometric objects (GEOMETRIC ENTITY NODE). We have copied this part of the structure onto the TTRS structure already presented on several occasions which you will be able to find in previous articles.

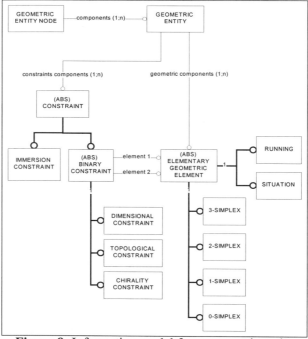

Figure 8: Information model for a geometric entity

However, we have put new entities in place:

The (CONSTRAINT) entity is subdivided into 4 sub-types which are the immersion, dimensional, topological and chiral constraints. In this article we are presenting dimensional constraints, without dwelling on the other 3 types of constraints, due to lack of space. The (ELEMENTARY GEOMETRIC ENTITY) entity is sub-divided into 4 sub-types which are the 0-simplex, the 1-simplex, the 2-simplex and the 3-simplex, which corresponds to an association of 1, 2, 3 or 4 points. Parallel to this first subdivision, the (ELEMENTARY GEOMETRIC ENTITY) entity is specialised in two sub-types which enable us to stipulate whether the (ELEMENTARY GEOMETRIC ENTITY) geometric entity is a **situation** element (these are EGRMs) or a **running** element belonging, for example to the surface of the geometric object represented.

This specialisation allows the tangency constraints between two surfaces to be defined by using the same geometric constraints as those used to define the relative position of MRGE between them. As an example, in order to specify that two circles C1 and C2 are tangents, we write a first coincidence constraint between the running point of C1 and the running point of C2, a second coincidence constraint between the tangent plane of the running point of C1 and the tangent plane of the running point of C2. See the illustration below.

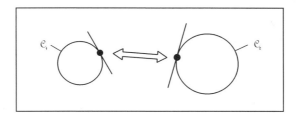

Figure 9: tangency constraint between two circles C_1 and C_2

6.5 General expression of a geometric tolerancing specification:

The objective of this paragraph is to demonstrate up our capacity to describe the form, dimensioning and tolerancing of a geometric entity using only our 13 constraints.

To do this, we have taken as a demonstration example a complex geometric form of the Béziers square type, positioned in relation to a reference system composed of 2 perpendicular planes that are the subject of a localisation tolerance in relation to this datum.

The first stage in this declaration consists of defining the toleranced surface itself. To do this, we will use the De Casteljau method.

Let P_0, P_1, P_2, and P_3 be the 4 poles of the curve; their relative positions in relation to the reference system are assumed to be known.

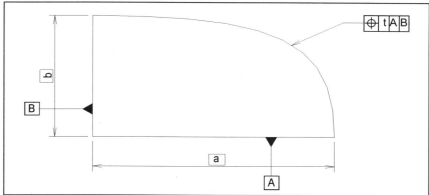

Figure 10: Specification of geometric tolerancing

We then simply have to declare the position of the sub-poles P_4, P_5, P_6, P_7, P_8 and, finally, of the point M which is the « running point » of the curve.

To do this, we only need to use one type of constraint the C2 distance point $\leftrightarrow$ point constraint,

For example, the relative declaration at point P_4 is as follows:

- C2 $\rightarrow$ d(P_0,P_1) = 1
- C2 $\rightarrow$ d(P_0,P_4) = x
- C2 $\rightarrow$ d(P_4,P_1) = 1 - x

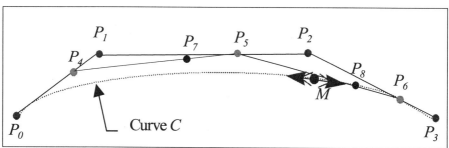

Figure 11: De Casteljau construction

The same type of declaration is used for all the other points P5, P6....M.

The second stage of this declaration consists of defining the relative position of the 2 reference surfaces, together with the position of the toleranced surface in relation to the reference system.

To do this, we have to:

- declare a constraint C7 with an angle $\alpha = \pi/2$ between the 2 planes A and B,
- declare 2 C3 distance point $\leftrightarrow$ plane constraints between each of the poles of the curve and the 2 planes of the reference system.

Finally, the third and last stage of this declaration consists of defining the boundaries of the tolerance zone by using 2 additional running points M_1 and M_2, by declaring 2 type C2 constraints, such that:

$$d(M_1,M) = d(M_2,M) = t/2$$

together with a perpendicularity constraint C13 between the line passing through M1 and M2 and the line passing through P_7 and P_8.

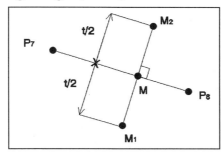

Figure 12: Declaration of the tolerance zone

7. CONCLUSION

In this article, we have tried to state the theoretical causes for globally inconsistent dimensioning and/or tolerancing of a mechanical part.

We have then developed a general and consistent method of dimensional specification, based on the use of the metric tensor, together with related properties and operations.

Finally, we have proposed a general structure for an information model relative to the geometric entity.

We consider this set of results as a sufficiently firm theoretical basis for the development of a new CAD/CAM geometric model that guarantees the global consistency of dimensioning and tolerancing.

8. BIBLIOGRAPHY

[BJO, 1995] BJORKE, O. ; "Manufacturing System Theory" ; TAPIR PUBLISHER 1995, N-7034, Trondheim, Norvège

[BOU et al., 1995]   BOUMA W., FUDOS I., HOFFMANN C. M., CAI J., PAIGE R., « Geometric constraint solver », Computer-Aided Design, Vol. 27, N°6, pp. 487-501, 1995.

[CAP et al., 1996]CAPOYLEAS V., CHEN X., HOFFMANN C. M., « Generic Naming in Generative, Constraint-Based Design », Computer-Aided Design, Vol. 28, N°1, pp. 17-26, 1996.

[CAR, 1925] CARTAN, E. ; "Leçons sur la géométrie des espaces de Riemann" ; 2° Edition Gauthier-Villars PARIS 1946 ; Ré-Edition. Jacques GABAY, PARIS 1988. ISSN 0989-0602.

[CLE et al., 1994] CLEMENT, A. ; RIVIERE, A. ; TEMERMAN, M. ; "Cotation Tridimensionnelle des Systèmes Mécaniques" ; HERMES, 1994.

[CLE, 1997] CLEMENT, A. ; "Livre blanc sur la conception routinière" ; Revue internationale de CFAO et d'informatique graphique, Vol. 12 ; N°5 pages 471 à 496 - octobre 1997.

[CON, 1996] CONSTANT, D. ; "Contribution à la spécification d'un modèle fonctionnel de produits pour la conception intégrée des systèmes mécaniques" ; Thèse Doctorat. Université J Fourier. Grenoble. France. Sept 1996.

[HAR, 1987] HAREL, D. ; "Statecharts : a visual formalism for complex systems" ; Journal of computer programming, vol. 8, n°1, pp 231-274, 1987.

[KOT et al., 1991] KOTA, S. ; WARD, A.C. ; "Functions, structures and constraints in conceptual design" ; Design Laboratory, Department of Mechanical Engineering and Applied Mechanics, University of Michigan, Ann Arbor.

[MAN, 1993] MANDORLI, F. ; "A reference kernel model for feature-based CAD systems supported by conditional attributed rewrite systems" ; 2nd ACM Solid Modeling 1993 Montréal Canada

[ODE et al., 1997] ODELL, J. ; RAMACKERS, G. ; "Toward a formalization of OO analysis", Journal of Object-Oriented programming, vol. 10, n° 4, pp 64-68, 1997.

[PAI, 1995] PAIREL, J. ; "Métrologie fonctionnelle par calibre virtuel sur MMT" ; Université de Savoie. Nov 1995.

[RAN et al., 1996] RANTA, M. ; MANTYLA, M. ; UMEDA, Y. ; TOMIYAMA, T. ; "Integration of functional and feature-based product modelling - the IMS/GNOSIS experience", Computer-Aided Design, Vol 28, N° 5, pp.371-381, 1996.

[SAU, 1997] SAUCIER, A. ; "Un modèle multi-vues du produit pour le développement et l'utilisation des systèmes d'aide à la conception en ingénierie mécanique" ; Thèse de doctorat de l'Ecole Normale Supérieure de Cachan, Juin 1997.

[SCH et al., 1994] SCHENCK, D. ; WILSON, P. ; "Information Modelling : the Express way" Oxford University Press 1994.

[SMY, 1977] SMYTH, M. ; "Effectively given domains" ; Theoritical Computer Science, 5, pp 257-274, 1977.

[TIC, 1994] TICHKIEVITCH, S. ; "De la CFAO à la conception intégrée" ; International Journal of CADCAM and Computer Graphics, HERMES, Vol, 9, n°5, 1994.

[TOM et al., 1993] TOMIYAMA, T. ; UMEDA, Y. ; YOSHIKAWA, H. ; "A CAD for Functional Design", Annals of the CIRP, Vol.42/1,pp143-146, 1993.

Optimising Tolerance Allocation for Mechanical Assemblies Considering Geometric Tolerances Based on a Simplified Algorithm

Ming-Shang Chen and Ken W. Young
School of Engineering
University of Warwick, Coventry CV4 7AL, UK
M-S.Chen@warwick.ac.uk

Abstract: In order to solve tolerance allocation optimisation problems involving geometric tolerances, almost all existing studies use feature-based representation techniques and apply genetic algorithms. This paper proposes an alternative, gradient-based optimisation technique which applies a simplified algorithm. Through the implementation of an illustrated example, it is shown that the efficiency is remarkably high. This approach is also effective for the general cases of nonlinear and large-scale tolerance allocation optimisation problems.
Keywords: tolerance allocation, geometric tolerance, nonlinear programming

1. INTRODUCTION

Tolerance allocation plays a very important role in the design and manufacture phases of a product life-cycle. It accommodates the function and quality of products to their economically efficient production. In the last three decades, numerous studies have been conducted into the area of tolerance allocation techniques. Most of them have used various mathematical programming methods to solve the optimal tolerance allocation problems for dimensional tolerances by minimising the manufacturing cost. In these approaches, gradient-based optimisation techniques are used to solve the optimisation problems. On the other hand, less research has been devoted to optimising tolerance allocation for geometric tolerances. Attempts to optimally allocate geometric tolerances have tended to employ genetic algorithms because of the difficulty involved in evaluating the gradients for the optimisation solution as required in mathematical programming methods [Nassef et al., 1993, 1995, 1997; Kanai et al., 1995]. Other researchers [Lu et al., 1991; Willhelm et al., 1992] have used the concept of tolerance primitives to specify relationships between features and to allocate geometric tolerances.

Thus far, gradient-based mathematical programming approaches have not generally been applied for the optimal allocation of both dimensional and geometric tolerances in mechanical assemblies. When geometric tolerances are involved in tolerance allocation problems, feature-based representation techniques are always used to derive the

27

constraint conditions, and then genetic algorithms are employed to solve the constrained optimisation problems. Although genetic algorithms have the advantage of being able to solve problems without calculating gradients (as is necessary for cost-tolerance optimisation problems), the computation of solutions is considerably time-consuming, which means that the efficiency is much lower than that of mathematical programming. Moreover, current studies have contributed to solving optimisation problems which satisfy only one or two functional requirements. More complicated mechanical assemblies involve perhaps dozens of functional requirements which must be converted into constraint functions for formulating tolerance allocation optimisation problems. This raises serious questions about the efficiency of applying genetic algorithms in such cases.

The present paper proposes an approach to tolerance allocation problems involving geometric tolerances with the use of a simplified algorithm. In essence, the algorithm is based on Lagrangian necessary conditions and can obtain the global optimum solution very efficiently. In order to derive the constraint functions of the optimisation problems, geometric tolerances are modeled according to current standards [BS 308, 1990] and are based on two-dimensional environments. In this paper, the geometric derivations, like the dimensions, are expressed by chain links. Consequently, the geometric tolerance constraint conditions can easily be derived as those of dimensional tolerances. An example is used to illustrate the optimal tolerance allocation for this problem.

2. GENERALISED TOLERANCE ALLOCATION OPTIMISATION PROBLEMS

For a mechanical assembly that has m assembly dimensions (sum dimensions) and n component dimensions, there will be m constraint equations including p equality and $m-p$ inequality ones for tolerance allocation optimisation problems. Let x_i and $Z_j(\mathbf{x})$ denote the component dimensions and their constituted assembly dimensions respectively, where $\mathbf{x} = [x_1, x_2, ..., x_n]^T$, $i=1,2,...,n$ and $j=1,2,...,m$. Based on minimising the total manufacturing cost $C(\mathbf{T}) = \sum_{i=1}^{n} C_i(T_{x_i})$, where $C_i(T_{x_i})$ and T_{x_i} are the manufacturing cost and tolerance of x_i, respectively, a constrained optimisation problem of tolerance allocation can be generalised as follows (Chen, 1995).

$$Min\ C(\mathbf{T}) \tag{1}$$

subject to

$$F_j(\mathbf{x}) = Z_j(\mathbf{x}) - \delta_j = 0 \qquad\qquad j = 1,2,...,p \tag{2}$$

$$F_j(\mathbf{x}) = \begin{cases} Z_j(\mathbf{x}) - \delta_j \geq 0 & if\ Z_j(\mathbf{x}) \geq \delta_j \\ \delta_j - Z_j(\mathbf{x}) \geq 0 & if\ Z_j(\mathbf{x}) \leq \delta_j \end{cases} \qquad j = p+1,p+2,...,m \tag{3}$$

$$T_{x_i} > 0 \qquad\qquad i = 1,2,...,n \tag{4}$$

where δ_j is the given design value of the assembly dimension and $F_j(\mathbf{x})$ represents the jth constraint function, which may be linear or nonlinear.

The above optimisation problem can be transformed into a quadratic form in terms of standard deviations $\boldsymbol{\sigma_x} = [\sigma_{x_1}, \sigma_{x_2}, \cdots, \sigma_{x_n}]^T$, where σ_{x_i} is the standard deviation of x_i. Based on the assumption of random assembly, all component dimensions are independent. By expanding the first order Taylor series of $Z_j(\mathbf{x})$ about the mean value $\overline{\mathbf{x}}$ and in turn taking the expectation and variance of $Z_j(\mathbf{x})$, the transformed optimisation problem is of the following form:

$$Min\ C(\boldsymbol{\sigma_x}) \tag{5}$$

subject to

$$f_j(\boldsymbol{\sigma_x}) = \sum_{i=1}^{n} [\frac{\partial Z_j(\mathbf{x})}{\partial x_i}]_{\overline{x}_i}^2 \sigma_{x_i}^2 - \sigma_{Z_j}^2 = 0 \qquad\qquad j = 1,2,...,m \tag{6}$$

$$\sigma_{x_i} > 0 \qquad\qquad i = 1,2,...,n \tag{7}$$

Where σ_{Z_j} is the standard deviation of $Z_j(\mathbf{x})$ and $f_j(\boldsymbol{\sigma_x})$ is the jth transformed constraint function.

Equations (5)-(7) are available not only for dimensional tolerances but also for geometric tolerances in mechanical assemblies. However, the aspects of the accumulation of geometric tolerances should be different from those of dimensional tolerances. It is necessary to model the accumulation of geometric tolerances for the derivation of constraint functions in the optimisation problems.

3.   THE ACCUMULATION OF GEOMETRIC TOLERANCES

Besides size, the geometric controls for features in mechanical components include form, orientation, location and run-out, etc. In addition to dimensional tolerances, the major deviations in geometry, such as parallelism, perpendicularity, position, coaxiality and run-out, will be taken into account in this paper.

3.1 Orientation Tolerances

Figure 1(a) shows a simple assembly consisting of two components. It is clear that the assembly dimension (sum dimension) Z_2 is an arithmetic sum of component dimensions x_3 and x_4. On the other hand, the assembly parallelism (sum parallelism) Z_1 can be derived according to the geometric relations between features shown in Figure 1(b), and is the parallelism of the top surface with respect to the datum surface S_1 after assembly. The parallelism of each feature is regarded as a design variable such as x_1 and x_2. Thus, the assembly parallelism can be derived according to the following linear sum.

$$Z_1 = \ell_2 \sin[\tan^{-1}(x_1 / \ell_1)] + x_2 \cos[\tan^{-1}(x_1 / \ell_1)] \tag{8}$$

$$\approx (\ell_2 / \ell_1)x_1 + x_2 \tag{9}$$

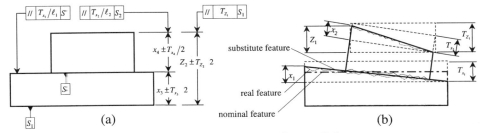

Figure 1; *Modeling of assembly parallelism*

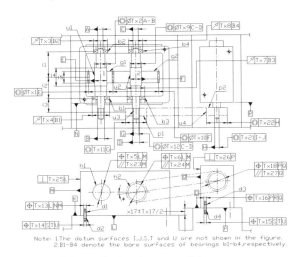

Note: 1.The datum surfaces I,J,S,T and U are not shown in the figure.
2.B1–B4 denote the bore surfaces of bearings b1–b4,respectively.

Figure 2; *A schematic drive system*

Similarly, for an n-component assembly, the assembly parallelism Z is:

$$Z = \sum_{i=1}^{n-1} \left(\ell_n / \ell_i \right) x_i + x_n \tag{10}$$

Equation (10) can also be applied to evaluate the assembly parallelism and perpendicularity when there are hybrid geometric relations of parallelism and perpendicularity between features. Based on Equation (10), all the assembly parallelism and perpendicularity in mechanical assemblies can be evaluated, and in turn the constraint functions can be derived to formulate the optimisation problem. In general, orientation tolerances in mechanical components would be manufactured to approximate the triangular distribution. Moreover, dimensional tolerances are considered to be normally distributed in mass production.

3.2 Location and Run-Out Tolerances
Figure 2 shows a schematic drive system consisting of a gear box and a spindle unit which are mounted on different bases and located with dowel pins. This simple drive system is

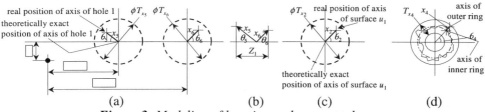

***Figure 3**; Modeling of location and run-out tolerances*

used to model location and run-out tolerances for optimising tolerance allocation. In addition, two-dimensional views are used to interpret the tolerances defined in the current standards [BS 308, 1990]. As shown in Figure 3(a), the positional deviations of the axes of the two housing holes h_1 and h_2 in the gear box shown in Figure 2 can be regarded as eccentricity variables which deviate from the theoretically exact positions of the axes but are limited to the tolerance zones. The behaviour of eccentricity caused by positional deviations has already been described in detail [Bjorke, 1989]. An important conclusion, drawn by Bjorke, is that an eccentricity variable generally follows Rayleigh distribution, which will be applied later. Figure 3(b) shows the deviation of distance of the two axes, which is a nonlinear sum as follows:

$$Z_1 = x_5 \cos\theta_5 + x_6 \cos\theta_6 \tag{11}$$

The above discussion holds for the dowel-pin holes (e.g. d1-d4 in Figure 2) and also for the coaxiality and run-out tolerances. For example, Figure 3(c) shows the model of coaxiality tolerance of the cylindrical surface u_1 on shaft 1, fitting with the bore of gear g_1 as shown in Figure 2. The circular run-out tolerance of bearing b_1 is modeled as shown in Figure 3(d). Furthermore, in the case of a bearing, the run-out tolerance of the outer ring with respect to the axis of the inner ring is regarded as twice the axis deviation of the two rings. Following the established models, the nonlinear sum of location and run-out deviations can then be derived.

3.3 Deviation of Distance and Parallelism of Two Axes

In general, it is necessary to derive constraint functions case by case for different mechanical assemblies. However, this derivation should be easy and regular if the functional requirements are focused on controlling the orientation and location deviations for mechanical assemblies, such as drive systems, which are definitely needed in machines. This paper illustrates this particular case, but the arguments are also applicable to more general cases. For the drive system shown in Figure 2, it is necessary to control the design functions, such as the backlash and contact of gear teeth, in order to maintain good driving capability and achieve an acceptable noise level. Accordingly, the functional requirements of this system are the deviation of distance between the two pitch cylinder axes of gears g_1 and g_2, and the parallelism of the axis of gear g_2 with respect to that of gear g_1. In addition, the same conditions required for pulleys p_1 and p_2 are also considered as part of the functional requirements.

In order to formulate the assembly tolerances based on the above functional requirements, chain links are introduced to represent the variables, including scalar and vector links. Figure 4(a) shows the chain of deviation of distance between the two pitch cylinder axes of gears g_1 and g_2, and Figure 4(b) shows that between pulleys p_1 and p_2, where all are vector links except the scalar link x_{17} in the chains. This scalar link comes from the dimensional tolerance controlling the different sizes of the two bases shown in Figure 2. Assuming that the dowel pins are tightly fitted with holes, and neglecting the deviations of clearances between gears and shafts in Figure 2, the deviations of distance $Z_1(\mathbf{x})$ in Figure 4(a) and $Z_2(\mathbf{x})$ in Figure 4(b) are derived as follows.

$$Z_1(\mathbf{x}) = \sum_{i=1}^{10} A_i x_i \cos \theta_i \tag{12}$$

$$Z_2(\mathbf{x}) = \sum_{i=6}^{8} A_i x_i \cos \theta_i + \sum_{i=11}^{12} A_i x_i \cos \theta_i + \sum_{i=18}^{22} A_i x_i \cos \theta_i +$$
$$\left[\sum_{i=13}^{16} A_i x_i \cos \theta_i \right] \cos \phi + [A_{17} x_{17}] \sin \phi \tag{13}$$

Similarly, the assembly parallelisms $Z_3(\mathbf{x})$ in Figure 4(c) and $Z_4(\mathbf{x})$ in Figure 4(d) are:

$$Z_3(\mathbf{x}) = \sum_{i=1}^{4} A_i x_i \cos \theta_i + \sum_{i=7}^{10} A_i x_i \cos \theta_i + \sum_{i=23}^{24} A_i x_i \tag{14}$$

$$Z_4(\mathbf{x}) = \sum_{i=7}^{8} A_i x_i \cos \theta_i + \sum_{i=11}^{12} A_i x_i \cos \theta_i + \sum_{i=19}^{22} A_i x_i \cos \theta_i + \left[\sum_{i=24}^{27} A_i x_i \right] \cos \phi \tag{15}$$

The coefficients A_i in the above equations signify how the geometric deviation of an assembly feature is related to that of the individual feature in a chain. They will be evaluated according to the location or size of a feature in an assembly. For example, if the tolerance value of the assembly parallelism in Equation (14) is applied to the width of gear g_1 in Figure 2, then the coefficient A_2 is ℓ_5/ℓ_4, but it is equal to 1 in Equation (12). In addition, A_7 in Equation (12) is equal to $\ell_1/(\ell_1 + \ell_2)$, but it is equal to $(\ell_1 + \ell_2 + \ell_3)/(\ell_1 + \ell_2)$ in Equation (13).

4. OPTIMISATION PROBLEMS WITH GEOMETRIC TOLERANCE CONSTRAINTS

4.1 Formulation of the Optimisation Problem

Based on the above modeling, a generalised assembly geometric deviation of features in a mechanical assembly can be expressed as:

$$Z_j(\mathbf{x}) = \sum_{i=1}^{q} A_i x_i + \sum_{i=q+1}^{r} A_i x_i \cos \theta_i + Y_j(x_{r+1}, x_{r+2}, ..., x_n) \tag{16}$$

where q denotes the number of dimension and orientation deviations and r is the number of location and run-out deviations. Y_j represent the nonlinear deviations caused by the deviations of features which do not belong to those in the first two terms of the equation.

Generally, the assembly geometric deviation $Z_j(\mathbf{x})$ in Equation (16) can be substituted into Equation (6) to obtain a transformed constraint function. Note that

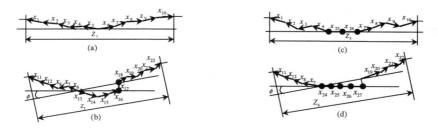

Figure 4; *Chain links of geometric deviations*

Equation (6) is derived from the first order Taylor approximation. It is not suitable for the substitution of the second term in Equation (16) because the term contains a product of stochastic variables x_i and $\cos 6_i$, and 6_i is uniformly distributed in the range between 0 and 2π [Bjorke, 1989]. The expectation and variance of $\cos 6_i$ have been demonstrated to be equal to 0 and 0.5, respectively. Therefore, the transformed quadratic constraint function is derived as:

$$f_j(\boldsymbol{\sigma}_x) = \sum_{i=1}^{q} A_i^2 \sigma_{x_i}^2 + \sum_{i=q+1}^{r} A_i^2 B_i^2 \sigma_{x_i}^2 + \sum_{i=r+1}^{n} (\frac{\partial Y_j}{\partial x_i})_{\bar{x}_i}^2 \sigma_{x_i}^2 - \sigma_{Z_j}^2 = 0 \qquad (17)$$

where $B_i = \{0.5[1 + \gamma_{x_i}^2 (1 + \alpha_{x_i})^2]\}^{1/2}$. γ_{x_i} and α_{x_i} denote the confidence coefficient and shift factor of Rayleigh distribution, respectively. Since the interval between the upper and lower limits of a distribution is considered a confidence interval, the tolerance T_{x_i} can be expressed as $T_{x_i} = 2\gamma_{x_i}\sigma_{x_i}$. For Rayleigh distribution with confidence level of 99.73%, the following data can be obtained: $\gamma_{x_i} = 2.625$, $\alpha_{x_i} = -0.271$, and $B_i = 1.527$.

The optimisation problem with geometric tolerance constraints can thus be formulated as follows.

$$Min\ C(\boldsymbol{\sigma}_x) \qquad (18)$$

subject to

$$f_j(\boldsymbol{\sigma}_x) = \sum_{i=1}^{q} A_i^2 \sigma_{x_i}^2 + \sum_{i=q+1}^{r} A_i^2 B_i^2 \sigma_{x_i}^2 + \sum_{i=r+1}^{n} (\frac{\partial Y_j}{\partial x_i})_{\bar{x}_i}^2 \sigma_{x_i}^2 - \sigma_{Z_j}^2 = 0$$

$$j = 1,2,...,m \qquad (19)$$

$$\sigma_{x_i} > 0 \qquad\qquad i = 1,2,...,n \qquad (20)$$

A number of cost-tolerance models have been proposed [Dong et al., 1994]. In this paper, the manufacturing cost is considered to be related to design tolerances and will be estimated based on the final manufacturing process with the use of the empirical cost curves plotted in [Dong et al., 1994]. The subtotal cost of the processes prior to the final one is regarded as being a fixed cost in the total cost. In fact, as described in [Kanai et al., 1995], every manufacturer has a set of standard costs which can be used to fit the cost curves for a real design.

4.2 Simplified Algorithm

For the tolerance allocation optimisation problem in Equations (18)-(20), it is apparent that the objective and constraint functions are convex. As a result, it can be concluded that the solution of the optimisation problem exists and is the global optimum. A simplified algorithm for the solution is used to solve the problem, which may be described briefly as follows. For Equations (18)-(20), the Kuhn-Tucker necessary conditions for a feasible point (σ_x^*, λ^*) can be simplified as:

$$\nabla_{\sigma_x} L(\sigma_x^*, \lambda^*) = 0 \tag{21}$$

$$\mathbf{f}(\sigma_x^*, \lambda^*) = 0 \tag{22}$$

$$\sigma_x^* > 0 \tag{23}$$

where $\mathbf{f} = [f_1, f_2, ..., f_n]^T$, $\mathbf{0} = [0,0,...,0]^T$ and the Lagrangian $L(\sigma_x, \lambda) = C(\sigma_x) + (\lambda)^T \mathbf{f}(\sigma_x)$. In the algorithm, Equation (23) is excluded so that the Kuhn-Tucker necessary conditions become Lagrangian necessary conditions. These conditions construct $n+m$ nonlinear simultaneous equations. The system of equations can be solved using Newton's method. The iterative update solution in the numerical algorithm is as follows:

$$\begin{bmatrix} \sigma_x^{(k+1)} \\ \lambda^{(k+1)} \end{bmatrix} = \begin{bmatrix} \sigma_x^{(k)} \\ \lambda^{(k)} \end{bmatrix} - \begin{bmatrix} \nabla^2 L & (\nabla \mathbf{f})^T \\ \nabla \mathbf{f} & 0 \end{bmatrix} \begin{bmatrix} \nabla L^{(k)} \\ \mathbf{f}^{(k)} \end{bmatrix} \tag{24}$$

5. EXAMPLE

A drive system shown schematically in Figure 5 is used to illustrate the optimal tolerance allocation. It includes a drive unit, a gear box and a spindle unit which are mounted on different base blocks and located with dowel pins. In total, there are 59 geometric controls for component features, a part of which are listed in Table I. For the constraints, 13 functional requirements are considered in this system. They are:

1. deviations of distance between the pitch cylinder axes of gear-gear pairs ($Z_1 \sim Z_4$) and those of gear-pulley pairs (Z_5 and Z_6);
2. parallelism of the axes of gears with reference to the datum axes of gears ($Z_7 \sim Z_{10}$) and those of pulleys with reference to the datum axes of gears (Z_{11} and Z_{12});
3. circular run-out of the spindle of the spindle unit (Z_{13}).

   Suppose the following limits of the above deviations are required: $T_{x_1} = T_{x_2} = T_{x_3} = T_{x_4} = 0.08$, $T_{x_5} = T_{x_6} = T_{x_7} = T_{x_8} = 0.04$, $T_{Z_9} = T_{Z_{10}} = 0.3$, $T_{Z_{11}} = T_{Z_{12}} = 0.2$ and $T_{Z_{13}} = 0.02$. On the other hand, the manufacturing costs for the features are based on the following cost-tolerance curves, as explained above.

Model 1 $\quad C_i(T_{x_i}) = 40 + 0.6095 / T_{x_i}^2$ $\tag{25}$

Model 2 $\quad C_i(T_{x_i}) = 50 + 5.682 / T_{x_i}$ $\tag{26}$

Model 3 $\quad C_i(T_{x_i}) = 150 + 27.84 e^{-3.661 T_{x_i}}$ $\tag{27}$

Model 4 $\quad C_i(T_{x_i}) = 250 + 98.87 e^{-18.39 T_{x_i}}$ $\tag{28}$

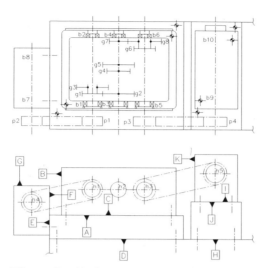

Figure 5; The drive system in the example

Table I; Related data of the example

Tolerance	Type	Feature	Datum	A_i		B_i	C-T model No	Optimal tolerance value
Tx_1	⊙	axis of pitch cylinder of g_1	bore of gear g_1	1	(in Z_1)	1.527	4	0.0594
Tx_2	⊙	shaft surface of gear g_1 for fitting	shaft surface for installing b_1 & b_2	1	(in Z_1)	1.527	5	0.0104
Tx_3	↗	outer surface of bearing b_2	bore of bearing b_2	0.266	(in Z_1)	1.527	4	0.1178
Tx_4	↗	outer surface of bearing b_1	bore of bearing b_1	0.873	(in Z_1)	1.527	4	0.0658
Tx_5	⊕	axis of hole h_1	surfaces A, B	1	(in Z_1)	1.527	3	0.0036
Tx_6	⊕	axis of hole h_2	surfaces A, B	1	(in Z_1)	1.527	3	0.0022
Tx_7	↗	outer surface of bearing b_3	bore of bearing b_3	0.873	(in Z_1)	1.527	4	0.0532
Tx_8	↗	outer surface of bearing b_4	bore of bearing b_4	0.266	(in Z_1)	1.527	4	0.0906
...	...	...	...	...		...	...	...
Tx_{23}	⊙	shaft surface of gear g_6 for fitting g_8	bore of gear g_6	1	(in Z_4)	1.527	5	0.0222
Tx_{24}	⊙	axis of pitch cylinder of gear g_8	bore of gear g_8	1	(in Z_4)	1.527	4	0.0328
Tx_{25}	//	axis of hole h_1	surface B	0.400	(in Z_5)	-	3	0.0116
Tx_{28}	⊙	axis of pitch cylinder of pulley p_2	bore of pulley p_2	1	(in Z_9)	1.527	2	0.1380
...	...	...	...	...		...	...	...
Tx_{48}	size	height between surfaces C and D	-	0.342	(in Z_{10})	-	1	0.3389
Tx_{49}	size	height between surfaces I and H	-	0.342	(in Z_{10})	-	1	0.3389
Tx_{52}	↗	outer surface of bearing b_{10}	bore of bearing b_{10}	1.291	(in Z_{11})	1.527	4	0.0094
...	...	...	...	...		...	...	...
Tx_{58}	⊥	surface K	surface J	0.376	(in Z_{12})	-	1	0.1983
Tx_{59}	//	axis of hole h_5	surface K	0.376	(in Z_{12})	-	3	0.0922

Model 5 $C_i(T_{x_i}) = 220 + 71.24e^{-4.348T_{x_i}}$ (29)

By applying the simplified algorithm, the solution of the optimisation problem is thus obtained. A part of the optimal tolerance values are listed in Table I. The total manufacturing cost is 13091.25. The solution needs 54 iterations and takes 4.69 seconds (CPU time) on a Pentium-Pro 233 PC. This shows that the efficiency is very high when the simplified algorithm approach is used to solve the tolerance allocation optimisation problem involving geometric tolerances.

6. CONCLUSION

In this paper, in addition to size, three major geometric controls - orientation, location and run-out - have been modeled on the basis of two-dimensional environments to derive the constraint functions for formulating the tolerance allocation optimisation problem. By applying the simplified algorithm, the solution of the optimisation problem is obtained and is the global optimum. Compared with the application of genetic algorithms, the proposed approach proves to be much more efficient. It is also very effective for solving nonlinear and large-scale optimisation problems constructed from more complicated mechanical assemblies. The approach thus offers an alternative solution for tolerance allocation optimisation problems involving geometric tolerances, and points to a useful direction for further development.

REFERENCES

[Bjorke, 1989] Bjorke, O.; *Computer-Aided Tolerancing*, Tapir Publishers, Trondheim, Norway

[BS 308, 1990] BS 308-1990; *Engineering drawing practice Part 3: Recommendations for geometrical tolerancing*; British Standards Institution, UK

[Chen, 1995] Chen, M.S.; "Optimizing tolerance allocation for a mechanical assembly with nonlinear multiple constraints"; *Journal of the Chinese Society of Mechanical Engineers*, Vol. 16, No. 4, pp. 349-361

[Dong et al., 1994] Dong, Z.; Hu, W.; Xue, D.; "New production cost-tolerance models for tolerance synthesis"; *Journal of Engineering for Industry*, Vol. 116, pp. 199-206

[Kanai et al., 1995] Kanai, S.; Onozuka, M.; Takahashi, H.; "Optimal tolerance synthesis by genetic algorithm under the machining and assembling constraints"; *Computer Aided Design*, Editor: Kimura, F., Chapman & Hall, London, pp. 235-251

[Lu et al., 1991] Lu, S.C.-Y.; Willhelm, R.G.; "Automatic tolerance synthesis: a framework and tools"; *Journal of Manufacturing Systems*, Vol. 10, No. 4, pp. 279-296

[Nassef et al., 1993] Nassef, A.O.; El Maraghy, H. A.; "Allocation of tolerance types and values using genetic algorithm"; *Proceedings of the 3rd CIRP Seminar on Computer Aided Tolerancing*, Cachan, France

[Nassef et al. 1995] Nassef, A.O.; El Maraghy, H. A.; "Statistical analysis and optimal allocation of geometric tolerances"; *Proceedings of the Computers in Engineering Conference and the Engineering Data base Symposium*, ASME, pp. 817-824

[Nassef et al., 1997] Nassef, A.O.; El Maraghy, H. A.; "Allocation of geometric tolerances: new criterion and methodology"; *Annals of the CIRP*, Vol. 46, No. 1, pp. 101-106

[Willhelm et al., 1992] Willhelm, R.G.; Lu, S.C.-Y.; "Tolerance synthesis to support concurrent engineering"; *Annals of the CIRP*, Vol. 41, No. 1, pp. 197-200

Functional and product specification
by Gauge with Internal Mobilities (G.I.M.)

DANTAN Jean-Yves, BALLU Alex
LMP – Laboratoire de Mécanique Physique
UPRES A 5469 CNRS – Université Bordeaux 1
351, Cours de la Libération
33405 TALENCE Cedex, France
Email: dantan@lmp.u-bordeaux.fr, ballu@lmp.u-bordeaux.fr

Abstract: With the systematic use of computers at all steps of product life cycle, especially at the level of design and manufacturing, a coherent model of product functional geometry is required. In this paper, we propose a model making it possible to build a virtual gauge with internal mobilities, and to specify a part, a mechanism or a manufacturing process.

The approach by gauge with internal mobilities establishes the domain of functionally acceptable variations of the non-ideal geometry, starting from three concepts:

- the interface gauge / part defining the relative position between the ideal features of the gauge and the non ideal features of the part,
- the gauge structure modeling the environment of the part, the environment is the parts influencing on the functional requirement,
- the functional characterization optimizing the geometrical characteristic related to the function.

Keywords: Tolerance specification, Functional specification, Product specification Tolerance modelling, Virtual Gauge.

1. INTRODUCTION

With the systematic use of computers at all steps of product life cycle, especially at the level of design and manufacturing, a coherent model of product functional geometry is required. The imperfections, inherent in the manufacturing processes, involve a degradation of the functional characteristics, and thus of the quality of the product. The part is seen as non-ideal surface, which is the physical interface of the part with its environment.

Our objective is to define the permissible geometrical variations of the parts. The permissible geometrical variations ensure a certain level of quality, which is defined by functional geometrical requirements. We will define the semantics of the specification model.

2. CLASSICAL GAUGE

The gauges of control, employed since nearly one century, have the advantage to control the relatively complex functional requirements like the assembly of parts between them.

The standards approach the geometrical tolerancing by gauge by defining three requirements. These requirements are the means of increasing the tolerances of form, orientation or position according to dimensions of the features concerned.

2.1. Envelope requirement

The envelope requirement [ISO 8015] exclusively expresses the assembly requirement of a shaft in a boring.

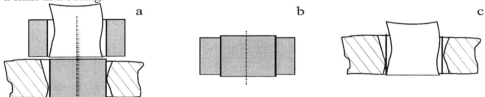

Figure 1: *Envelope requirement.*

It is evidenced by a relation between the size and the form of a feature. The non-ideal feature must fit in a half space that is delimited by the virtual perfect form at the maximum material (Fig.1a).

The assembly requirement between the two gauges (Fig.1b) ensures the assembly requirement between the two parts (Fig.1c).

For a single feature, either a cylindrical surface or a feature established by two parallel plane surfaces, the envelope requirement may be applied. For the more complex assemblies, the requirement will be done by the maximum material requirement.

2.2. Maximum material requirement

The maximum material requirement [ISO 2692] expresses the assembly requirement of two parts.

It is evidenced by the establishment of a relation between the size of a feature and its orientation or its position.

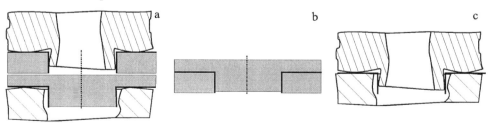

Figure 2: *Maximum material requirement.*

The assembly will be possible if the two parts must fit in a half space that is delimited by the virtual perfect form at the maximum material (Fig.2a). The assembly requirement is simulated by the worst parts (the state at maximum material). The assembly requirement between the two gauges (Fig.2b) ensures the assembly requirement between the two parts (Fig.2c).

2.3. Least material requirement

Least material requirement [ANSI Y 14.5] expresses the limitation of the gap between two parts. It is similar to maximum material requirement, simply we consider the state at least material. The limitation of the gap makes it possible for example to ensure a requirement of precision. It is necessary that whatever the relative position of these two parts, the requirement is respected. This limitation is simulated by the worst part (the state at least material).

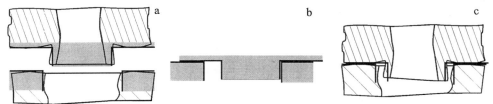

Figure 3: *Least material requirement.*

The gap will be limited if the two parts must belong to a half space that is delimited by the virtual perfect form at the least material (Fig.3a). The gap between the two gauges (Fig.3b) maximizes the gap between the two parts (Fig.3c).

2.4. Requirement formalization

The Virtual Boundary Requirements [Srinivassan, Jayaraman, 1989] generalise the maximum material requirement and the least material requirement. The surfaces must belong to the half spaces defined by the Virtual Boundary Requirement. The association of a half space and of an offset operation at each virtual boundary, generalizes the requirements. A mathematical formalism is presented with this generalization. This model finds like direct functional application the cases of the assembly and the limitation of the gap between two parts [Robinson, 1997].

Anselmetti illustrated the translation of the functional requirements of an elementary mechanism in geometrical specifications. The assembly requirement and the precision requirement are analyzed to define the maximum material requirement and the least material requirement using virtual boundary, location tolerance and projected tolerance zone [Anselmetti, 1995].

3. GAUGE WITH INTERNAL MOBILITIES (G.I.M.)

The main idea of the conventional gauge is to limit the geometrical variations of the part compared to its worst geometry.

The research of Anselmetti [Anselmetti, 1995], Ballu and Mathieu [Ballu, Mathieu,1997] is a first step towards more functionality.

In this paper, we propose a new concept of gauge allowing a geometrical specification, which give a better approach of the design intent. A mechanism must satisfy functions. These functions are provided by kinematic joints. Each part of the mechanism takes part in one or more functions through its structural and functional relations with the other parts of the mechanism [Constant, 1996].

We call environment of a part the set of the other parts of the mechanism.

In order to specify a part, its influence on the functional requirement must be limited. The main idea of the gauge with internal mobilities (G.I.M.) is to limit the geometrical variations of the part compared to the worst geometry of its environment.

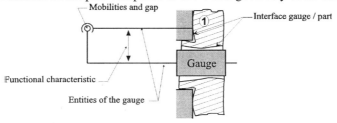

Figure 4: gauge with internal mobilities (G.I.M.).

For that, three new concepts are introduced, the concept of interface gauge / part, the concept of internal mobilities, and the concept of functional characteristic.

- The concept of the **interface gauge / part** define the relative position between the ideal features of the gauge and the non-ideal features of the part, simulating the contacts between the parts.
- The concept of internal mobilities modelizes the environment of the part. The environment is simulated by a **gauge structure** composed of several entities having mobilities and gap between them.
- The concept of **functional characteristic** gives a criteria of appreciation of the part with respect to the function.

The following paragraphs present these three concepts.

A mathematical formalism is presented in the form of geometrical constraints on the relative position between the non-ideal features of the part and the ideal features of the gauge [Ballu, Mathieu, 1993].

3.1. Interface gauge / part

The interface gauge / part simulates the contact with the close parts. We distinguish four types of contact: fixed contact, floating, quasi floating and slipping.

3.1.1. Fixed contact

A contact is regarded as fixed when the two parts in contact do not have a possible relative movement. The relative position of the parts can be constrained technologically in a given configuration. The contacting surfaces ensure a reference. This reference is adjusted on the non-ideal surface of the parts according to the technological constraints. An association between the ideal surface of the gauge and the non-ideal surface of the part modelizes this adjustment. Association operation identifies one or more ideal features from one or more non-ideal features. Association depends on criteria which are function of the technological constraints [Ballu, Mathieu, 1993].

Examples: A planar contact pair (Fig.5), a hooped assembly between a shaft and a boring (Fig.6).

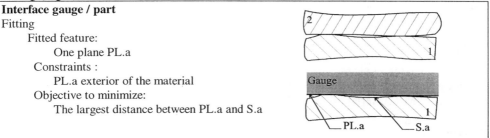

Interface gauge / part
Fitting
    Fitted feature:
        One plane PL.a
    Constraints :
        PL.a exterior of the material
    Objective to minimize:
        The largest distance between PL.a and S.a

Figure 5: *Fitting operation of a plane in the case of a planar contact pair.*

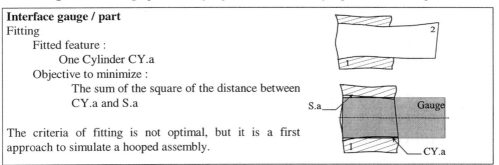

Interface gauge / part
Fitting
    Fitted feature :
        One Cylinder CY.a
    Objective to minimize :
        The sum of the square of the distance between
        CY.a and S.a

The criteria of fitting is not optimal, but it is a first approach to simulate a hooped assembly.

Figure 6: *Fitting operation of a cylinder in the case of hooping.*

3.1.2. Floating contact

A contact is regarded as floating when relative movements of the two parts allow a separation of surfaces. The gap must be limited in order to satisfy the functional requirement. Virtual boundary defines the worst limit of these gap. From where, virtual boundary between the ideal surface of the gauge and the non-ideal surface of the part modelize this type of contact. The virtual boundary constrained the non-ideal surface in a half space [Srinavassan, Jayaraman, 1989].

- For a assembly requirement or a minimal clearance requirement, the non-ideal geometry of the part is limited by the worst geometry of its environment, the state at the maximum material of the close parts. It is a limiting gauge [Pairel, 1995].

    Example : A bilateral planar contact pair (Fig.7)

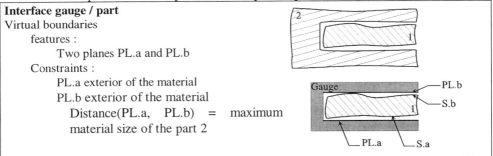

Interface gauge / part
Virtual boundaries
    features :
        Two planes PL.a and PL.b
    Constraints :
        PL.a exterior of the material
        PL.b exterior of the material
        Distance(PL.a, PL.b) = maximum material size of the part 2

Figure 7: Virtual boundaries, maximum material condition.

- For a maximum clearance requirement or maximum deviation requirement, the non ideal geometry of the part is limited by the worst geometry of its environment, the state at least material of the close parts [Ballu, Mathieu, 1997].

    Example: a cylindrical pair (Fig.8)

Interface gauge / part
Virtual boundary
    feature :
        One cylinder CY.a
    Constraints :
        CY.a exterior of the material
        Radius of CY.a = least material radius of the part 2

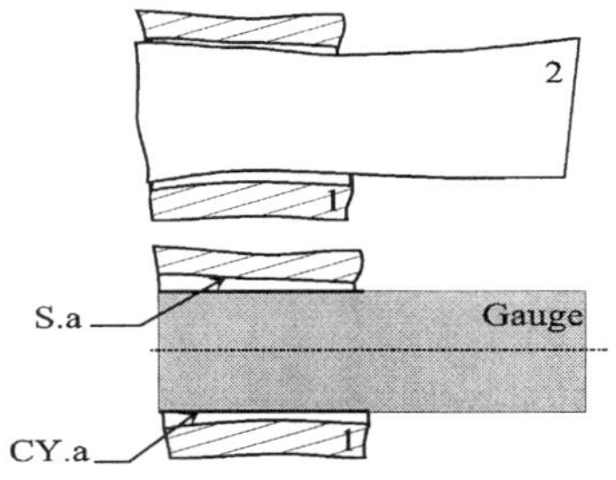

Figure 8: Virtual boundary, least material condition.

3.1.3. Quasi-floating contact

During its life cycle, a mechanism can have several operating modes. A floating contact can be constrained technologically in a configuration. It is a quasi-floating contact. A contact is regarded as quasi floating between two non-ideal surfaces when no relative movement is possible, but that the relative position depends on a gap.

The approach for this type of contact is similar to the floating contact, except for a requirement of maximum or minimal clearance where it is not constrained.

3.1.4. Slipping contact

A contact is regarded as slipping between two non-ideal surfaces when there can be only tangential relative movement and not normal relative movement. The contact of the two parts is constrained technologically by a mechanical action. A fitting between the ideal surface of the gauge and the non-ideal surface of the part modelizes this type of contact. The criteria of fitting is function of the mechanical action.

Example: The contact of a roller on a cam (Fig.9)

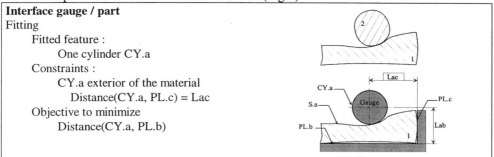

Interface gauge / part
Fitting
Fitted feature :
One cylinder CY.a
Constraints :
CY.a exterior of the material
Distance(CY.a, PL.c) = Lac
Objective to minimize
Distance(CY.a, PL.b)

Figure 9: *Fitting in the case of a cam.*

3.2. Gauge structure

The gauge structure modelizes the environment of the part. The environment of the part is composed of the parts influencing the functional requirement. These parts have mobilities between them; the gauge structure is composed of ideal geometrical entities also having mobilities with gap.

- With a part of the environment an entity of the gauge corresponds. The entity of the gauge is composed of:
    - TTRS [Clement et al., 1997],
    - ideal surfaces for the entities in contact with the specified part.

    The relative position of the geometrical ideal element is defined mathematically by constraints on the situation characteristics.

    It is possible to include position deviations and orientation deviations between geometrical ideal element.

- Mobilities between entities of the gauge, ensured by a TTRS, correspond to mobilities between the parts of the environment.

- Gap between entities of the gauge, i.e. between the geometrical elements, correspond to the worst gap between the parts. They represent the gap between the virtual boundaries of the parts.

    - For a requirement of assembly or minimal clearance, we generally modelize by the states at the maximum material, least gap.

    - For a requirement of maximum clearance or maximum variation, we generally modelize by the states at least material, maximum gap.

3.3. Functional characteristic

In the case of the classical gauge, the respect of the specification by a part characterizes its requirement like a Boolean. However, in design, it is essential to have coherence between what the product must be able to realize and what it realizes.

What it must be able to realize, is defined by a functional requirement. A functional requirement is an interval in which the value of a characteristic must be.

What it realizes, is defined by the value of a characteristic. This characteristic and its interval must be defined by the specification.

We integrate this concept in the functional specification by G.I.M. by defining a condition and a function objective.

Example: an angular precision (Fig.10)

Interface gauge / part	Gauge structure
Fitting	Gauge entity 2:
Fitted feature :	Features:
One plane PL.2b	One plane PL.2b
Constraints :	One cylinder CY.2a
PL.2b exterior of the material	One point PT.2c
Objective to minimize :	Constraints :
The largest distance between PL.2b and S.1b	Angle (PL.2b, CY.2a)=90°
	Distance (PT.2c, axis CY.2a)=L1
Virtual boundary	Distance (PT.2c, PL.2b)=L2
feature:	Gauge entity 3:
One cylinder CY.3e	Features:
Constraints :	One cylindre CY.3^e
CY.3e exterior of the material	One point PT.3c
Radius of CY.3e = $\phi D'$ mini	Constraint:
Virtual boundary	Distance (PT.3c, axis CY.3e)=L1
feature:	Mobilities and gap
One cylinder CY.2a	Constraint:
Constraints :	Distance (PT.3c, PT.2c)=J1
CY.2a exterior of the material	
Radius of CY.2a = ϕD mini	
Functional characteristic	
Fitting:	
Interface constraints and structure constraints	
Function objective to minimize :	
Angle(CY.2a, CY.3e)	
Condition :	
Angle(CY.2a, CY.3^e) $\leq \alpha$	

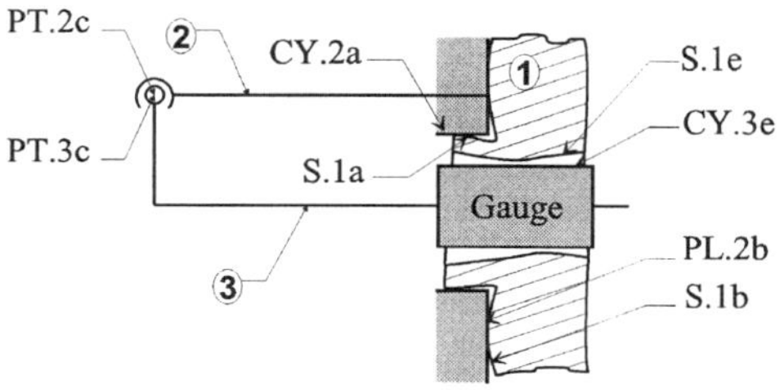

Figure 10: Functional characterization of an angular precision.

4. EXAMPLE

The example is the EAP part of acceleration to gunpowder of ARIANE 5. The structure of the EAP is mainly constituted of:

• The front skirt, (1)

- The motor to propergol, (4)
- The EAP is linked up to the main part (3) by two systems.

The main function of the EAP part is to propel ARIANE 5 in a direction. This function is expressed by a functional geometrical requirement: the angular deviation between the axis of the motor (4) and the axis of ARIANE 5 must be lower than α (Fig.11).

We are interested in the geometrical specification of the part 1:

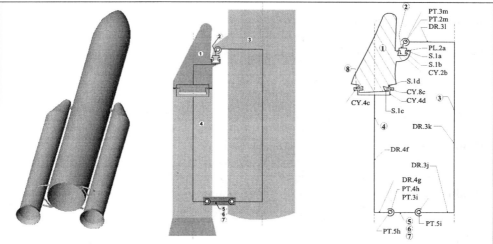

Interface gauge / part
Interface between the entity 2 and the part 1
> Planar contact pair, fixed contact between PL.2a and S.1a (fitting: exterior of the materiel and minimize the largest distance)
> Cylindrical pair, quasi-floating contact between CY.2b and S1.b (Virtual boundary, least material form of the part 2)

Interface between the entity 4 and the part 1
> Cylindrical pair, quasi-floating contact between CY.4d and S1.d (Virtual boundary, least material form of the part 4)

Interface between the entities 8 (180 radial pawns of positioning) and the part 1
> Cylindrical pair, quasi-floating contact between CY.8c and S1.c (Virtual boundary, least material form of the parts 8)

Gauge structure
The gauge structure is constituted of seven entities (2 – 8) with mobilities and gaps

Functional characteristic
Fitting:
> Interface constraints and structure constraints
> Function objective to minimize : Angle(DR.4f, DR.3k)

Condition :
> Angle(DR.4f, DR.3k) ≤ α

Figure 11: *Geometrical specification of the part 1.*

5. CONCLUSION

The geometrical specification by G.I.M. corresponds directly to the function of the part and can directly be controlled and be measured. During measurement the value of the functional characteristic can be quantified.

This approach is applicable to the manufacturing specification, it makes it possible to validate the process and to impose the optimal tolerances.

For the integration of process knowledge into design, this model can be easily used. It makes it possible to express geometrical specifications following various points of view: functional, metrological and manufacturing. The intents of the various actors can be taken into account by this model.

REFERENCES

[Anselmetti, 1995]   Anselmetti B.; Tolerancing method for function and manufacturing ; In: *Proc. Of ILCE 95*; Paris; France; Fevruary 1-3; 1995

[Ballu, Mathieu, 1993]   Ballu A. and Mathieu L., Analysis of dimensionnal and geometric specifications : Standard and model; In: *Proc. Of 3rd CIRP Seminar on Computer Aided Tolerancing*; Cachan; France; April 27-28; 1993

[Ballu, Mathieu, 1997]   Ballu A. and Mathieu L., Virtual gauge with internal mobilities for verification of functional specifications; In: *Proc. Of 5rd CIRP Seminar on Computer Aided Tolerancing*; Toronto; Canada; April 27-29; 1997

[Clement et al., 1997] Clement A., Riviere A., Serre P., Valade C., The TTRS: 13 constraints for dimensioning and tolerancing; In: *Proc. Of 5rd CIRP Seminar on Computer Aided Tolerancing*; Toronto; Canada; April 27-29; 1997

[Constant, 1996] Constant D., Contribution a la specification d'un modele fonctionnel de produits pour la conception integree de systemes mecaniques; Thesis; Universite Grenoble 1; France; October 3; 1996

[Pairel, 1996]   Pairel E., Metrologie fonctionnelle par calibre virtuel sur machine a mesurer tridimensionnelle; Thesis; Universite de Savoie; France; December 20; 1995

[Robinson, 1997] Robinson D.M., Geometric tolerancing for assembly with maximum material parts; In: *Proc. Of 5rd CIRP Seminar on Computer Aided Tolerancing*; Toronto, Canada; April 27-29; 1997

[Srinivassan, Jayaraman, 1989]   Srinivassan V. and Jayaraman R., Conditional tolerances; In: *IBM Journal of Research and Develpment*; Vol. 33; n° 2; pp 105-124; 1989

Development of a Computer-Aided Tolerancing System in CAD Environment

Jhy-Cherng Tsai and Wen-Wei Wang
Department of Mechanical Engineering
National Chung-Hsing University
250 Kuokuang Road
Taichung City, Taiwan 40227
The Republic of China
jctsai@mail.nchu.edu.tw

Abstract: Although mechanical designers can take the advantage of computer-aided tools to assist their tasks, yet tolerance design is still based on experience or trial and error methods. This paper is an effort to develop a computer-aided tolerancing system with systematic analysis in a CAD (Computer-Aided Design) environment. The developed software contains a geometric dimensioning and tolerancing (GD&T) database and five major modules that are (1) feature extraction module, (2) GD&T specification extraction module, (3) fitting assignment module, (4) tolerance network construction module, and (5) interface for tolerance analysis. The developed system assists the designer to evaluate tolerance accumulation of an assembly that provides a reference for tolerance refinement. The spindle assembly of a CNC turning center is also used as an example to illustrate the tolerancing system.
Keywords: computer-aided tolerancing, CAD, dimensioning and tolerancing, tolerance analysis.

1. INTRODUCTION

Tolerancing is one of the indices to evaluate product quality and cost. It plays a key role in product design and manufacturing. As tight tolerance assignment ensures design requirements in terms of function, quality and part exchange, it also induces more requirements in manufacturing, inspection and service that results in higher production cost. Tolerance design hence is usually a trade-off among activities involved in product life cycle.

Although today's designers can take the advantage of computer-aided tools to assist their tasks, yet tolerance design is still based on experience or trial and error methods. There is an immediate requirement to have computer-aided tools for tolerance design with systematical analysis. This paper is an effort to address this issue and developed a tolerancing software in a CAD (Computer-Aided Design) environment to assist designers in their tolerance assignment tasks.

Because dimensional tolerances are intuitive in tolerance assignment, linear

47

dimensional tolerance stack-up and computer-aided tools have been developed in early days such as n [Knappe, 1963] and in [Hillyard & Braid, 1978]. These tools, including linear tolerance stack-up and dimensional sensitivity analysis, are now also available either in individual software package or in many modern CAD systems [Turner & Gangoiti, 1991; ElMaraghy et al., 1991]. As geometric tolerances become more critical in modern product design [ISO, 1983], their analysis and synthesis, however, are still not available in these CAD systems. In this paper, we stress on handling the geometric dimensioning and tolerancing (GD&T) information in a CAD system and the usage of the information for tolerance analysis.

2. SYSTEM ARCHITECTURE

The first consideration to develop the tolerancing system is to select a proper CAD system that is commonly used with an open architecture and good application programming interface (API). Among many CAD systems we evaluated, AutoCAD and Solidworks were selected because the first one is widely used and the other one has an open architecture and reasonable API. The tolerancing system based on AutoCAD has been developed and reported [Tsai & Chang, 1997]. While AutoCAD supports limited solid modeling, we also developed a tolerancing system based on Solidworks. The developed system is composed of five major modules, *i.e.*, the feature extraction module, the GD&T specification extraction module, the fitting assignment module, the tolerance network construction module, and an interface for tolerance analysis.

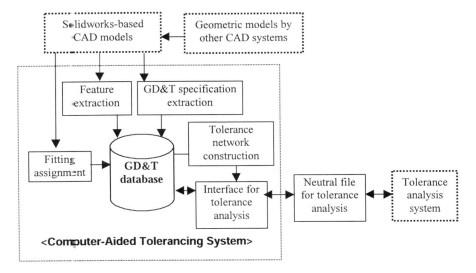

Figure 1; *System architecture of the developed computer-aided tolerancing system.*

Figure 1 shows the system architecture of the developed system where the CAD-based tolerancing system consists of five modules and the GD&T database. Geometric models developed in other CAD systems can be transformed into Solidworks for feature extraction and GD&T assignment. The feature extraction module extracts information of geometric features (simplified as features) from a boundary-representation-based geometric model. It also determines the coordinate associated with each feature and detects feature types for appropriate tolerance assignment. The GD&T specification extraction module deals with GD&T information of a design, including both dimensional tolerances and geometric tolerances. The fitting assignment module provides graphic user interface to assist the designer to assign fitting and tolerancing attributes for assembled parts. The tolerance network construction module constructs the tolerance network [Tsai & Cutkosky, 1997] of a design based on features and GD&T and fitting information. This network is intended for tolerance analysis supported by an existing software. The interface for tolerance analysis is designed to transform data structures of the network into the required format thus kinematic errors of a design can be analyzed at design time. It assists the designer to evaluate the effects of tolerance accumulation and provides a reference for tolerance refinement. In the following sections, we briefly discuss functions of the five modules and the GD&T database.

3. SYSTEM MODULES

3.1 Feature extraction module

Because GD&Ts are assigned associated with features, the developed system first extracts the information of features. Although it is still arguing of the classification of features, this system deals with certain features that are commonly used in GD&T assignment [ISO, 1983]. As shown in Table I, five types of features and their associated attributes are extracted from the embedded CAD system [Qamhiyah et al., 1996]. These extracted features are stored in a feature attribute database that is used in conjunction with the GD&T database to be discussed in the following section.

Table I; Types of feature used in the system.

Type of feature	Attributes extracted associate with the feature
Plane	feature coordinate and the normal vector of each plane
Cylinder	feature coordinate, the axis and radius of the cylinder
Cone	feature coordinate, axis of the cone, inclined angle, and the radius of the bottom surface
Sphere	feature coordinate and the radius of the sphere
Torus	feature coordinate, axis, and the inner and the outer radii

3.2 GD&T specification extraction module

GD&T specifications assigned to a part are often treated as annotations or graphic symbols in current CAD systems. To use these GD&T specifications in design analysis and synthesis, manufacturing, and inspection, one needs to extract such information. Our intention is to store the information in a database such that it can be reused for further use. To extract tolerancing information, we noticed that the representations for dimensional tolerances and geometric tolerances are different. Dimensional tolerances, including size tolerances such as diameter, radius and angular tolerances and distance tolerances, specified the allowances of size or distance variation. Geometric tolerances, on the other hand, limit the variations of relative location or shape of features. As a result, geometric tolerances can be classified as cross-referenced tolerances (CRTs) and self-referenced tolerances (SRTs). A dimensional tolerance, based on such classification, can be treated as an SRT. The data structures for storing CRTs and SRTs thus are designed differently. As GD&T specifications applies to features, these GD&T-related information are also extracted and stored.

Figure 2 illustrates the extraction and storage of the GD&T information. The perpendicularity shown in the figure is a CRT which specifies the allowable variation in perpendicularity of the planar surface with respect to datum surface A. The datum surface A is located with respect to the datum reference frame (DRF) of the part. As a result, the perpendicularity tolerance and its value 0.05mm are stored in the GD&T database with a pointer linked to the datum feature A. The plane feature, the datum feature A, the DRF, and frames associated with features are extracted and stored in the feature attribute database. The perpendicularity information in the GD&T database, in turn, is pointed back to the plane feature that recorded in the feature attribute database.

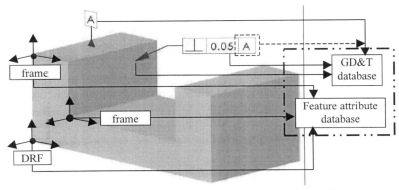

***Figure 2**; Extraction of GD&T specifications from CAD model.*

3.3 Fitting assignment module

In addition to dimensional and geometric tolerancing assignment, fitting and tolerancing assignment is close related to function requirement in design and therefore should be handled in the tolerancing system. Fitting assignment involves mating features between parts, *i.e.*, mating features should be of the same type for consistency. Parameters of the mating condition depend on the type of mating. As an example, figure 3 shows the mating features and required parameters for a rotational mating. Dimensional tolerances associated with each mating condition are extracted and stored in the GD&T database when a mating condition is assigned.

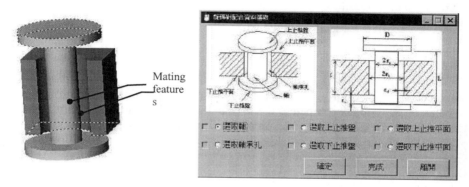

Figure 3; Illustration of mating features and parameters of rotational mating in fitting and tolerancing assignment.

3.4 Tolerance network construction module and interface for tolerance analysis

As one of the goals of the tolerancing system is to take GD&T information for tolerance analysis, we developed the tolerance network (TN) construction module and an interface for tolerance analysis. The TN is a graph with features and feature frames as nodes and GD&Ts as links [Tsai & Cutkosky, 1997]. The network is a representation that can be used for tolerance analysis as tolerance stack-up, including dimensional and geometric tolerances as well as fitting tolerances, can be calculated based on the propagation path in the network. Figure 4 is an illustration of a TN where nodes and links are constructed based on information from the GD&T database and the feature attribute database.

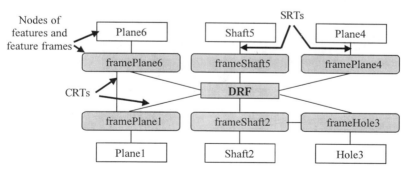

Figure 4; An example of a tolerance network for a part.

52

Mating condition between parts corresponds to a link between mating features with GD&T information associated with the mating condition. The constructed TN is then translated into a neutral file for a tolerance analysis software via the developed interface. Tolerance propagation paths are analyzed and kinematic error accumulated along each path due to tolerancing specifications is then calculated based on error models of tolerances and fittings [Tsai et al., 1995].

4. IMPLEMENTATION AND AN APPLICATION EXAMPLE

The tolerancing system is developed in Visual Basic as it supports different interfaces for databases and OLE automation under Windows operation system. The system is running in a personal computer in MS-Windows with Solidworks as the embedded CAD system. To discuss the function of the developed tolerancing system, the spindle of a turning center as shown in figure 5 is used for illustration in this section.

Geometric model of the spindle is first constructed in the embedded CAD system. GD&T specifications and fitting conditions are then assigned. Such specifications are extracted and stored in the GD&T and the feature attribute databases via the feature extraction module, the GD&T specification extraction module, and the fitting assignment module. The corresponding TN of the spindle is constructed by the TN construction module. Figure 6 shows part of the constructed TN in a table where the first two fields are nodes and the third field is the link between them. In this design we are concerning about the relative positioning error of the spindle with respect to the machine base caused by tolerancing. Because the spindle consists of more than 40 parts, there exist many possible paths for evaluating the relative positioning error. One of the critical paths is the one starting from the spindle cover to the rear ring via the front ring, the roller bearing, the spacer, and the thrust bearing.

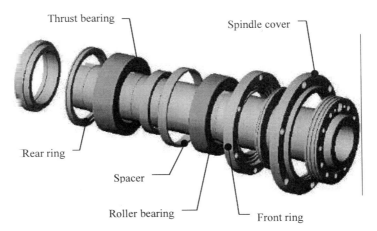

Figure 5; The spindle example of a turning center.

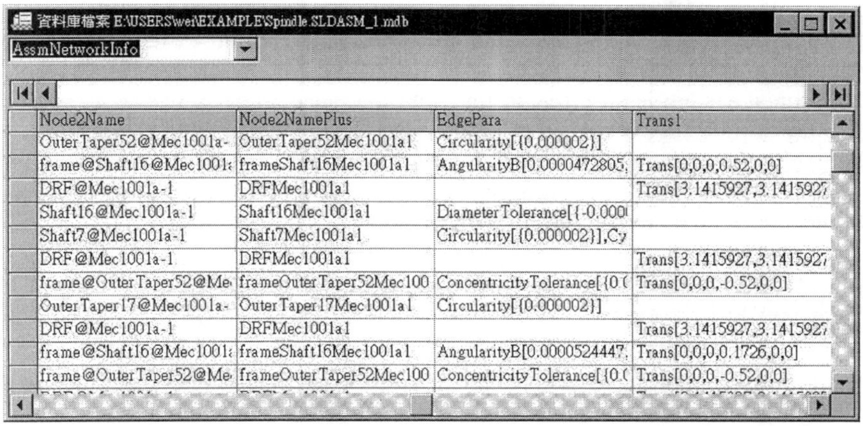

Node2Name	Node2NamePlus	EdgePara	Trans1
OuterTaper52@Mec1001a-	OuterTaper52Mec1001a1	Circularity[{0.000002}]	
frame@Shaft16@Mec1001	frameShaft16Mec1001a1	AngularityB[0.0000472805,	Trans[0,0,0,0.52,0,0]
DRF@Mec1001a-1	DRFMec1001a1		Trans[3.1415927,3.1415927
Shaft16@Mec1001a-1	Shaft16Mec1001a1	DiameterTolerance[{-0.000(	
Shaft7@Mec1001a-1	Shaft7Mec1001a1	Circularity[{0.000002}],C9	
DRF@Mec1001a-1	DRFMec1001a1		Trans[3.1415927,3.1415927
frame@OuterTaper52@Me	frameOuterTaper52Mec100	ConcentricityTolerance[{0(	Trans[0,0,0,-0.52,0,0]
OuterTaper17@Mec1001a-	OuterTaper17Mec1001a1	Circularity[{0.000002}]	
DRF@Mec1001a-1	DRFMec1001a1		Trans[3.1415927,3.1415927
frame@Shaft16@Mec1001	frameShaft16Mec1001a1	AngularityB[0.0000524447,	Trans[0,0,0,0.1726,0,0]
frame@OuterTaper52@Me	frameOuterTaper52Mec100	ConcentricityTolerance[{0.(	Trans[0,0,0,-0.52,0,0]

Figure 6; *Part of the tolerance network constructed from the GD&T database.*

To calculate tolerance accumulation along this path, GD&T specifications and corresponding error models are retrieved from the GD&T and feature attribute databases. The specifications are further translated into a neutral file corresponding to the path with associated error models by the interface for tolerance analysis. The resultant error by statistical tolerancing analysis is shown in figure 7 where the bonding box represents the worst error range.

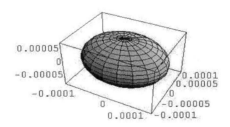

Figure 7; *Resultant error range caused by tolerances along a specified path of the spindle example.*

5. CONCLUSION AND DISCUSSIONS

We discussed the development of a tolerancing system in a CAD environment and demonstrated its function in this paper. The developed system consists of five modules for extraction and storage of feature information, GD&T specifications, and fitting specifications. The extracted specifications are stored in the GD&T database and the feature attribute database. A TN construction module and an interface for tolerance analysis are developed to construct the corresponding TN and to translate the GD&T specifications for tolerance analysis.

The developed system is an effort to integrate tolerance analysis with CAD systems in order to assist mechanical designers in tolerance design. Tolerancing-induced problems can be evaluated and avoided at design time before a design is sent to manufacturing. The system, however, is a concept-proof prototype. Types of features and mating conditions are limited and need further expansion.

ACKNOWLEDGMENT

This work was supported by the National Science Council of the Republic of China under contracts NSC 87-2212-E-005-001 and NSC 88-2212-E-005-026.

REFERENCES

[ISO, 1983] ISO 1101; *Technical drawings - Geometrical tolerancing - Tolerance characteristics and symbols.* International Standard Organization, Switzerland.

[ElMaraghy et al., 1991] ElMaraghy, H. A.; Wu, Z.; ElMaraghy, W. H.; A tolerance system for mechanical design. In: *Proceedings of the 1991 ASME Intelligent Design and Manufacturing for Prototyping,* PED-Vol. 50, pp.113-123.

[Hillyard & Braid, 1978] Hillyard, R.C. and Braid, I.C.; Analysis of dimensions and tolerances in computer-aided design. *Computer-Aided Design,* 10(3), pp.161-166.

[Knappe, 1963] Knappe, L.F.; A technique for analyzing mechanism tolerances. *Machine Design,* pp.155-157.

[Qamhiyah et al., 1996] Qamhiyah, A. Z.; Venter, R. D.; Benhabib, B.; A generalized method for the classification and extraction of form feature. In: *Proceedings of the 1996 ASME Design Engineering Technical Conference and Computers in Engineering Conference,* paper 96-DETC/CIE-1319.

[Tsai & Cutkosky, 1997] Tsai, J.-C. and Cutkosky, M.R.; Representation and reasoning of geometric tolerance in design. *Artificial Intelligence in Design, Analysis and Manufacturing,* 11(4), pp.325-341.

[Tsai et al., 1995] Tsai, J.-C.; Guo, D.-N.; Cheng, K.-D.; Variational kinematic models for geometric tolerances and fittings. In: *Proceedings of the 4th National Applied Mechanisms and Robotics Conference,* Vol. II, paper AMR95-107.

[Tsai & Chang, 1997] Tsai, J.-C. and Chang, J.-S.; Development of an AutoCAD-based geometric tolerancing system. In: *Proceedings of the 1997 ASME Design Engineering Technical Conferences,* paper DETC97/DAC-3998.

[Turner & Gangoiti, 1991] Turner, J.U.; Gangoiti, A.B.; Commercial software for tolerance analysis. In: *Proceedings of the 1991 ASME Computer in Engineering Conference,* Vol. 1, pp.485-503.

TOWARDS REALISTIC SIMULATION OF MACHINING PROCESSES

Patrick Le Pivert and Alain Rivière
Groupe de Recherche en Ingénierie Intégrée des Ensembles Mécatroniques
3, rue Fernand Hainaud, 93407 Saint-Ouen, FRANCE
Tel.: 33 1 49 45 29 20   Fax: 33 1 49 45 29 29
email : ariviere@ismcm-cesti.fr
patricklepivert@ismcm-cesti.fr

Abstract: The purpose of realistic simulation of machining processes is to compare geometric design specifications to specifications derived from manufacture. The specifications may be of a different nature: «worst case » or « statistical ». This article poses the problem of writing specifications for phase contracts and tolerance transfer, and presents a principle for solution using the « resolution tensor ».

Keywords: simulation, process plan, tolerance transfer, specification for manufacturing.

1. INTRODUCTION

The designing of mechanical assemblies requires the definition of the nominal geometry of its component parts. In addition, there are inaccuracies in manufacturing processes which result in dimensional and geometric variations in machined parts.

The purpose of definition design tolerancing, therefore, is to take these process errors into account by limiting surface position variations of the part in relation to the nominal geometry.

When a manufacturing process is being set up, it is essential to verify the geometric consistency between specifications which are admissible as regards operation and specifications achievable in manufacturing.

The aim of realistic simulation is to anticipate the behaviour of the manufacturing process with respect to its capacity to obtain parts in conformity with drawing office specifications [*Tolerancing specification for design §2*]. Conformity of parts being manufactured is verified by means of tolerancing specifications for manufacturing [*Tolerancing specification for manufacturing: §3*].

Passing from one to the other of these types of tolerancing specification entails a specification transfer operation [*Specification transfer: §4*] which involves seeking the tolerancing specification for manufacturing [*Tolerancing specification for manufacturing §4.1*] and a method of resolution [*Resolution principle: §4.2*]. Simulation is completed by comparing specifications for design and specifications resulting from manufacturing [*Comparison of specifications: §5*].

2. TOLERANCING SPECIFICATION FOR DESIGN

Tolerancing specification for design is written in accordance with ISO - standards by defining the position and size of the tolerance zones within a reference system [fig1].

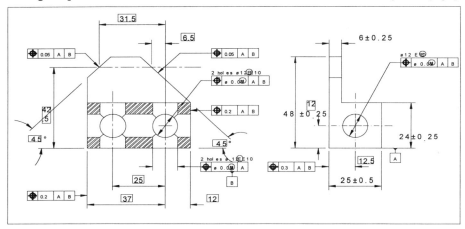

Figure 1 : Dimensioning and tolerancing specifications on a design draft

The reference systems and tolerance zones are defined by operations on the "Skin model" [**ISO/CD17450**] and, whilst statistical approaches [**SRI, 1997**] have been made in this domain, the tolerancing model generally adopted by drawing offices is a « worst case » model. Only the position defined by the dimensions in boxes and the domain boundaries have any significance.

In this type of specification, the distribution of the parts is not specified, only the characteristics of the individual part count. This information is of interest for isolated parts but does not provide any constraints for the series to be produced, in particular with respect to the average position, distribution or accepted risk.

3. TOLERANCING SPECIFICATION FOR MANUFACTURING

In order to comply with the specifications established by the design draft, the methods engineer designs a process that can contain several phases: the manufacturing process plan.

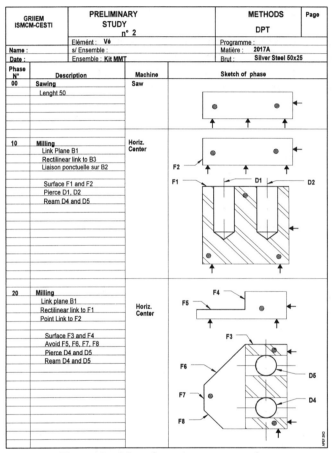

GRIIEM ISMCM-CESTI	PRELIMINARY STUDY n° 2		METHODS DPT	Page
	Elémént : V6		Programme :	
Name :	s/ Ensemble :		Matière : 2017A	
Date :	Ensemble : Kit MMT		Brut : Silver Steel 50x25	

Phase N°	Description	Machine	Sketch of phase
00	**Sawing** Lenght 50	Saw	
10	**Milling** Link Plane B1 Rectilinear link to B3 Liaison ponctuelle sur B2 Surface F1 and F2 Pierce D1, D2 Ream D4 and D5	Horiz. Center	F2 F1 D1 D2
20	**Milling** Link plane B1 Rectilinear link to F1 Point Link to F2 Surface F3 and F4 Avoid F5, F6, F7, F8 Pierce D4 and D5 Ream D4 and D5	Horiz. Center	F4 F5 F3 F6 D5 F7 D4 F8

Figure 2 : Manufacturing process plan

Each phase can then be considered as a specific design draft with only the positioning and machined surfaces being concerned by the specification. This specification may have several forms dependent upon whether it is the machined part or the process that is being specified. These two specifications can be seen as the attribute of isolated manufacture ('worst case' specification) or the attribute of a series (statistical specification).

58

Specifying the process or the manufactured part affects the means and methods used for verification. In a « part-oriented verification », the aim is to ensure that, at the end of each phase, the parts in their intermediary state are acceptable for the next phase. In a « verification-oriented process », verification ensures that the process does not present any dispersion beyond that used to validate it during simulation. First and foremost, the causes should be recorded and verified one by one: at the very least, the position of the tool and the position of the part in relation to the positioning surfaces of the part holder.

Simplification of industrial procedures leads naturally to the first solution. The manufacture of the inclined, toleranced surface F8 [fig 1] can lead to the specification for the next manufacturing process [Fig2]. This presentation associates the tool position error in relation to the part holder and the error between the reference system built onto the part and another system which nominally coincides on the part holder.

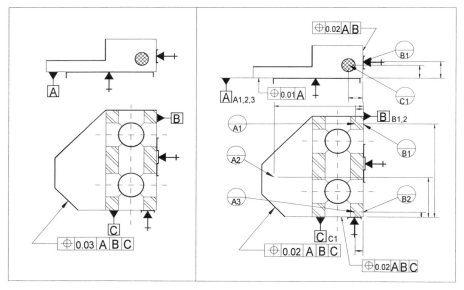

Figure 2
Tolerancing specification
in manufacturing

Figure 3
Tolerancing specification in manufacturing
with partial references

Tolerancing specification in manufacturing, which allows the behaviour of the process to be monitored, leads to the definition of a reference system determined from partial references modelling the positioning. The machined surfaces and the placement surfaces are positioned within this reference system [fig 3]

Irrespective of the adequacy of these indications, they are often expressed by applying a 'worst case' principle, and inadequately reflect the needs for verification during manufacture, which has to be done by sampling. Verification results are, therefore, usually expressed in terms of average value, variation type and capability.

Proposals for statistical specification have already been presented [**SRI, 1997**]. They are particularly suited to tolerancing specification for manufacturing. The previous specification thus takes the following form:

The indication ST indicates that this relates to statistical tolerancing. Cp and Cpk are capability indices. Therefore, this is 'worst case' tolerancing for design and statistical tolerancing which is adequate for mass production.

4. SPECIFICATION TRANSFER

Specification transfer must meet two requirements:
– In terms of quality, it must express the tolerancing specification for manufacturing needed to monitor the process which makes specifications for design easier to obtain,
– It must allow comparison of the specifications for design with the resulting specifications for manufacture

4.1. Searching for tolerancing specifications for manufacturing

The strategy for searching for specifications to be included in the phase contract is supported by the structure of specifications for design and process planning.

For direct specifications for design, that is to say, relating to surfaces which are active in the same phase (i.e. machined surfaces or surfaces used for positioning in this phase), the structure of specifications for manufacture is identical to those for design.

Generally-speaking, in cases where specifications are not direct, an approach that takes the process into account is preferred. This time, the machined surfaces will be positioned in relation to the part holder referential and the positioning surfaces will also be under constraint in their position as regards this referential. The following graphical representations can be obtained by using the graphic illustration of the TTRS model [**CLE et al. 1994**],:

$$DESIGN\ TTRS : (A \cup (D1 \cup D2)) \cup F8$$

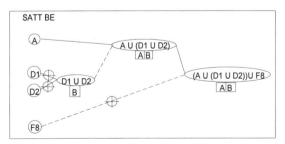

In phase 10 :

MACHINING TTRS : $\left(B1_{1,2,3} \cup B1\right)$

MACHINING TTRS : $\left(\left(\left(B1_{1,2,3} \cup B3_{1,2}\right) \cup B3\right)\right)$

MACHINING TTRS : $\left(\left(\left(B1_{1,2,3} \cup B3_{1,2}\right) \cup B2_1\right) \cup B2\right)$

MACHINING TTRS : $\left(\left(\left(B1_{1,2,3} \cup B3_{1,2}\right) \cup B2_1\right) \cup F1\right)$

MACHINING TTRS : $\left(\left(\left(B1_{1,2,3} \cup B3_{1,2}\right) \cup B2_1\right) \cup F2\right)$

MACHINING TTRS : $\left(D1 \cup D2\right)$

MACHINING TTRS : $\left(\left(\left(F1 \cup F2\right) \cup B1_1\right) \cup \left(D1 \cup D2\right)\right)$

In phase 20

MACHINING TTRS : $\left(B1_{1,2,3} \cup B1\right)$

MACHINING TTRS : $\left(\left(\left(B1_{1,2,3} \cup F1_{1,2}\right) \cup F1\right)\right)$

MACHINING TTRS : $\left(\left(\left(B1_{1,2,3} \cup F1_{1,2}\right) \cup F2_1\right) \cup F2\right)$

MACHINING TTRS : $\left(\left(\left(B1_{1,2,3} \cup F1_{1,2}\right) \cup F2_1\right) \cup F8\right)$

4.2. Resolution principle

The objective is to obtain a result with a specification resulting from manufacturing. The surfaces active in this specification may or may not belong to the same phase; they may be machined surfaces or positioning surfaces.

Resolution involves two stages:

– Firstly, uncertainties at a certain number of points on the active surfaces are sought – reference elements and toleranced element - for the specification in question.

– then a quantity characteristic of the variation in the position of certain points of the toleranced element is calculated by means of a virtual metrology operation.

First stage:

Based on the hypothesis of small displacement around the nominal position, due to uncertainty of positioning and dimensionally stable solids, it can be demonstrated [CLE, 1991][LEP, 1998] that the variance in the displacement of a point of a solid in a given direction can be calculated by:

$$\left| VAR(M)_n = [n]^T * \overline{\overline{R}} * [n] \right|$$

where

- [n] is the single column matrix of the co-ordinates of the unit vector defining the calculation direction

- $\overline{\overline{R}}$ is the 3x3 matrix of the resolution tensor at point M for a given position, the components of the tensor depending on the co-ordinates of M, the co-ordinates of the points of contact, the contact normals and the variation of the position at points of contact.

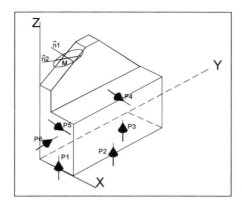

Figure 3 : Illustration of the calculation by the tensor

These results are true if the points of the solids are centred on the average position or are offset in relation to this position.

By considering the variance of the position of the points of a machined surface, it is possible, by combination, to express the variation in a phase of the position of points of surfaces:

✓ machined in this phase
✓ used as a positioning surface in this phase
✓ machined in a previous phase
✓ used as a positioning surface in a previous phase

Second stage:

The previous calculations enable the displacement variance of any point of the solid to be determined in any direction. This is particularly true for points chosen as reference elements of the specification for design. This work allows the resolution tensor to be instanced at any points of the toleranced element. Calculation of the displacement variance of the toleranced element, in relation to the reference system in the direction defined by the tolerance zone is immediate.

To some extent, this is a virtual measurement operation which provides a result in the form of variance.

5. COMPARISON OF SPECIFICATIONS

Comparison of a specification for manufacturing recorded in statistical form cannot be directly compared with a « worst case » specification. The transfer from one to the other can only be done by calculating the average zone overflow risk [**SRI et al., 1997**]. This risk calculation takes the distribution of the variable into account (normal, triangular or equiprobable) as well as the position of its average value.

The values of Cp and Cpk are determined for a given risk and the tolerance value provided by the design specification, which allow an assessment to be made as to whether or not the process is capable.

The variance equations formulated to determine the variances are still true, irrespective of whether the average value is centred or not. However, complete verification will only be achieved by also verifying the position of the average value.

6. CONCLUSION

We have seen that design and manufacture specifications are not recorded in the same manner. Statistical writing of tolerancing specifications for manufacturing is justified because it corresponds to the information processed by the verification.

The resolution tensor provides an indication of population dispersion. This result, in the form of variance, allows us to say whether or not the process is capable, as long as we can always adjust it to an average position. To take into consideration the possibility of a deviation in relation to the average position, an additional calculation based on small displacement must then be carried out..

7. BIBLIOGRAPHY

[CLE, 1991] A.CLEMENT, « The Resolution of positioning solids ». CIRP Stanford Vol 40/1

[CLE et al., 1994] A. CLEMENT, A. RIVIERE ,M. TEMMERMAN « Cotation tridimensionnelle des systèmes mécaniques »(3-D specification of mechanical systems), PYC Editions, 1994

[LEP. 1998] P. LE PIVERT , « Contribution à la modélisation et à la simulation réaliste des procédés d'usinage » (Contribution to modelling and realistic simulation of machining processes), Doctorate Thesis, Ecole Centrale de Paris, GRIEEM-LISMMA, 1998

[SRI et al., 1997] V. SRINIVASAN M.A. O'CONNOR, F.W. SCHOLZ , « Techniques for Composing a Class of Statistical Tolerance Zones », in advanced Tolerancing Techniques, pp.139-165, Editor ; H.C. Zhang, John Wiley & Sons, NEW York, 1997

[SRI, 1997] V. H., SRINIVASAN « ISO Deliberates Statistical Tolerancing », Proceedings of 5th CIRP International Seminar on Computer-Aided Tolerancing, Canada, pp25-36, April 1997,

[ISO/CD17450] Dimensional and geometrical product specifications and verification

ESTIMATING POSITION TOLERANCE PROCESS CAPABILITIES

Dr. Leonard E. Farmer
School of Mechanical and Manufacturing Engineering
The University of New South Wales
Sydney NSW 2052
Australia
Email: L.Farmer@unsw.edu.au

ABSTRACT: The nature of position variations is discussed and this is related to the form of process capability data for positioning machining processes. Strategies are then developed for estimating the position tolerances for single features and groups of features. The focus of these strategies is to provide position tolerances that are compatible with the positioning capabilities of processes. A curve has been derived from which adjustment factors can be determined for all three positioning tolerance cases discussed in the paper. The procedures assume the position deviation distributions of processes are normal and the results are for the commonly used ±3 standard deviation limits.
Keywords: Position Tolerances, Dimensioning, Tolerancing, Process Capabilities, Geometry Tolerances

1. INTRODUCTION

A continuing problem that confronts product designers is the assignment of appropriate tolerances to the dimensions that specify designs in order that the functional requirements of the product are satisfied and the product can be produced at optimum cost. Much research effort has been expended on this subject, particularly with regard to optimising costs/tolerances. Papers on cost/tolerance modelling date back to the 1950s, whilst more recent work is described by [Gadallah et al., 1998]and [Dong et al., 1998]. A practical difficulty usually associated with the application of cost/tolerance models, is the availability of accurate cost data. The situation is improving with the increasing use of computer based data logging systems and the adoption of more appropriate cost systems such as Activity Based Costing.

It would appear that this difficulty with applying cost/tolerance models is likely to continue for some time and the pragmatic, informal optimisation procedure often followed by designers will be the environment at which this paper is directed. Briefly, the procedure comprises of the following steps. First, a method or process is selected for producing a dimension of a product design. The selection of the process is based on some generally accepted opinion on; *What would be the most economically desirable method for producing the dimension.* Second, the tolerance specified for this dimension will be derived from the expected capability of the process. Third, when all processes and tolerances have been selected, an analysis is made to assess how well the functional requirements of the product will be achieved. This procedure is continued until it is agreed that the best set of methods or processes and tolerances have been chosen.

Arguably, the two most prevalent types of dimensions that appear on product design drawings are related to the size of features and the location or position of features relative to each other. Specifications for form, orientation and surface texture appear less frequently and are usually expressed in general tolerance notes.

Process capability information is generally available for the size of features, such as hole diameters and the distance between planes, produced by common machining, moulding, casting and forming methods. This is not the case for dimensions that locate or position features. Methods for arriving at appropriate tolerances for these types of dimensions will be the focus of this paper.

2.   NATURE OF POSITION VARIATION

The nature of the variability in the position of features is illustrated in the accuracy specifications of machine tools. The general form of specifications for position are set out in national and machine tool industry or builders standards or described in books such as Schlesinger's [Schlesinger, 1979]. Specifications for position accuracy usually refer to maximum deviations from set distances of movement along a machine tool axis. The typical form of a positional accuracy specification for a CMM is;

$$\pm p_{xa} \quad = \quad C_{1x} + C_{2x} X \quad\quad\quad (1)$$

where

$$\pm p_{xa} \quad = \quad \text{position accuracy for a set position ,}$$
$$C_{1x} \quad = \quad \text{intercept constant,}$$
$$C_{2x} \quad = \quad \text{gradient constant and}$$
$$X \quad = \quad \text{magnitude of the position setting dimension.}$$

The form of this function is illustrated in Figure 1. It has a value of C_{1x} when $X = 0$ and a constant gradient of C_{2x}. Position accuracy is specified as a bilateral deviation from a set position. The deviations from a set position will be distributed as shown in Figure 1, and the limits given by equation (1) are usually chosen to contain 95% of the deviations. The distribution of the deviations is often assumed to be normally distributed.

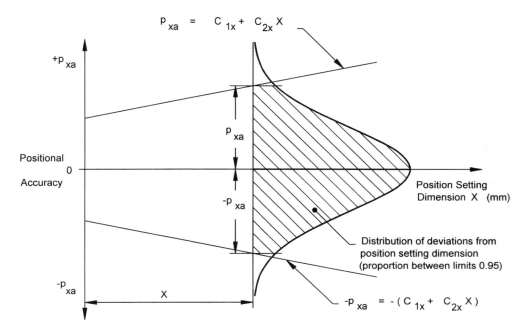

Figure 1; *Form of positional accuracy deviations along an axis of a Coordinate Measuring Machine*

3. POSITION PROCESS CAPABILITY

The nature of the variability in the positioning of features has been shown to be a function of the magnitude of the positioning dimension. This is compatible with the family of standard tolerance grade curves that are part of the ISO system of Limits and Fits [ISO, 1988]. These curves have been adapted for position tolerances, as shown in Figure 2, for tolerance grades IT7, IT8, IT9 and IT10. The predicted magnitude of a position tolerance p_x, is a function of the magnitude of the dimension X, and the tolerance grade number. Whilst the gradient of these curves decreases with increasing X, it is essentially constant for the larger values of X and, for these values, are similar to the positioning accuracy function in Figure 1. Thus it is reasonable and convenient to express position tolerance process capabilities in terms of standard tolerance grade curves.

There is little information published on the position process capabilities, however, Bjorke [Bjorke, 1978] does provide information for the process of Boring and this is shown in TABLE I. A typical application for the data given under *The distance from a*

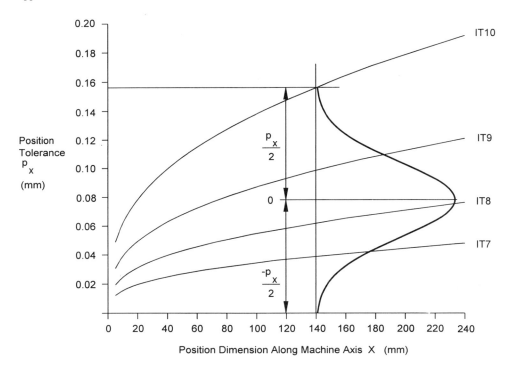

Figure 2; *Standard tolerance grade curves for IT7 to IT10. The Normal distribution approximates position variations for the process of Boring a dimension of the type 'The distance between two centrelines'.*

centreline to a reference plane is illustrated in Figure 3 with the ⌀15 hole, whilst the dimensions specifying the distance between the two ⌀20 holes illustrates, *The distance between centrelines*. The position tolerance p_x, for both dimension types, is covered by the range of tolerance grades IT7 to IT10. The selection of the appropriate tolerance grade from this range would be arrived at after considering factors such as, the material to be bored, the condition of the boring machine, the shape and size of the workpiece, skill of the machine operator, etc. Thus, if the tolerance grade of IT10 was selected and the position dimension was 140 along a boring machine axis then the position tolerance, p_x, would be approximately 0.155, as shown in Figure 2. The terms Es, and Var s, in the table are the expected values and variances, respectively, of the Unit Beta distributions for the position variations. The limits of these distributions are 0 and 1. The four extreme combinations of Es and Var s for *The distance between centrelines* dimension type is shown in Figure 4. These distributions vary from symmetrical at one limit to being moderately negatively skewed at the other. They tend to be flat or platykurtic, relative to normal distributions. For the purposes of comparison, a normal distribution is

TABLE I; *Expected Position Tolerances and Parameters of Unit Distributions for the Process of BORING. [Bjorke, 1978 - Appendix E]*

DIMENSION TYPE	p_x		Es		Var s	
	min	max	min	max	min	max
The distance from a centreline to a reference plane.	IT7	IT10	0.48	0.52	0.028	0.040
The distance between centrelines.	IT7	IT10	0.50	0.55	0.028	0.047

The above data applies to dimensions in the range 5 to 500mm. for parts with masses less than 100Kg.

superimposed on Figure 4, that has its ±3 standard deviation limits corresponding to the limits of 0 and 1 for the Unit Beta distributions. Given the computational difficulties of using Beta distributions and the uncertainties when defining the limits of Beta distributions, it is perhaps reasonable to assume that the distribution of position variations for a given length and tolerance grade can be represented satisfactorily by a normal distribution. This, at least appears to be the case for the Boring process and it is illustrated in Figure 2.

4.   POSITION TOLERANCE DIMENSIONING - CASE 1

The first case of position tolerance dimensioning is illustrated by the position tolerance for the ∅15 hole in Figure 3. Two dimensions specify the position of a feature, a hole in this case, relative to existing or reference features, faces A, B and C. The position tolerance only applies to the feature. To determine this tolerance first nominate a manufacturing process for producing the feature. Then select the appropriate tolerance grade for positioning the feature from process capability data. The tolerance is then calculated from Standard Tolerance grade curves, as shown in Figure 2, or from the equivalent table in the ISO System of Limits and Fits [ISO, 1988]. That is, assume that the ∅15 hole in Figure 3 is to be produced on a Boring machine. The dimension type is the distance from *a centreline to a reference plane* in Table I. Further, assume that the operation is of average difficulty and the hole will be positioned with a tolerance grade of IT9. From Figure 2 it can be seen that the position tolerances for distances of 12 and 50 are 0.04 and 0.07 wide respectively and they are illustrated in Figure 5.

   These tolerances were obtained from data that appears to be applicable to position deviations along the axis of movement of a Boring machine and the deviations along two axes have now been brought together. If it is assumed that the distributions of

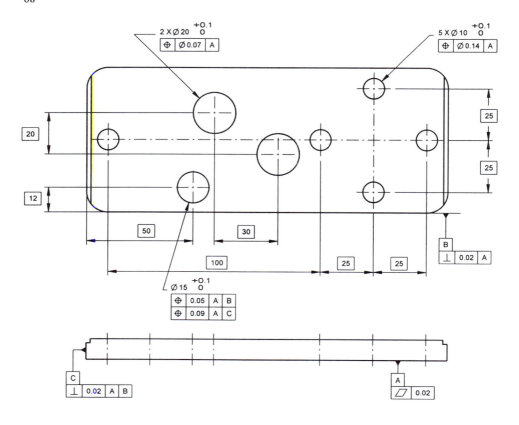

Figure 3; *Partially dimensioned workpiece that illustrates the three cases of position tolerance specifications.*

position deviations for both axes of movement are normal and the position deviations on each axis are independent of the other, then the deviations from the true position of the hole follow a bivariate normal distribution. If it is assumed that the tolerances obtained from Table I and Figure 2 represent the ±3 standard deviation limits of the normal position deviation distribution along a machine axis, then the ±3 standard deviation zone for the bivariate distribution will be elliptical [Paradine et al., 1968], as shown in Figure 5. This zone is larger in some areas than the individual ±3 standard deviation limits for the axes.

Two strategies are suggested for obtaining appropriate position tolerance values that observe the ±3 standard deviation requirements. Where width position tolerances are to be specified, multiply the machine axis position deviations by 1.16. The value of 1.16 was obtained from the probability density function for the bivariate normal distribution [Paradine et al., 1968],. Thus, for the ∅15 hole the position tolerances would be at least

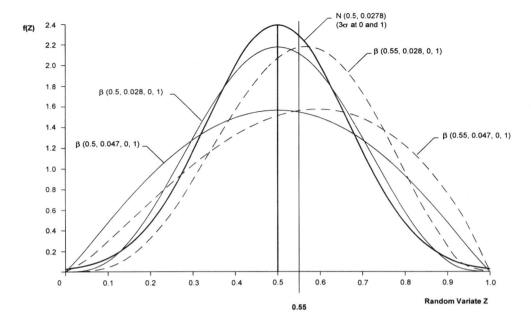

Figure 4; *Unit Beta Distributions for extreme combinations of parameters Es and*
Var s for the dimension type 'The distance between two centrelines'.
The Normal distribution has its ±3 standard deviation limits corresponding
to the limits of 0 and 1 for the Unit Beta distributions.

0.046 and 0.081, as shown in Figure 5. For the case where a cylindrical position
tolerance specification is required, the diameter of this tolerance is obtained by
multiplying the larger of the two width position tolerances by 1.16. That is, for the ⌀15
hole the position tolerance would be at least ⌀0.081, as shown in Figure 5.

5.   POSITION TOLERANCE DIMENSIONING - CASE 2

The second case of position tolerance dimensioning is illustrated by the position
tolerance for the two ⌀20 holes in Figure 3. This case differs from the first in the sense
that the position of two features are involved and the position requirement has to be
satisfied simultaneously for both features. The process capabilities given in Table I for
the distance between centrelines dimension type, are applicable to this case and they are
assumed to apply to each feature. That is, if we once again assume a position process
capability of IT9 for producing the holes. From Figure 2, the centre distance dimensions

70

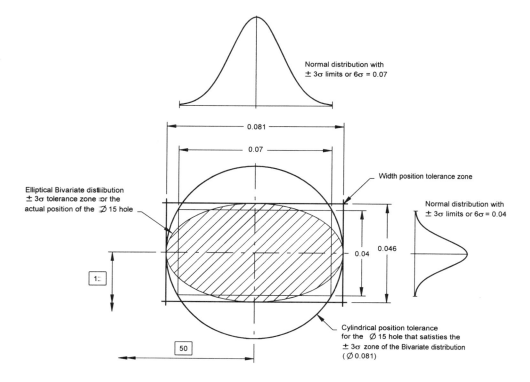

Figure 5; *Position tolerance specification and interpretation for ⌀15 hole in Figure 3*

of 20 and 30 have variations in position of at least 0.05 and 0.055 for each hole. Once again we encounter the bivariate distribution situation and the largest position tolerance (0.055), multiplied by 1.16 is used to obtain the diameter of the position tolerance (⌀0.064). It should also be noted that this specification includes two features and a method for taking this into account when deciding upon a minimum position tolerance is discussed in Case 3.

6. POSITION TOLERANCE DIMENSIONING - CASE 3

The third case of position tolerance dimensioning relates to the situation where two or more features are contained in a position tolerance specification. This is illustrated in Figure 3 with the position tolerance specification for the group of five, ⌀10 holes. The question here is; *how does the number of features in a group effect manufacturing difficulties and to what extent do position tolerances need to be varied to offset these*

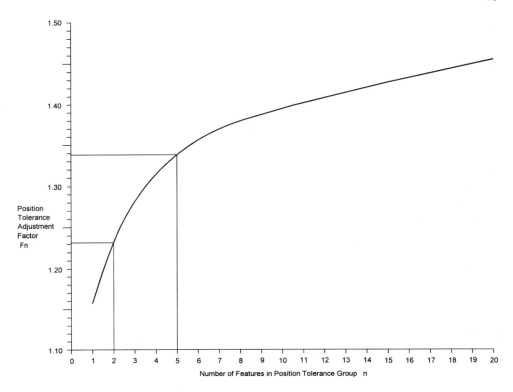

Figure 6; Position tolerance adjustment factor for groups of n features

difficulties? The above has been recognised for some time and the Interservices Committee [HMSO, 1948] suggested the following: *"production difficulties may be regarded as being inversely proportional to the position tolerance, and directly proportional to the square of the number of features in a group"*. This statement has many interesting implications, however, for the present the interest will be confined to the task of specifying position tolerances that satisfy the probability (0.9973), associated with the ±3 standard deviation limits, of process positioning capabilities.

The same procedure is used, as described for Cases 1 and 2, for obtaining the position tolerance for one feature in the group. The maximum dimension between the centres of the features is used as the positioning dimension. This would be 150 for the five holes in Figure 3 and the tolerance from, Figure 2, would be 0.1. If the probability of an acceptable feature is p, and for a group of n, features is p_{an}, then $p_{an} = p^n$. If the probability of p_{an}, is known (0.9973 for ±3 standard deviation limits), then p, can be calculated for values of n. Assuming the position distributions are normal the position tolerance can be calculated for the features in the group. The curve in Figure 6 has been derived for the purpose of calculating group position tolerances. The curve includes the adjustment for the bivariate normal distribution, described in Case 1, as well as the number of features in the group. That is, for the example of five holes, the adjustment factor obtained from Figure 6 is 1.34 and the minimum position tolerance for the holes is 0.134.

7. CONCLUDING REMARKS

Strategies have been developed for estimating the position tolerances for single and groups of features that are compatible with the positioning capabilities of processes. A curve has been derived from which adjustment factors can be determined for all three positioning tolerance cases.

REFERENCES

[Bjorke, 1978] Bjorke, O.; "Computer-Aided Tolerancing"; Tapir Publishers

[Dong et al., 1998] Dong, Z.; Wang, G.G.; "Automated Cost Modeling for Tolerance Synthesis Using Manufacturing Process Data, Knowledge Reasoning and Optimization"; In: *Geometric Design Tolerancing: Theories, Standards and Applications*; By: ElMaraghy, H.A.; Chapman and Hall; ISBN 0412-83000-0

[Gadallah et al., 1998] Gadallah, M.H.; ElMaraghy, H.A.; "A New Algorithm for Combinatorial Optimization: Application to Tolerance Synthesis with Optimum Process Selection"; In: *Geometric Design Tolerancing: Theories, Standards and Applications*; By: ElMaraghy, H.A.; Chapman and Hall; ISBN 0412-83000-0

[HMSO, 1948] Interservices Committee for Dimensioning and Tolerancing of Drawings; Dimensional analysis of Engineering Designs, Volume I - Components; London: His Majesty's Stationary Office

[ISO, 1988] ISO 286-2:1988; "ISO System of Limits and Fits; Part 2: Tables of standard tolerance grades and limit deviations for holes and shafts"; International Standards Organisation

[Paradine et al., 1968] Paradine,C.G.; Rivett, B.H.P.; "Statistical Methods for Technologists"; English Universities Press

[Schlesinger, 1979] Schlesinger, G; "Testing Machine Tools", 8th Ed.; Pergamon Press

The validation of a process plan by propagated dispersion zones

Pierre Gaudet, Guy Cloutier, Clément Fortin,
École Polytechnique de Montréal
Département de génie mécanique, Section fabrication,
Case postale 6079, Succursale Centre-ville
Montréal (Québec)
H3C 3A7
gaudet@meca.polymtl.ca

Abstract: Simulating a process plan is of great assistance to obtain a virtual product using the known capabilities of the manufacturing processes to produce a part. A common practice in industry to evaluate the capabilities of a manufacturing process is to quantify them using standard specifications as per tolerancing standards, and to perform statistical analysis to obtain the means and variances. This expression of the capability is therefore obtained at the price of aggregating several dispersions into a hopefully meaningful quantity called the dispersion range that can be compared to a tolerance zone. Following this synthesis, as aggregated measures lose track of independent causes, some level of detail is inevitably lost. Regenerating first level data to perform consistent simulations is thus complicated by the otherwise convenient synthesis of information. To avoid inconsistencies, we introduce the concept of *"coherence filters"* and apply them to simulate a population of virtual parts. The validation of a process plan is then performed using a known metrological algorithm to analyze each simulated feature.

Keywords: Jacobian, vectorial tolerancing, dispersion twists, Monte-Carlo simulation.

1. INTRODUCTION

In concurrent engineering, the production engineer's role is crucial as he is involved early during the design phase of a new product to validate fast and economic manufacturing tactics and to evaluate the expertise and equipment needs (fixtures, tools and gauges). One aspect of validating a process plan is to ensure the dimensional control of the product. This must be based on a certain number of facts and data, related to several aspects of manufacturing, some of these being process sequencing, capabilities, cost and the like. However the coupling within and between data sets rapidly increases the complexity of documenting and forecasting the performance of the process plan. At all times, the production engineer needs to justify his decisions and this could be an easier task if he relies on efficient tools to verify various process plans and to document the expertise involved. The tolerance charting technique was

73

developed to help the production planner to deal with dispersions' stack-ups. The manufacturing analysis purpose is variable and although it is not only limited to dispersions' stack-up analysis, the main objective of this work is dedicated to this subject.

2. BACKGROUND

The term *tolerance* is often used to express a manufacturing *dispersion* range. By abstraction, both expressions can be considered equivalent in the manner they are usually computed in worst-case stack-ups using their limit boundaries. However, the *tolerance* concept refers to the engineering boundary within which a feature of a part must reside. On the other hand, the *dispersion* ranges are better characterized using statistical distribution on a population of events of a feature. It is thus natural to review the methods developed for either tolerance or dispersion stack-ups.

Over the years, many tools were developed as researchers focused on solving specific tolerance analysis and synthesis problems, giving birth to a variety of tools. The small displacement theory (twist) is used to solve analytically the tolerance allocation in a chain which is now well understood to be a tri-dimensional problem. The twist is a convenient way to characterize the possible degrees of freedom between pairs of part features. The TTRS concept [Clément et al., 1997] was proposed to help the semantic organization of data required to solve tolerance stack-ups in computer-aided design software [Gaunet, 1994][Ballot, 1995]. When coupled to twist theory, this technique can solve generic examples. Tolerance specification using vectorial tolerancing is similar to twist theory and some works were dedicated to create a manufacturing feedback method to solve different tolerance allocation problems [Martinsen, 1995]. Also similar in concept, the kinematic tolerancing was demonstrated on tolerance transfer problems useful to compute equivalent re-expression of engineering tolerances with a different reference frame for drafting and process planning purposes [Rivest, 1993]. The method consists of regenerating the tolerance zones by incrementally sweeping the boundaries of a tolerance. The tolerance allocation problem then amounts to analyze clouds of valid points. If the zones are assumed to be canonic volumes as per tolerancing standard, the tri-dimensional analysis can suitably be simplified to a two-dimension analysis. With long accumulation chains, since the zones may not be simple volumes, such simplifications always impose excessive restrictions to the tolerance allocation when pursuing zero-defect production. Recent work by Bennis [Bennis et al., 1998] also deals with a simplified approach of kinematic tolerancing.

Another approach uses a technique, based on Monte-Carlo simulation [Ianuzzi, 1995][Nassef, 1993] which determines valid values for a set of variables within known ranges with statistical distribution functions on each tolerance. A genetic algorithm then contributes to compute the most economical solution to stack-up chains. The method was demonstrated on industrial examples using a standalone subroutine. A Monte-Carlo simulation was also used in conjunction with the TTRS technique to create the "resolution tensor" concept [Le Pivert, 1998]. The application method is shown to be conservative in some cases and it must be concluded that derivation of statistical distribution functions on variables must be carried out with care when treating critical cases where marginal conditions exist.

The trend in research efforts is now to combine different concept theories that have evolved separately. In that perspective, it is proposed to use a vectorial tolerance specification and a kinematic propagator using a Monte-Carlo generator of valid parts.

3. THE PROCESS PLANNING SIMULATION

While the metrological control of a part is based on the principle that tolerances are independent, the simulation of some manufacturing process needs to satisfy simultaneously all tolerances applying on the same features. The simulation tool also

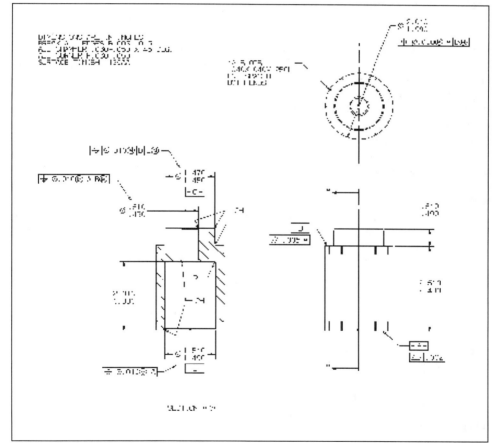

Figure 1: A simple generic example

needs to account for all possible sources of a cumulative dispersion zone. The example presented in figure 1, a cylindrical part which could serve as a spacer, is used to demonstrate the technique.

When validating a new part concept, multiple points of view can address a mix of design and manufacturing intents such as the following objectives.

1- Assess the performance of one or more process plans to manufacture parts.

2- Assess assemblability of two identical parts by engaging diameter C on one part in diameter B of a second part, with predominant contact between planes D and A.

3- Compute the cumulative dispersions with different configurations of an N part assembly with known dispersions on each part.

The relevant dimensional constraints are generally obtained from drawings under a tolerancing standard like ASME Y 14.5M-1994. The standard defines the tolerances to be applied to position, form, orientation and size of a feature explicitly described by tolerance zones which all points of a surface or the axis of a feature must reside into. The simulation solution to the first objective of the example part is detailed using the situation dispersions (position and orientation dispersions) of plane D with respect to A.

3.1 Composition method of the manufacturing twists.

The part is produced with a rapid prototyping technique. All part's features are produced when the part is set up on plane A and diameter B. The manufacturing dispersions of each feature are independent from any other part's features dispersions. By using twist theory, the specification of three manufacturing twist domains is required between the set-up and the manufactured features. The twist specification is by its theoretical reference frame situation using the following chaining.

RF_0[1] is aligned on plane A of the part, RF_1 is aligned on the fixture plane of the set up. RF_2 is aligned on the theoretical origin of plane D. RF_3 is aligned on the manufactured plane D. All reference frames are co-parallel after aligning their Z-axis with the part axis.

The three resultant twist domains to produce face D are built as follows[2]:

$$^{0,0}\tau_1 = \begin{bmatrix} ^{0,0}P_1 \\ \hline ^{0,0}O_1 \end{bmatrix} \text{Microscopic twist describing set-up dispersions from } RF_0 \text{ to } RF_1$$

$$^{1,1}\tau_2 = \begin{bmatrix} ^{1,1}P_2 \\ \hline ^{1,1}O_2 \end{bmatrix} \text{Macroscopic twist describing the transport from } RF_1 \text{ to } RF_2$$

$$^{2,2}\tau_3 = \begin{bmatrix} ^{2,2}P_3 \\ \hline ^{2,2}O_3 \end{bmatrix} \equiv \begin{bmatrix} ^{2,2}\vec{p}_3 + (^{0,0}\vec{O}_1 \times \vec{T}_1) + (^{2,2}\vec{O}_3 \times \vec{T}_2) \\ \hline ^{2,2}O_3 \end{bmatrix} \text{Microscopic twist of face D}$$

---

[1]RF_n means: Reference Frame in this text; subscript n refers to the RF number.

[2]Dispersion twists are specified by $^{a,b}\tau_c$ denoting "the situation (position & orientation) of RF_c from RF_b, expressed in RF_a".

In theory, each of the six dispersion parameters composing a twist must take into account position and orientation elementary dispersions. At one point, on a surface, the influence of these parameters can be superimposed but each parameter is bounded by a plus and minus dispersion value boundary. If the orientation dispersions of a feature are specified using a parallelism specification on a standard surface with known dimensions, the size of the setup and working faces are required to compute, to scale down the orientation dispersions angles and to build the required coherence filters.

Plane A size: $\vec{T}_1 \equiv \begin{bmatrix} T_{x1} & T_{y1} & T_{z1} \end{bmatrix}^T = \begin{bmatrix} 1" & 1" & 0 \end{bmatrix}^T$; plane D size: $\vec{T}_2 \equiv \begin{bmatrix} T_{x2} & T_{y2} & T_{z2} \end{bmatrix}^T = \begin{bmatrix} 1" & 1" & 0 \end{bmatrix}^T$;

Position dispersion:

$$^{0,0}\vec{p}_1 \equiv \begin{bmatrix} x_{p1} \\ y_{p1} \\ z_{p1} \end{bmatrix} = \begin{bmatrix} \varsigma_{x1} \cdot \delta_{x1} \\ \varsigma_{y1} \cdot \delta_{y1} \\ \varsigma_{z1} \cdot \delta_{z1} \end{bmatrix} = \begin{bmatrix} \varsigma_{x1} . 005" \\ \varsigma_{y1} . 005" \\ \varsigma_{z1} . 001" \end{bmatrix} ; \quad ^{2,2}\vec{p}_3 \equiv \begin{bmatrix} x_{p2} \\ y_{p2} \\ z_{p2} \end{bmatrix} = \begin{bmatrix} \varsigma_{x2} \cdot \delta_{x2} \\ \varsigma_{y2} \cdot \delta_{y2} \\ \varsigma_{z2} \cdot \delta_{z2} \end{bmatrix} = \begin{bmatrix} \varsigma_{x2} . 001" \\ \varsigma_{y2} . 001" \\ \varsigma_{z2} . 010" \end{bmatrix} ;$$

Orientation dispersion

$$^{0,0}\vec{O}_1 \equiv \begin{bmatrix} \varepsilon_{a1} \cdot t_x / T_{y1} \\ \varepsilon_{b1} \cdot t_y / T_{x1} \\ \varepsilon_{c1} \cdot t_z \end{bmatrix} = \begin{bmatrix} \varepsilon_{a1} . 001" \\ \varepsilon_{b1} . 001" \\ \varepsilon_{c1} . 0" \end{bmatrix} ; \quad ^{2,2}\vec{O}_3 \equiv \begin{bmatrix} \varepsilon_{a2} \cdot t_x / T_{y2} \\ \varepsilon_{b2} \cdot t_y / T_{x2} \\ \varepsilon_{c2} \cdot t_z \end{bmatrix} = \begin{bmatrix} \varepsilon_{a2} . 001" \\ \varepsilon_{b2} . 001" \\ \varepsilon_{c2} . 0" \end{bmatrix} ;$$

By hypothesis the domain of Monte-Carlo scalar parameters H and γ are $[-1,1]$. Each of the above numeric parameters is obtained when the contributions of the others are null. The dispersions' values are fed to a simulator for demonstration purposes.

Coherence filters

The role of coherence filters is to validate if a set of randomly generated parameters meets a dispersion range specification. When a generated set does not meet the criteria, it is rejected and another set is generated. The coherence filter concept is similar to the formulas proposed by Bialas [Bialas et al., 1997] when discussing conversion problems from standard to vectorial tolerance specifications. In the present model, two hypotheses are used with the coherence filter concept: 1- a surface contour is approximated by a circumscribing rectangular perimeter. 2- coherent sets of dispersion parameters are applied to perfect form simulated features.

In coherent twists, the sum of position and orientation dispersions at any point of a feature must be limited axially using a constraint similar to the following formula:

$$\left| \sum z \right| \le .010"$$

The z component of the vector $({}^{0,0}\vec{O}_1 \times \vec{T}_1) + ({}^{2,2}\vec{O}_3 \times \vec{T}_2)$ in ${}^{2,2}\tau_3$ represents the parallelism dispersion between RF$_1$ and RF$_3$. This must be submitted to a second coherence filter:

$$\left| ({}^{0,0}\vec{O}_1 \times \vec{T}_1) + ({}^{2,2}\vec{O}_3 \times \vec{T}_2) \right| z \le .001"$$

3.2 Propagation of simulated dispersions.

Two kinematic propagation models are well known. The multiplication of homogeneous transforms representing individually one displacement between 2 RFs, enables the computation of an exact solution to propagation [Craig, 1986]. The kinematic tolerancing theory from Rivest [Rivest, 1993] uses this approach. Simulation is performed by iteration to obtain a series of events which are resulting from multiplying each time the homogeneous transforms. In this method, the dispersion contributors from a chain are lost once the homogeneous transforms are multiplied.

To the first order, an approximation can be obtained by using the Jacobian matrix; its description can be found in [Paul, 1981]. This method, as demonstrated in [Lapperrière et al, 1998] enables computation of a resultant dispersion using a resolution system expressed as follows:

$$\underset{(6\times1)}{^{n,0}\tau_n} = \underset{(6\times6(n+1))}{J} * \underset{(6*(n+1)\times1)}{\left[^{0,0}\tau_0 \quad ^{1,0}\tau_1 \quad \ldots \quad ^{n,n-1}\tau_n \right]^T}$$

In our example, and because of the mechanical nature of the problem, the Jacobian is constructed by stacking a number of sub-matrices (6x6) equivalent to the number of RFs in the manufacturing chain built to analyze a feature. Each of the sub-matrices in the Jacobian is resulting from the inverse product of kinematic transforms, between an intermediate RF and the final RF. In a general case, of n RFs we have:

Mechanical chain: $\underbrace{^1_0C \rightarrow ^2_1C \rightarrow ^3_2C}_{Op\acute{e}ration1} \rightarrow \underbrace{^4_3C \rightarrow ^5_4C \rightarrow ^6_5C}_{Operation2} \rightarrow \ldots \rightarrow ^n_{n-1}C$

$Jacobian = \left[^n_0C \quad ^n_1C \quad ^n_2C \quad \ldots \quad ^n_nC \right]$ where $^n_0C = ^n_{n-1}C .. ^3_2C ^2_1C ^1_0C, \; \ldots$

3.3 Dispersion zones generation.

Once the manufacturing chaining of RFs is determined for the simulation objective, the Jacobian matrix can be built and Monte-Carlo simulation can begin. The simulator generates a given number of coherent dispersions twists which can be stacked as columns in an input matrix. Because the Jacobian never changes for a given manufacturing process plan on a part or for a given assembly, the input matrix can be pre multiplied by the Jacobian matrix to obtain the cumulative twists' matrix. Each column of the cumulative twist matrix then expresses one possible origin situation of the observed feature RF from its theoretical origin situation and expressed in an analysis RF.

If the dispersion zone is about an axis or a volume, all the cumulative twists can be transported a number of times to simulate the entire zone, by sweeping either randomly or incrementally the size parameters about the relevant entity RF origin.

3.4 Dispersion zones validation.

The validation of the simulated dispersion zones can be performed by analysing a population of coherent twists generated to simulate the manufacturing of plane A, plane D, diameters B and C; refer to figure 1 to identify the planes and diameters.

The position D/A is directly obtained from the min-max z-component of simulated plane D twist events. The parallelism dispersion of plane D/A is obtained by converting the maximal orientation dispersions of a population about the x and y axis to a parallelism specification. The position zone of diameter C with respect to plane A and diameter B is obtained by simulation of the manufacturing process. Each population of twists are obtained using different dimension chains and are expressed in the main reference frame. To verify the true position tolerance of diameter C with respect to plane D and diameter B, the two independent chains must be combined into a single chain, by inverting the plane D orientation dispersion events and by appending them the position dispersion events of diameter C; by doing so, the common set-up dispersions to each direct chains will vanish in the combined chain. By analysing the resultant events in the plane D RF, the position C/D dispersion zone is obtained. The zone sizes of position C/D are smaller than position C/A because the set-up dispersions are contributing to the latter and also because the results are expressed in a different RF. To compute zone sizes, a least square algorithm is used to compute the maximum radius of a cylindrical zone about the least-square mean orientation from the population of simulated twist events. The size of a simulated dispersion zone depends on the number of events generated, as shown in table 1 and the coverage of a zone depends on the number of simulated contributors required to represent a feature. When compared to the drawing specification zone size computed as per a common worst case tolerance stack-up at MMC, any of the simulated dispersion zone diameter is included in the specified tolerance zone.

Population size	Position D/A	Parallelism D/A	Position B/A	Position C/A	Position C/D	Minimum-fit zone
10	±.0077"	.0014"	∅.0049"	∅.0066"	∅.0057"	∅.0082"
100	±.0085"	.0016"	∅.0050"	∅.0099"	∅.0079"	∅.0062"
1000	±.0095"	.0017"	∅.0050"	∅.0124"	∅.0086"	∅.0057"
5000	±.0096"	.0019"	∅.0050"	∅.0129"	∅.0084"	∅.0051"
Dwg spec.	±0.010"	0.005"	∅.0100"	∅.0163"*	∅.0100"	∅.0000"*

*Table 1: Dispersion zone size with number of simulated events.(*computed by stack-up)*

3.5 Assess assemblability

The available play, between diameters C and B of two mating parts, depends primarily on the difference of size between the mating diameters which can be expressed as a cylindrical dispersion zone between features. In the example, the size of this zone is calculated to be ɩ.020"-.060" x .5"L when there are no orientation dispersions of the mating diameters. Then all orientation dispersions from any mating features involved & orientation of planes or perpendicularity of diameters to planes & must be subtracted from the zone diameter to get the minimum fit zone. Any position dispersion will not alter the assemblability when the problem is reduced to diameters insertion as in the example. For a simulation purpose, the problem results in simulating the orientation dispersion contribution to compute the minimum fit zone from in table 1. By worst stack-up calculation, the fit zone diameter would thus be reduced to .000"-.060". This result also implies that there is no alignment optimization per rotation of the parts about their z-axis; such an optimization would be required when the computed assemblability zone would become negative. In such a case, the mating is then restricted by the diameters fit at MMC condition and a different dimensional chaining would be required or the size of the diameters should be modified to obtain a positive zone fit. From table 1, the position C/D zone value is always positive ensuring a plane-to-plane predominant contact with any population size.

3.6 Simulating an assembly of 3 parts.

After verifying by simulation the aptitude of a process plan to produce good parts, the same simulation technique can be used to model an assembly of three parts stacked on top of each other. The objectives are then to compute resultant parallelism of plane D_3 at one end of the assembly, on the 3rd part, with respect to plane A_1 at the other end of assembly, on the first part. Also, the resultant position dispersion zone of diameter C_3 with respect to A_1 and B_1 can be simulated. Figure 2 presents an image of uniformly distributed coherent events simulating dispersion zones without position optimization of the parts in the assembly, of plane D_3 and diameter C_3 at MMC by randomly sweeping ten times the normal to each plane D_3 events, located at 7.5" from plane A_1, and diameter C_3 axis within a .5" length from plane D_3. By analysing 1000 simulated assembly events, the position 7.5" ±.0245" and parallelism .0046" of plane D_3 to plane A_1 are obtained. The minimal dispersion zone of diameter C_3 is equivalent to a positioning zone 1.038". This simulated minimal dispersion zone help to compute the minimal fit zone of each part mating or the co-axiality of diameters B_1 and C_3.

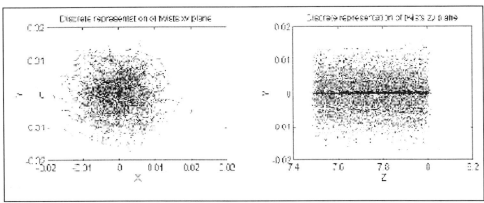

Figure 2: Discrete representation of dispersion zones of D_3,C_3 from A_1, B_1.

4. CONCLUSION AND FUTURE DEVELOPMENT.

The quality of any simulation tool relies on its replication fidelity when compared to the genuine process. To validate a process plan simulator the following two issues must be addressed: to ensure the correctness of the coherence filters and to propose a method to delineate any manufacturing process aptitude to produce a feature. The first issue can be dealt with by comparing the simulation results with metrological results on a population of real parts; the second issue can be solved by using the metrological results from a population of real parts to refine the coherence filters. However, analysis results must then cope with any form defect of the real features. In a simulation tool, the exact representation of form defects represents an entire research field and is, most of the time, unnecessary when the dispersion domains are bounded with representative values describing any possible mating contact. These issues will be addressed by using a rapid prototyping technology to produce a population of real parts.

As demonstrated, the coherence filters subordinate dispersion parameters of one or more input twists to linear or quadratic geometrical constraints. Quadratic constraints are not easy to consider in an analytical resolution system of the tolerance allocation problem. Simulation of features followed by metrological analysis techniques could then be preferred to replicate the inherent statistical distribution of a cumulative dispersion chain. Another interesting application of simulation is to replicate known deviations of a part with representative coherence filters to review the deviation severity of discrepant parts or on how to rework them. This implies to replace the Monte-Carlo simulator with a deterministic simulator.

The simulation of a tolerance zone is a different problem from the dispersion zone simulation. A full 3-D tolerance zone simulator tool needs to generate each cumulative tolerance zone limit by finding the worst combination of tolerances in any spatial direction. The generation of a zone from discrete points can help to define general validity domains of a feature. Using the Jacobian technique, this is feasible for short propagation chains. However with long propagation chains, the possible situation domain of the observed feature involves too many tolerances to extract efficiently, the bounding combination of tolerances. This issue is now being addressed.

Finally, the process plan simulator also finds an important outcome to provide a good basis for statistical tolerance allocation per ASME Y-14.5 standard. Once the manufacturing processes capabilities are known on each part, it was shown that a virtual assembly simulation enables more complex analysis using known techniques. It is believed this could result in important savings when designing complex assemblies made using hundreds of manufacturing processes.

REFERENCES:

[Ballot, 1995] Ballot, É., "Lois de comportement géométrique des mécanismes pour le tolérancement", *Thèse de doctorat*, École normale supérieure de Cachan, 1995, 247p.

[Bennis et al., 1998] Bennis, F.; Pino, L.; Fortin,C.; "Geometric tolerance transfer for manufacturing by an algebraic method", *Proceedings of IDMME'98 conference*, Compiègne, France, may 27-29 1998, pp. 713-720.

[Bialas et al, 1997] Bialas, S., Humienny, Z., Kiszka, K., "Relations between ISO 1011 geometrical tolerances & vectorial tolerances - conversion problems", *5th CIRP conference on CAT*, Toronto, 1997, pp. 37-48.

[Clément et al, 1997] Clément, A., Rivière, A., Serré, P., Valade, C., "The TTRS: 13 constraints for dimensioning and tolerancing", *5th CIRP conference on CAT*, Toronto, 1997, pp. 73-82.

[Craig, 1986] Craig, J.J.; "Introduction to robotics, mechanics & control", *Addison-Wesley publishing company*, 1986, 303p.

[Cubberly et al., 1989] Cubberly, W.H., Bakerjian, R., "Tool and Manufacturing Engineers Handbook, Desk Edition", *Society of Manufacturing Engineers*, 1989.

[Gaunet, 1994] Gaunet, D., "Modèle formel de tolérancement de position. Contributions à l'aide au tolérancement des mécanismes en CFAO.", *Thèse de Doctorat*, Laboratoire de mecatronic de L'ISMCM-École Normale Supérieure de Cachan, France, 1994, 339 p.

[Ianuzzi, 1995] Ianuzzi, M.P., "Tolerance optimisation using genetic algorithms: benchmarking with manual analysis", *Proceedings of the 4th CIRP design seminar*, University of Tokyo, Japan, april 5-6 1995, pp. 219-234.

[Le Pivert, 1998] Le Pivert, P.; "Contribution à la modelisation et à la simulation réaliste des processus d'usinage", *Thèse de doctorat*, École centrale de Paris, 1998, 164 p.

[Lapperrière et al., 1998] Laperrière, L., Lafond, P., "Identification of dispersions affecting predefined functional requirements of mechanical assemblies", *Proceedings of IDMME'98 conference*, Compiègne, 1998, pp. 721-728.

[Martinsen, 1995] Martinsen, K., "Vectorial tolerancing in manufacturing systems", *Dr. Ing. Thesis*, Department for production and quality engineering, Norwegian Institute of technology, University of Trondheim, Norway, 1995, 139 pp.

[Nassef et al, 1993] Nassef, A.O., El Maraghy, H.A., "Probabilistic analysis of geometric tolerances", *4th CIRP Seminar on CAT*, Tokyo, 1993, pp. 187-203.

[Paul, 1981] Paul, R. P., "Robot manipulators: Mathematics, programming, and manipulators", *The MIT press*, Cambridge, Massachusets, 1981, 279p.

[Rivest, 1993] Rivest, L., "Modélisation et analyse tridimensionnelles des tolérances dimensionnelles et géométriques", *Thèse de doctorat*, November 1993, 185 p.

[Wirtz, 1991] Wirtz, A.; "Vectorial tolerancing for production quality control and functional analysis in design", *CIRP International working seminar on computer-aided tolerancing*, Penn State university, 16-17 may 1991, pp. 77-84.

A three dimensional tolerance transfer methodology

Dr. Alain Desrochers and Sander Verheul (U. of Twente)
École de technologie supérieure
Department of Automation and Production Engineering
1100, Notre-Dame West, Montréal, H3C 1K3, Canada
Telephone : (514) 396-8930, Fax : (514) 396-8595, E-Mail : alain@gpa.etsmtl.ca

Abstract: This paper presents the adaptation of tolerance transfer techniques to a model called TTRS for Technologically and Topologically Related Surfaces. According to this model, design and process plan specifications can both be represented as a succession of surface associations on a combined representation tree. This tree can be enriched by the addition of technological attributes such as dimensional and geometrical tolerances. This provides the basis of a proper framework for the computation of tolerance transfer. Rules are indeed established to simulate tolerance chains or stack up which are generated according to paths or loops on the TTRS tree. Sets of equations are therefore generated and solved for each of these loop.

The principles presented above, are validated on a realistic example.

Keywords: CAD/CAM; tolerance; surfaces; stack up; transfer; TTRS

1. INTRODUCTION

Modern CAD/CAM systems allow three dimensional modelling of mechanical parts and can simulate their machining on the screen, while generating the machine codes. Nevertheless, a large gap still exists between the "design" (CAD) and "manufacturing" (CAM) part of these software packages. This gap should be filled by completing the geometrical model through the addition and integration of additional technological information.

1.1 Introduction to the method

The method presented in this paper combines the design specifications and the unknown process plan specifications in one three dimensional model.

The model applies the TTRS (Technologically and Topologically Related Surfaces) approach, which is based on the successive association of elementary surfaces. Using this approach, the design specifications and process plan specifications can be modelled by either separate or combined trees or graphs. Traditional tolerance transfer techniques can then be applied to solve the unknown process plan specifications.

The TTRS model applies to three dimensional parts and has been the topic of several papers [Desrochers et al., 1994, 1995].

1.2 Bibliographical review

The domain of tolerance transfer has been the topic of active research in recent years. Numerous graphical approaches have been used for the representation of tolerance chains. One, is the graph representation [Irani et al., 1989] for identifying the dimensional chains. Another is the matrix-tree-chain [Tang et al., 1988] which uses a matrix to represent blueprint dimensions and working dimensions. A third approach, based on directed trees has been used [Ji et al., 1996] to represent working dimensions, blueprint dimensions and stock removal. Other original graphical approaches have been proposed: the first is based on the construction of imbedded "windows" directly from the process plan [Ngoi et al., 1996] while the second, called the "maze chart" method, uses a pie-chart to represent the process links [Ngoi et al., 1997]. All methods presented mainly address the representation and identification of dimensional chains.

2.   THE TTRS MODEL

In this section the TTRS approach will be explained and applied to represent design specifications as well as process plan specifications.

2.1 Definitions

The TTRS model is based on a recursive binary association process applied to elementary surfaces forming TTRS objects. These associations are represented by partial or complete datum systems called Minimum Geometric Datum Elements, or MGDE. Three *basic* MGDE, which are the plane, the line and the point, are combined and used to construct those theoretical reference frames. More specifically, primary datum will be represented by a plane, secondary datum by a line and tertiary datum by a point. The association process can be represented by a hierarchical tree or graph.

In addition to the resulting MGDE, the TTRS objects bear a content which is both topological and technical. The topological aspect includes the geometry, and relative position of the surfaces or TTRS involved in the association. The technological aspect can include optional aspects such as surface finish, orientation, position and dimensional tolerances.

2.2 Product representations

A product is initially represented by its design or blueprint specifications. When machining however, different dimensions and tolerances will be required for the process plan representation. The TTRS method can be applied to both representations.

In the case of design specifications, a TTRS will be generated by an association between a datum and a surface (or TTRS). Design dimensions and tolerances can be added to the TTRS object as attributes. Functional features can also be defined by the TTRS method and then be included as such in the final TTRS representing the part (figure 3).

In the case of process plan specifications, a TTRS will be generated by an association between a setup and a surface which is to be machined. Process plan dimensions and tolerances can also be added to the TTRS objects. Moreover, machining features or multiple surfaces created in a single machining operation, can be modelled and added directly to the process plan representation tree (figure 3).

In both the design representation and process plan representation tree, the MGDE of the last datum system or setup has no influence and can therefore be discarded. It should be mentioned though, that it does represent the reference frame of the part. The design representation tree and the process plan representation tree can be combined as they share the same elementary surfaces (figure 3).

3. TOLERANCE TRANSFER WITH THE TTRS METHOD

Tolerance transfers are required whenever design specifications can not be achieved directly in one single machining operation. In those cases, computations have to be carried out to evaluate the machining precision required in order to reach the desired tolerance on the blueprint. In terms of TTRS representation, tolerance transfer will ensue from the difference between the TTRS design and process plan trees.

3.1 Tolerance stack up in general

Tolerance stack up or tolerancing chain is a well known method for tolerance transfer which can be described by two principles :

The first principle : 'every dimension on the blueprint can be expressed as the *signed* summation of the dimensions extracted from the process plan'

$$\text{Design dimension} = \sum \text{Process plan dimensions}$$

The second principle : 'every dimensional tolerance on the blueprint can be expressed as the *unsigned* summation of the required machining tolerances'.

$$|\text{Dimensional tolerance}| = \sum |\text{Machining tolerances}|$$

Solving the equations produced by these principles will yield the process plan specifications.

3.2 Rules for tolerance specification and transfer with the TTRS method

Dimensioning and tolerancing with the TTRS method implies that a set of relations between the basic MGDE's of the associated surfaces or TTRS are derived. This is discussed in detail in [Desrochers et al., 1994].

The application of these principles is illustrated in figure 4 for the example that will be presented in the following section.

In addition to the conventional stack up principles, rules will now be stated for deriving the stack up equations from the combined representation tree.

Rule 1: The process plan representation tree must be a valid TTRS tree. This implies that the machining sequence in the tree may differ slightly from the original process plan sequence *within a same setup*.

Rule 2 : For each design specification, one equation must be extracted from the process plan representation tree.

Rule 3 : The equations are generated by following the loop from the base of the process plan representation tree from one basic MGDE to another, adding the appropriate dimensions and tolerances in the loop according to the stack up principles.

Rule 4 : A dimension is negative in an association if the associated surface has a negative distance from the current surface, according to a given reference frame

Rule 5 : The dimensional and tolerancing information related to a TTRS object should not be included in the equation if the corresponding MGDE has not evolved from the previous association. However, the information related to the last TTRS in the loop should always be added.

In the process plan representation tree, the MGDE's represent the setup used in the next machining operation. This implies that the corresponding TTRS representing a setup should always be generated just prior to using it. This could cause some minor difference with the actual process plan. This is the basis for rule 1.

Rules 2, 3 and 4 result from the stack up principles stated in section 3.1 when these are applied to the tree structure.

In the process plan representation, only the dimensions and tolerances which are related to the design specifications should be taken into account for a given setup. The application of rule 5 will eliminate all irrelevant dimensional and tolerancing information from the loop. The association sequence should be followed to ensure consistency.

Also, when working dimensions in three dimensions (3-D), it has proven practical when these are oriented along the reference frame axis. For instance, a distance should be represented by its components parallel to the axis of the global reference frame.

3.3 Geometrical tolerance transfer

Traditionally, positional tolerance transfer is computed by applying the same calculations as for dimensional tolerances. To enable this, the three dimensional (3-D) tolerance zone must be split in three unidirectional (1-D) cases. Only translations within the tolerance zone are considered and possible rotation are therefore discarded. When this approach is applied, the rules and principles presented above can be used for positional as well as for dimensional tolerances.

The influence of rotational tolerances will increase as the number of tolerances in the tolerance stack up increases. If the rotations are taken into account in a stack up, the accumulation of the angles will result in a real 3-D representation of the corresponding tolerance zone. However possible, this will result in computational complexity which can be dealt with, using homogeneous transformations and screw parameters [Desrochers et al., 1997 a) and b)].

Angular deviations are mostly caused by geometrical machine uncertainties, which the machine operator cannot control. Translational uncertainties however, results from a deviation from the exact value of a machine setting. This can be controlled by the operator. As the machine is made very precise, the translational uncertainties will usually overrule the rotational defects. As the scope of this paper is to present a method that supports traditional stack up methods, rotational tolerance transfer will not be taken into account here.

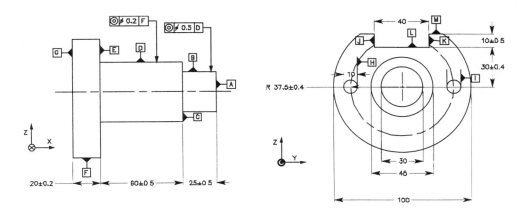

Figure 1; *Design specifications of the part*

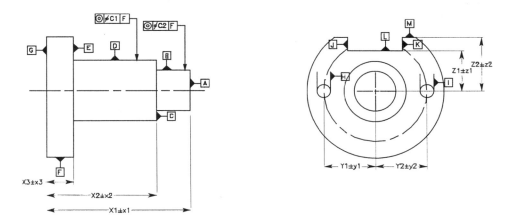

Figure 2; *Process plan specifications of the part*

4. APPLICATION TO A 3-D TEST CASE

The principles and rules presented in previous sections will now be applied to a three dimensional example including both dimensional and positional tolerance stack up.

4.1 The part

The design specifications of the part are presented in figure 1 while its machining specifications are shown in figure 2 for the corresponding process plan:

With FG as setup : machining of A, DE, BC

With DE as setup : machining of H, I, JKL, M

Surfaces B,C and D,E can be combined in two 'shoulder feature', with the surfaces C and E acting as shoulders while surfaces J, K and L can be combined in a 'keyway feature'. These features are illustrated in figure 3. The corresponding TTRS for these features can be directly added to the combined representation tree, which is presented in the same figure.

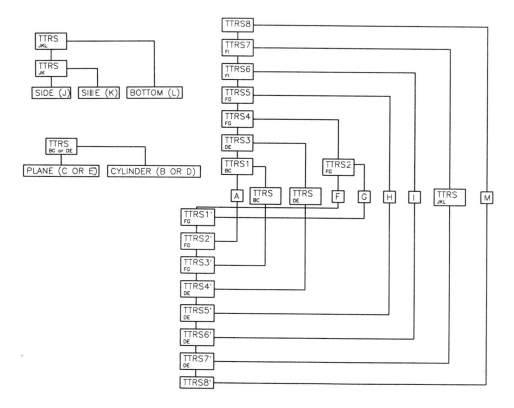

Figure 3; *TTRS representation tree for the part*

It can be seen, in the process plan side of the representation tree, that BC is associated before DE while in the original process plan, DE is machined first. This results from the fact that DE is being used as the set up in the following machining sequence. This is taken into account in rule 1 which in turns has no influence on the outcome of the tolerance transfer.

4.2 Tolerance stack up calculations

The tolerance stack up calculations will now be performed for dimensional and positional design specifications according to the principles and rules stated in section 3.1 and 3.2. For calculating $X1 \pm x1$, $X2 \pm x2$ and $X3 \pm x3$ the set of equations will be derived from the process plan tree.

$X3 \pm x3$ is the dimension and tolerance between surface G and E and is contained in TTRS4. A loop between these surfaces must be found in the process plan representation tree. Applying rule 5 from section 3.3, the appropriate set of TTRS objects can be derived, which is shown in the first row of figure 4. The arrows indicate the direction the loop passes through the TTRS objects, in order to determine positive and negative dimensions in the stack-up equation. In the TTRS objects, the sequence of the letters represents the order of precedence of the corresponding basic MGDE's. For example, in TTRS4, D represents the line and E the point.

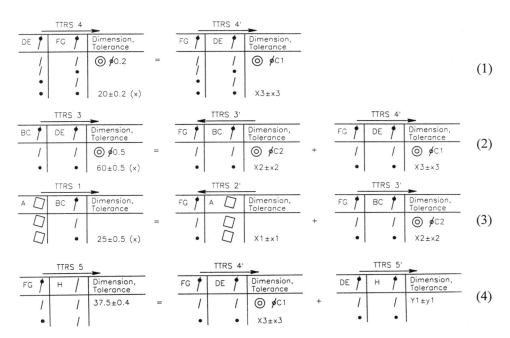

Figure 4; *TTRS equations derived from loops on the representation tree*

$X2 \pm x2$ represents the distance and tolerance between surface G and C while $X1 \pm x1$ represents the distance and tolerance between surface G and A. The appropriate set of TTRS objects are presented in rows 2 and 3 of figure 4. The irrelevant basic MGDE relations have been discarded. The equations for calculating the process plan specifications can now be derived from the TTRS objects :

TTRS4 $= -$TTRS4'
EG $= -$GE $=$ EG $\qquad$ (1)
$-(20 \pm 0.2) = -(X3 \pm x3)$

TTRS3 $=$ TTRS3' $+$ TTRS4'
CE $=$ CG $+$ GE $\qquad$ (2)
$-(60 \pm 0.5) = -(X2 \pm x2) + X3 \pm x3$

TTRS1 $=$ TTRS2' $+$ TTRS3'
AC $=$ AG $+$ GC $\qquad$ (3)
$-(25 \pm 0.5) = -(X1 \pm x1) + X2 \pm x2$

Solving these equations yield:
$X1 = 105$ and $x1 = 0.2$; $X2 = 80$ and $x2 = 0.3$; $X3 = 20$ and $x3 = 0.2$

Regarding the concentricity tolerances, C1 and C2 can be calculated using row 1 and 2 from figure 4. As only tolerances are calculated, only absolute values need to be considered.

TTRS4 $=$ TTRS4' and TTRS3 $=$ TTRS3' $+$ TTRS4'
DF $=$ FD $\qquad$ BD $=$ BF $+$ FD
$\varnothing 0.2$ $=$ $\varnothing C1$ $\qquad$ $\varnothing 0.5$ $=$ $\varnothing C1 + \varnothing C2$
Yielding : $\varnothing C1 = \varnothing 0.2$ and $\varnothing C2 = \varnothing 0.3$

$Y1 \pm y1$ (distance between D and H) will now be calculated using the last row of figure 4. The irrelevant basic MGDE relations have again been discarded.
$$37.5 \pm 0.4 = \varnothing C1 + Y1 \pm y1$$

$\varnothing C1$ was found to be 0.2 in the previous calculations. In this case it should be interpreted as ± 0.1 which yields $Y1 \pm y1 = 37.5 \pm 0.3$

The remaining tolerances can be calculated similarly, extracting the equations from the appropriate TTRS objects :
$Y2 \pm y2 = 37.5 \pm 0.3$ and $Z1 \pm z1 = 30 \pm 0.3$ and $Z2 \pm z2 = 10 \pm 0.2$

$Y2 \pm y2$ is equal to $Y1 \pm y1$, as could be expected. In the process plan representation tree however, H is associated to D and I is associated to D separately. This means that both dimensions and tolerances should be represented in the model.

All process plan dimensions and tolerances have now been calculated in all directions of the reference frame (x,y,z). The necessary equations have been derived from the three dimensional model in a systematic manner.

5.   CONCLUSION

In this paper a three dimensional model has been presented from which the process plan specifications can be calculated using the traditional tolerance stack up principles. The TTRS model has proven to be a valid approach for representing both the design specifications as well as the process plan specifications. It can be used in combination with appropriate rules to calculate the process plan specifications in a systematic manner, which enables the integration of the method in CAD/CAM software.

Future work
The presented method supports traditional tolerance transfer techniques for dimensional and positional tolerances. It enables these techniques to be applied to three dimensional cases. It would be desirable however to be able to calculate all process plan tolerances (perpendicularity etc.) in a systematic manner. In addition to translational tolerances, rotational tolerances would have to be supported. This introduces computational complexities as discussed in section 3.3. Work has been done however to represent tolerance zones three dimensionally with matrices and screw parameters [Desrochers et al., 1997 a) and b)]. These approaches could be used in conjunction with the presented method. Investigations in these directions have proven to be promising.

ACKNOWLEDGMENTS

The principal author wishes to acknowledge the support of the Natural Science and Engineering Research Council of Canada through its individual grant program.

REFERENCES

[Desrochers et al., 1994] Desrochers, A.; Clément, A.; "A dimensioning and tolerancing assistance model for CAD/CAM systems", *Int. J. of Advanced Manufacturing Technology*, Vol. 9, 352-361.

[Desrochers et al., 1995] Desrochers, A.; Maranzana, R.; "Constrained dimensioning and tolerancing assistance for mechanisms", *4th CIRP Seminar on Computer Aided Tolerancing*, The university of Tokyo, Tokyo, Japan.

[Desrochers et al., 1997 a)] Desrochers, A.; Rivière, A.; "A matrix approach to the representation of tolerance zones and clearances", *Int. J. of Advanced Manufacturing Technology*, Vol. 13, 630-636.

[Desrochers et al., 1997 b)] Desrochers, A.; Delbart, O.; "Determination of part position uncertainty within mechanical assembly using screw parameters", *5th CIRP Seminar on Computer Aided Tolerancing*, Toronto, Canada.

[Irani et al. 1989] Irani, S. A.; Mittal, R. O.; Lehtihet, E. A.; "Tolerance chart optimization", *International Journal of Production Research*, Vol. 27, No. 9, 1531-1552.

[Ji et al., 1996] Ji, P.; Fuh, J. Y. H.; Ahluwalia, R. S.; "A digraphic approach for dimensional chain identification in design and manufacturing", *Transaction of the ASME: Journal of Manufacturing Science and Engineering*, Vol. 118, 539-544,

[Ngoi et al., 1996] Ngoi, B. K. A.; Seow, M. S.; "A Window approach for tolerance charting", *International Journal of Production Research*, Vol. 34, No. 4, pp. 1093-1107.

[Ngoi et al., 1997] Ngoi, B. K. A.; Tan, C. S.; "Graphical approach to tolerance charting – A "Maze chart" method", *Int. J. of Advanced Manufacturing Technology*, Vol. 13, 282-289.

[Tang et al., 1988] Tang, X. Q.; Davies, B. J.; "Computer aided dimensional planning", *International Journal of Production Research*, Vol. 26, No. 2, 283-297.

Transforming ISO 1101 Tolerances into Vectorial Tolerance Representations – A CAD-Based Approach

Werner Britten, Christian Weber

University of the Saarland, Department of Engineering Design/CAD
P.O. Box 151150, D-66041 Saarbruecken, Germany
phone: +49 681 302-3607, fax: +49 681 302-4858, e-mail: britten@cad.uni-sb.de

Abstract: Computer aided tolerancing systems (CAT systems) have to represent tolerance information unambiguously in E^3 to perform 3D tolerance analysis. However, current international standards in the field of geometrical dimensioning and tolerancing (GD&T) often lead to misunderstandings concerning the 3D-consequences of tolerances and their combinations. In this paper a concept of a CAD system with extended tolerance support is outlined. As in commercial 2D / 3D-CAT systems a conversion of GD&T tolerances into 3D-representations is necessary. Therefore, a transformation procedure for integration into CAD systems is described. An angularity and position tolerance at a plane is given as an example. The contribution closes with a brief outlook at the validation of the concept.

Keywords: CAT, Geometrical Product Specification, Vectorial Tolerancing, CAD

1. INTRODUCTION

The tolerancing technique of current international tolerance standards evolved mainly from machine-shop and drafting room practice. So, standards like ISO 1101 or ISO 286 focus on the syntax of graphic tolerance representations which are mainly intended for human use. However, computer aided tolerance software requires clear input data and a mathematical sound theory to provide support to designers during the tolerance assignment and analysis process. Commercial CAT software vendors already claim that their systems could create unambiguous 3D-tolerance models out of given drawing indications according to ISO 1101. In our opinion further research in this field is necessary to bridge the gap between current geometrical dimensioning & tolerancing (GD&T) standards and an exact computer embedded tolerance representation.

Vectorial dimensioning and tolerancing (VD&T) has proved to be an unambiguous tolerance representation based upon the co-ordinate geometry and the co-ordinate measurement technique. Furthermore the VD&T data structure is in line with those computer internal models employed in CAD-, NC- and CMM-software and therefore suits well for the establishment of a CAD-NC-CMM control loop. However, the completely new kind of tolerance scheme, the necessity of an altered drawing indication and the great number of existing drawings (which comply with ISO-tolerance standards, GD&T) have prevented a broader use of VD&T in engineering practice. In the endeavour to cope within a VD&T environment especially with older drawings the idea of converting GD&T schemes into VD&T indications evolved a few years ago.

In our approach the CAD system with its powerful 3D-capability and its application interface is proposed to alleviate the problems resulting mostly out of the ambiguity of the current tolerance standards. So, the ISO-based tolerance user interface of the devised CAD-embedded system leads the designer through the tolerance assignment process. Thus unambiguous tolerance combinations or – in the sense of the transformation procedure – non-regular datum systems can be avoided. The transformation procedure will be performed semi-automatically with designer's interaction, e.g. for reference point definition. After the calculation of vectorial tolerance values a tolerance analysis or the visualisation of deviations can be performed within the CAD environment. The analysis functionality needs not to be as sophisticated as those provided by commercial CAT-Software. Instead, we focus on an easy-to-use system for frequent use by the designer who is not necessarily a tolerance expert.

2. CAD SYSTEM WITH EXTENDED TOLERANCE SUPPORT

The proposed concept of a CAD system with extended tolerance support and its integration into co-ordinate measurement technique is illustrated in Figure 1.

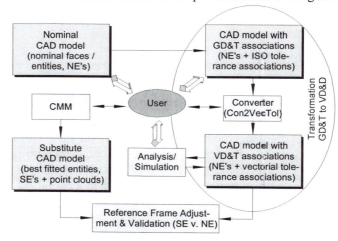

Figure 1; Concept of a CAD system with extended tolerance support.

After the nominal CAD model (upper left in Figure 1) of a part or an assembly is created, conventional tolerance associations are defined interactively by the user. The designer feels familiar with the conventional toleranced CAD-model. For tolerance analysis and simulation a VD&T-model is used to capture the tolerance information (lower right in Figure 1). The reason is that the CAD model with vectorial associations suits very well for tolerance validation. Besides the CAD-VD&T model another CAD model with substitute faces is necessary for reference frame adjustment and validation. Substitute elements result from best-fit operations applied on measured point clouds delivered by the co-ordinate measurement machine (CMM, left in Figure 1).

A converter, noted 'Con2VecTol' in Figure 1, is necessary to transform the interactively defined ISO tolerance scheme into the VD&T model which is used for analysis purposes internally. This step from the ISO-tolerance world into a mathematically sound representation in E^3 often causes trouble not only in our system but also in commercial CAT systems. Looking at Requicha's classification of tolerance models [Requicha, 1993] these problems arise because a tolerance model with tolerance *zone* semantics (like GD&T) is transformed into a system with *parametric* tolerance semantics (like VD&T).

3. TRANSFORMATION FROM GD&T TO VD&T

In this section two different transformation procedures are presented briefly. The first one devised by an ISO advisory group within the GPS (Geometrical Product Specification) initiative focuses on basic steps and restrictions necessary to perform a conversion. Additionally, our own proposal stresses the support which can be provided by a CAD system during the transformation process.

3.1 Transformation procedure devised by ISO / TC 213 advisory group 3
The relationship between both tolerancing systems has been examined recently and the layout of a transformation procedure was elaborated. The procedure conceived by Bialas in his ISO proposal [AG3-N27, 1997] is summarised in Figure 2.

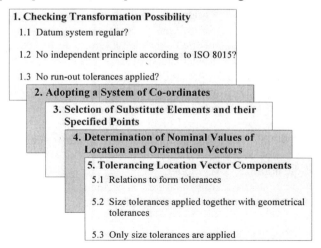

1. Checking Transformation Possibility
  1.1 Datum system regular?
  1.2 No independent principle according to ISO 8015?
  1.3 No run-out tolerances applied?

2. Adopting a System of Co-ordinates

3. Selction of Substitute Elements and their Specified Points

4. Determination of Nominal Values of Location and Orientation Vectors

5. Tolerancing Location Vector Components
  5.1 Relations to form tolerances
  5.2 Size tolerances applied together with geometrical tolerances
  5.3 Only size tolerances are applied

Figure 2; Transformation procedure from GD&T to VD&T [AG3-N27, 1997].

The procedure starts with the checking of the transformation possibility and comprises five major steps altogether. Since Bialas does not consider support provided by the CAD system steps like 'determination of nominal values of location and orientation vectors' are easier to perform in a CAD embedded system outlined in Figure 1.

3.2 CAD embedded transformation procedure

The following CAD embedded transformation procedure takes advantages out of the functionality provided by the CAD system. In Figure 3 it is symbolised as a bridge which fills the gap between the GD&T and the VD&T bank.

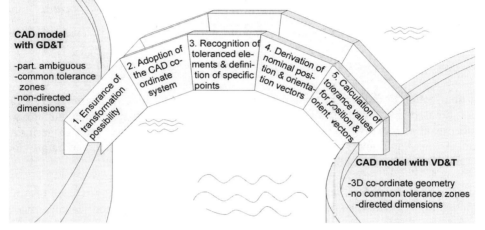

Figure 3; *CAD embedded transformation procedure from GD&T to VD&T.*

In the first step an interactive tolerance specification module within the embedded system ensures that only those ISO tolerances and their combinations are associated with nominal geometric elements of the CAD model which do not cause trouble during transformation. Changes like directed instead of undirected dimensions or restrictions concerning special kinds of ISO-tolerance combinations are necessary to provide an unambiguous tolerance scheme. Other changes within the tolerance input module called 'ConTol' issue from the face-oriented VD&T approach. Since the vectorial tolerance approach depends largely on the natural or model parametrisation of *faces*, line-based tolerances according to ISO 1101 (e.g. straightness or circularity) can not be represented in VD&T easily. So, not every tolerance provided by GD&T can be associated with every face type of the nominal CAD model.

It should be mentioned that the measures implemented in this stage may lead to losses concerning the traditional functional intend of some conventional tolerances or the traditional way of inspection. Until now tolerance practice has often denied adoptions of new interpretations as a consequence of innovations e.g. in the field of inspection technique. After the rejection of 'pure' vectorial tolerancing time will tell if an

ISO tolerance approach which is restricted in this way will survive. However, this transformation preparations within the conventional tolerance advisory module called 'ConTol' make a separate check after the assignment of tolerances redundant.

The adoption of the CAD model's co-ordinate system as the global tolerance frame of an assembly is performed in the second step. In the third phase geometric elements of the nominal CAD model associated with GD&T information are automatically recognised and the workpiece and element co-ordinate systems (WCS and ECS) are erected. Evaluation of datum frame information hold in the CAD-GD&T model results in the WCS. Face-inherent ECSs are stated either automatically on the basis of face properties (e.g. centre of gravity as reference point, main moments of inertia as reference axes) or can be declared interactively by the designer. CAD functionality is further employed in the fourth step when nominal vectors for the position or orientation of faces have to be established. Since nominal position vectors of a face are equal to the vectorial difference of ECS and WCS in the nominal model this type of information is inherent in the CAD model.

In the fifth step of the transformation procedure work is focused on a case sensitive calculation of vectorial tolerance values for the recognised elements. Despite lots of algorithms for standard tolerance combinations associated with the treated face types the system can not renounce on designer's support. User interaction in this phase is necessary e.g. to decide how a common tolerance zone has to be divided between the different vectorial tolerance components. This topic will be explained in detail by an example given below.

3.3 Tolerance zones as link between GD&T and VD&T

The transition from a tolerance system which comprises both 13 generic [Rivest et al., 1993] and an almost uncountable number of interfering tolerance zones to a parameter oriented tolerance system looks as follows:

After gathering all of the ISO tolerances connected to a nominal face of the CAD model a kind of 'common tolerance zone' for this face and its associated tolerances is elaborated. The determination of vectorial tolerance values begins thereafter with the decision what amount of the 'common tolerance zone' can be used by what kind of vectorial tolerance component of the surface. The distribution of the whole tolerance space on a face-specific set of vectorial parameters is controlled by the variable q. Variable q consists of a face-specific number of components corresponding to the degree of freedom and the number of size variables of the face. E.g. for planes there is one component controlling the position perpendicular to the plane (z-axis) and two others which bound the deviation of the plane's orientation vector around the x and y-axis.

This procedure is similar to the kinematic structure [Rivest et al., 1993] employ to solve 3D-tolerance transfer problems. The main difference to that very interesting approach is the use of the CAD system's functionality to calculate and to store tolerance values, which describe the allowed variations of the toleranced elements within the 'common tolerance zone'.

In Figure 4 the image of the bridge shown in Figure 3 is given in a more detailed top-view. In the middle of the bridge the different common tolerance zones are depicted. They serve as link between a great number of tolerances, toleranced and datum elements on the ISO side and – in this case – five surface types used at the vectorial bank.

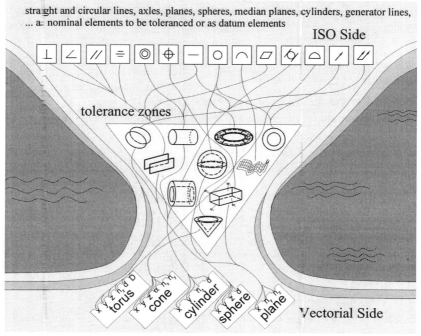

straight and circular lines, axles, planes, spheres, median planes, cylinders, generator lines, ... as nominal elements to be toleranced or as datum elements

Figure 4; Top-view of the bridge between GD&T and VD&T sides.

Though, the bridge as a symbol is not of constant width. With the smallest width at the level of the linking tolerance zones it expands in both directions – the ISO-bank and, to a smaller extent, also in the vectorial direction. That means that we have to access the best suited common tolerance zone for every single and combined tolerance which is possible on the ISO side. The lines which connect the ISO tolerance symbols with the tolerance zones in the middle of the bridge symbolise this step of the procedure. Afterwards the calculation of values of relevant vectorial face parameters starts. Since the tolerance zone has to be distributed among tolerance values of different face parameters, the width of the bridge increases from the middle to the vectorial bank, too.

4. EXAMPLE

In this section an example is presented to explain the procedure described above. The transformation of the angularity and a position tolerance attached to a plane – see

Figure 5 – is demonstrated. On the left in Figure 5 the graphic tolerance symbols of ISO 1101 are used to indicate the starting point of the transformation. Right in the same picture the plane is highlighted that should be associated with an appropriate vectorial tolerance scheme that is in line with the given ISO based tolerance input.

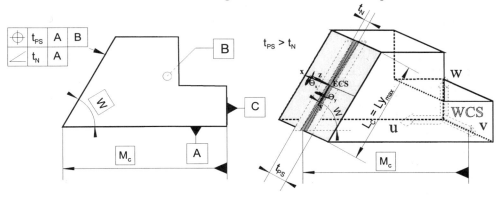

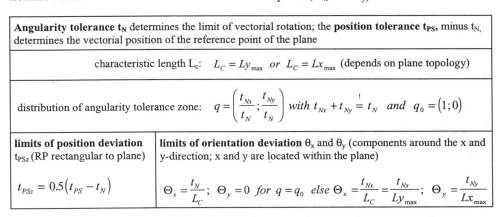

Figure 5; *Angularity and position tolerance at a plane.*

The workpiece co-ordinate system WCS can be adopted unchanged from the GD&T to the VD&T side. The plane's reference frame or element co-ordinate system ECS can be determined by face properties (centre-of-gravity as origin and moment-of-inertia-axes for directions) automatically. The relationships of the ECS related to the WCS and thereby the nominal position and orientation vector of the plane are part of the nominal CAD model and can be retrieved whenever it is necessary. In vectorial tolerancing terms a plane has three tolerance parameters corresponding to the natural or model parameters of the geometric element 'plane' (translation z along the plane's normal vector and the rotations around the x and y-axis, θ_x and θ_y).

Angularity tolerance t_N determines the limit of vectorial rotation; the **position tolerance t_{PS}**, minus t_N, determines the vectorial position of the reference point of the plane	
characteristic length L_c: $\quad L_C = Ly_{max} \ $ or $\ L_C = Lx_{max}$ (depends on plane topology)	
distribution of angularity tolerance zone: $\quad q = \left(\dfrac{t_{Nx}}{t_N} ; \dfrac{t_{Ny}}{t_N} \right)$ with $t_{Nx} + t_{Ny} \overset{!}{=} t_N \ $ and $\ q_0 = (1;0)$	
limits of position deviation t_{PSz} (RP rectangular to plane)	**limits of orientation deviation** θ_x and θ_y (components around the x and y-direction; x and y are located within the plane)
$t_{PSz} = 0.5(t_{PS} - t_N)$	$\Theta_x = \dfrac{t_N}{L_C}; \ \Theta_y = 0 \ for \ q = q_0 \ else \ \Theta_x = \dfrac{t_{Nx}}{L_C} = \dfrac{t_{Nx}}{Ly_{max}}; \ \Theta_y = \dfrac{t_{Ny}}{Lx_{max}}$

Table I; *Vectorial values for a plane with an angularity and a position tolerance.*

The values z, θ_x and θ_y have to be calculated and represent the tolerancing scheme after the conversion. The angularity tolerance t_N determines the limits of the vectorial rotation and the difference between the position tolerance t_{PS} and t_N determines the vectorial position of the plane's reference point (see Table I). A characteristic length L_c and the distribution 'vector' q are necessary to calculate the limits of rotation. L_c – the maximum of the planes extensions along the x and y-direction – can be assessed using CAD functionality. The default value of q is (1; 0) which says that the whole angularity tolerance can be used to calculate the rotation value θ_x. In this case the second rotation component θ_y equals zero (see Table I). As already mentioned, the limits of position deviation t_{PSz} depends on the difference between both conventional tolerances t_{PS} and t_N.

5. OUTLOOK

Based on a vectorial tolerancing prototype integrated into CoCreate's SolidDesigner [Weber et al., 1998] a validation of the outlined concept of a CAD system with extended tolerance support has already started. Yet, the conventional tolerancing module called 'ConTol' – it serves as tolerance advisor during the tolerance specification process – is being implemented. However, a great number of algorithms have to be elaborated for the numerous input configurations the 'Con2VecTol' module has to face. The current GD&T / 3D tolerance interfaces of commercial CAT software may profit from further research in that field of tolerance conversion.

REFERENCES

[AG3-N27, 1997] ISO/TC 213 AG3; Transformation of Geometrical Dimensioning and Tolerancing (GD&T) into Vectorial Dimensioning and Tolerancing (VD&T); *ISO document TC213 / AG3 N27* submitted by S. Bialas, Warsaw Univ. of Technology

[Requicha, 1993] Requicha, A.; Mathematical Meaning and Computational Representation of Tolerance Specifications; In: *Proceedings of the International Forum on Dimensional Tolerancing and Metrology* (ASME, CRTD-Vol. 27), pp. 61-68, Dearborn, ISBN 0-79180-697-9

[Rivest et al., 1993] Rivest, L.; Fortin, C.; Desrochers, A.; Tolerance Modeling for 3D Analysis: Presenting a kinematic formulation; In: *Proceedings of the 3rd CIRP Design Seminar on Computer Aided Tolerancing* (CAT 93), pp. 51-74, Cachan, ISBN 2-212-08779-9

[Weber et al., 1998] Weber, C.; Britten, W.; Thome, O.; Feature Based Computer Aided Tolerancing – A Step Towards Simultaneous Engineering; In: *Proceedings of the 31st International Symposium on Automotive Technology and Automation 1998 (ISATA 98)*, pp. 501-507; Duesseldorf, ISBN 0-95325-760-6

Industrial application of Vectorial Tolerancing to improve clamping of forged workpieces in machining

Oliver Krimmel
Norwegian University of Science and Technology,
Dept. of Production and Quality Engineering
N-7034 Trondheim, Norway
Oliver.Krimmel@protek.ntnu.no

Kristian Martinsen
Materials Technology, Raufoss ASA
N-2831 Raufoss, Norway
kristian.martinsen@raufossas.no

Abstract: Fixturing of forged workpieces in the succeeding machining process is troublesome due to complex shape, small areas suitable for clamping and geometrical deviations typically 10 times larger than the allowed tolerances on the machined surfaces. The deviations on the forged parts can cause randomly insufficient clamping and positioning which leads to random errors on the workpieces from the machining process. This paper describes how Vectorial Tolerancing is used to analyse the interface between the forging process and the machining process. The analysis gives statistics of 3D deviations in orientation, location and form. The Vectorial tolerancing principals are used in the design of the fixtures in order to get the best reproducibility of the workpiece orientation and location while clamped, and sufficient contact and clamping forces. Vectorial Tolerancing principals are also used to design measuring equipment used in the forging department to measure the clampability of the forged parts.
Keywords: Vectorial Tolerancing, Fixtures, Machining, Forging

1. INTRODUCTION

Raufoss ASA are using massive forging on many of their products for defence, water supply and automotive industry. Fixturing of forged workpieces in the succeeding machining process is, however, troublesome. The fixture grips on surfaces with large deviations compared to the tolerances of the finished part. The workpiece can have a complex shape with small areas suitable for clamping. This can result in insufficient clamping forces and difficulties in reproducing the orientation and location of the workpiece. These errors are often random and can be difficult to discover when statistically methods such as Statistical Process Control (SPC) are used. This is a problem especially for mass production, where forging and SPC are widely used.

This paper describes how Vectorial Tolerancing is used to analyse the interface between the forging process and the machining process. Using VT to analyse the process capability is very useful, because of the distinction between size, form, location and orientation. Experiments and analyses are performed on real workpieces in the production line of Raufoss ASA. The parts are forged and afterwards machined in multispindle chucking automatics. The workpieces are automatic or manually loaded, and fixtures in the machine tools are tailor-made for each type of workpiece. The fixtures are two jaws that have a shape as a negative to the workpiece. The part is supposed to be in contact with the jaws over wide areas of the entire surface.

1.1 Related work

Some of the reviewed papers describe flexible automated fixturing systems especially suitable for flexible manufacturing systems, one-of-a-kind and small batch production. [Tuffentsammer, 1981] describes in this keynote paper a numerical controlled fixture system. He defines four levels of automatisation of loading and fixturing. The level of automatisation in the workshop used in this study is level 2 or 3. [Shirinsadeh, 1996] describes various flexible and automated fixturing strategies. This will be strategies for reaching lever 4 of automatisation in [Tuffentsammer, 1981]. [Rong et al., 1997] describes an automated fixture configuration design of modular fixtures for FMS. [Pham et al., 1990] describes an expert CAD system called AutoFix, which configures complex fixtures from a database of modular elements. AutoFix also has a FEM module to simulate the deflexions of the workpiece. However, the methods described in these papers are not very suitable for large volume production.

[Hong et al., 1996] describes an analytical tool for fixture layout design using kinematic modelling. The Kinematic model is very similar to Vectorial Tolerancing. [Cecil et al., 1996] describes a three phased methodology for fixture design. In Phase II: Functional analysis, 3-2-1 locator point strategy is used to fix the position of prismatic workpieces. A similar approach is used in this paper regarding the use of locators and clamps. Cecil et al. have, however, focused on prismatic workpieces and milling and drilling operations. Neither Hong et al. nor Cecil et al. have taken frictional forces and elasticity into consideration. Hong et al. assumes the positionators are ideal points. However, Hong et al.'s model is useful to check if the fixture gives a totally kinematic constraint of the workpiece. [Xiuwen et al., 1996] and [Jeng et al., 1995] has developed models for clamping forces and frictional elastics in the clamping points. In the study described in this paper, only the geometric deviations on the workpiece are taken into account.

2.   VECTORIAL TOLERANCING

Vectorial Tolerancing (VT) was introduced by Wirtz [Wirtz 1988], [Wirtz 1990], [Wirtz 1991], [Wirtz 1993], [Wirtz, et al., 1993]. [Martinsen 1993] showed how VT can be used for all types of surfaces. VT is a three dimensional mathematically unambiguous model for describing geometry and tolerances using vectors. VT provides a clear distinction between the geometrical features size, form, location and orientation for each surface of a workpiece. Such distinction is important for both functional analysis and manufacturing process control [Martinsen, 1995]. The principle of 6 degrees of freedom (DOF) and surface position and deviation described by vectors in a 3D space is the basic concept of VT.

2.1 Coordinate system

To define a 3D coordinate system all 6 degrees of freedom must be fixed. A primary direction defined by a unit vector fix two rotational DOFs. The secondary direction is defined by a unit vector that is perpendicular (orthogonal coordinate system) to the primary direction. Now, all three rotational degrees of freedom are fixed. Finally, the origin of the coordinate system is defined by a point, thereby fixing the three translation degrees of freedom. To define a workpiece coordinate system (WCS), all 6 DOFs are defined by chosen surfaces or combinations of surfaces on the workpiece. There can be several coordinate systems on a workpiece forming a tree structure of coordinate systems. The WCS must be unambiguous and reproducible. The parameters used to define the coordinate system have by definition no deviation.

2.2 Vectorial surface description

An infinite plane can be translated in two directions and rotated in one without changing the location and orientation of the plane. This is called the open DOF for a plane. Similarly, an infinite cylinder can be translated in the axis direction, and rotated around the axis without changing the location and orientation of the cylinder. Hence, a cylinder has 2 open degrees of freedom. In VT, the position of a surface is described with a location vector:

$$\mathbf{P_0} = [X_0, Y_0, Z_0] \tag{1}$$

and a orientation vector:

$$\mathbf{E} = [E_x, E_y, E_z] \text{ where } |\mathbf{E}| = 1. \tag{2}$$

The location vector fixes the translation DOF of a surface and the orientation fixes two rotational DOF of a surface. For surfaces that do not have any well-defined centre point, the point on the surface or on an axis of the surface has to be chosen. The orientation vector of a cylinder is the direction of the axis and for a plane the normal vector. In addition to location and orientation, a surface has a form that are defined by the implicit equation $F(x,y,z) = 0$, or parametric equations $\mathbf{P}(u,v) = [X(u,v), Y(u,v), Z(u,v)]$ [Martinsen 1993], [Martinsen, 1995]. Sizes are constants in these equations, typically the radius of a cylinder.

2.3 Deviation vectors

On a real workpiece, the real location, orientation and size are defined by a geometrically ideal substitution surface calculated by the gaussian best fit algorithm from assessed measuring points. Since gaussian best-fit algorithm is used, the form deviations are separated from the other features an can be treated separately. The real location vector will have a deviation from nominal, and the location deviation can be written:

$$\delta P_0 = P_{0real} - P_{0nom} = [\ \delta x_0,\ \delta y_0,\ \delta z_0] \qquad (3)$$

Similar to the location deviation vector is the orientation deviation vector the nominal orientation vector minus the real orientation vector.

$$\delta E = E_{real} - E_{nom} = [\ \delta E_x,\ \delta E_y,\ \delta E_z] \qquad (4)$$

A deviation between two coordinate systems can be described this way too, however using a primary and a secondary orientation vector. [Martinsen, 1995]

3. METHOD FOR DESIGN OF FIXTURE

The method for design of fixture follows these steps:
1. Analysis of forging process variations using VT
2. Design of fixture based on the vectorial tolerances
3. Analysis of fixture behaviour
4. Design of measuring jig based on the design of fixture

3.1 Analysis of forging process variations using VT

The dies in the forging process studied in this paper, are made of an upper and a lower part, forming two halves of the workpiece. There are basically two types of deviations on the forged parts: Deviations that are dependent on accuracy in the match between the die parts, and deviations that are not. In the first case the deviations on the workpiece will be dependent on the orientation and location deviations between the die halves. Better knowledge of the 3D geometrical forging process capabilities will make it possible to design fixtures which take these into account.

A die coordinate system is defined for both halves; upper die coordinate system, and lower die coordinate system. Nominally, the die coordinate systems and the workpiece coordinate systems are identical. However, the real coordinate systems will have deviations in all 6 DOFs. By measuring the two corresponding halves on the workpiece as two different surface groups, the deviations from the forging process are measured, and the vectorial process variations can be found.

Deviations caused by mismatch between die halves:
- Location deviation between the halves in z-direction: This is the operation direction of the press, and deviations here will be because of deviations in the travel of the press. Typically insufficient closure of the die.

- Location deviation in x and y as well as all orientation deviations are caused by displacement of the die halves. This is due to deflections in the press and play in the centring mechanics.

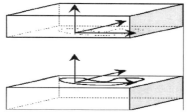

Figure 1; Upper and lower die coordinate systems

Deviations not caused by mismatch:
- Form and size deviations in one half are caused by wear of the dies and insufficient filling in the forging operation. The die wear gives of course a systematic trend for one specific die.

3.2 Fixture design requirements
The fixture should meet the following requirements:

General requirements:
- Minimum total costs
- Safe clamping of all workpieces within allowed deviations
- Minimum number of moving parts
- No problems from accumulation of chips on fixtures
- Minimum deformation of workpiece (sum of deviations from machining and deformation in clamping)
- Minimum fixture wear

Geometrical requirements:
- Accept the x,y,z -components of the location deviation vector $\delta \mathbf{P_0}$. The tolerance can be calculated from the following formula for the process capability, which is based on the normal distribution [Juran 1974]:

$$USL - LSL = c_p 6\sigma \qquad (5)$$

  c_p is chosen to 1,3 in the Raufoss Quality System.
- Accept the orientation deviation error of the upper half geometry in relation to the lower half geometry. The tolerances of the $\delta \mathbf{E}$ vector can also be calculated according to the specification for the location deviation error.

- Accept deviations in cylindricity. The tolerance for the cylindricity has to be calculated form the distribution of the test samples.
- Accept an error of all diameters, also calculated from the distribution.

3.3 Analysis of fixture behaviour

The fixture positionators will uniquely define the workpiece coordinate system if the positionators constraint the workpiece in all 6 DOF [Hong et al, 1996]. By superposing the USL and LSL for orientation, location, size and form, the capability of the fixture compared to the tolerances of the finished part can be calculated.

3.4 Fixture design guidelines

Generally, a fixture is composed of three basic elements: positionators, clamps and supports [Rong et al. 1997]. The positionators should ensure acceptable reproducibility in the positioning of the parts. This means the positionators should restrict the workpiece in all 6 DOFs. Clamps are for holding the workpiece firmly, preferably against positionators. Supports are sometimes applied to decrease deflection of the workpiece.

These are guidelines or "rule-of -thumbs" for the fixture design, derived form the vectorial process capability analyses that have been made. However for some workpieces the guidelines must be violated. These are special guidelines for the fixtures for forged parts. General guidelines for fixtures such as those described in [Cecil et al. 1996] are also valid.

1. The workpiece should be constraint in all 6 DOF. If possible, this should be obtained by the positionators. The positionators should be able to reproduce the workpiece coordinate system within the tolerances on location and orientation workpiece. This can be done by using the method for creating a coordinate system a described earlier. The method described in [Hong et al. 1996] can be used to mathematically check for kinematics constraint. The minimum number of points necessary to mathematically describe an ideal surface can be calculated from the following equation:

$$N_{min} = 6 - N_o + N_R \qquad (6)$$

where N_c is the number of open DOF on a surface, and N_R is the number of sizes.

2. If possible: The workpiece should be positioned using only one of the halves of the workpiece, and clamping on the other half.

3. The clamp(s) should be somewhat flexible in order to be able to accept deviations in the workpiece location and orientation deviations as described earlier in the paper. (often, this flexibility can be obtained from the hydraulic mechanism of the fixture).

3.5 Design of measuring jig based on design of fixture

A measuring jig for fast and simple measurements next to the forging process is required in order to perform statistical analysis of the process capability in terms of the clampability of the parts. This measuring jig should in general reproduce the workpiece coordinate system within allowed measuring uncertainties. A simple measurement can be taken by using a jig with measuring positionators (nominally) at same position as for the fixture for machining. The measuring gauges can be places at the same locations as the clamps.

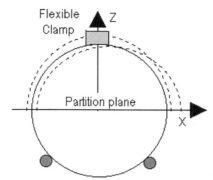

Figure 2; positionators and flexible clamp

4.   EXAMPLE

A series of forged workpieces as shown in Figure 3 have been measured on a coordinate measuring machine to test the fixture design method. The points P1 through P3 are used to illustrate the theoretical partition plane of the upper and lower die, i.e. the partition of the workpiece into two halves.

4.1 Step 1 :Analysis of the process variations
The location and orientation vectors between the lower half part and the upper half part of the workpiece has been measured. Their components are shown in Figure 5 and Figure 6. It can be seen, that the z-component of the location deviation vector is dominant. This can be explained by a gap between the upper and the lower die. Between sample 9 and 10, the press force has been increased, which led to a reduction of the gap. It can be seen, that the average of the z-component of the location deviation vector drops from 0,5 to 0,3mm. This however, does not lead to an alternation in the orientation deviation vector simultaneously.

The symmetry that can be observed in the orientation deviation vector is due to the rotation of the upper die relative to the lower die around the bisector of the angle between the x- and y-axis. The location deviation vector δP_0 is defined as following:

$$\delta P_0 = P_{0real} - [0,0,0] \tag{7}$$

The orientation deviation vector δE is calculated as following:

$$\delta E = E_{real} - [1,0,0] \tag{8}$$

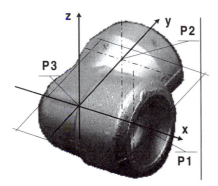

Figure 3; *Workpiece and orientation of the coordinate system*

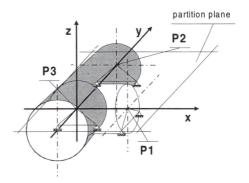

Figure 4; *Theoretical geometry*

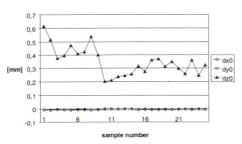

Figure 5; *Location deviation vector between the partition plane of the upper and lower half*

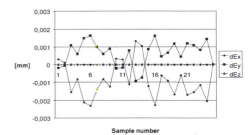

Figure 6; *x, y and z – component of the orientation deviation vector δE*

4.2 Step 2: Design of the fixture based on the vectorial tolerances

The positionators have been located at the points as illustrated in Figure 4.

mm	Positionator 1	Positionator 2	Positionator 3	Positionator 4	Positionator 5	Positionator 6
X	32,00	32,00	-16,62	16,62	-16,62	16,62
Y	-13,93	13,93	45,00	45,00	0,00	0,00
Z	-13,93	-13,93	-16,62	-16,62	-16,62	-16,62

Table 1; *Location of the positionators*

4.3 Step 3: Analysis of the fixture behaviour

The fixture behaviour is analysed by a location and orientation deviation vector between a theoretical plane in the fixture and the upper half part of the workpiece. The theoretical plane can be found, if an ideal workpiece would be clamped in the fixtures with positioners located as described in Table 1. Figure 4 shows a simplified workpiece which is compose out of two cylinders with the same diameters as the cylinders around the y and x axis in P1 and P2. This has been used as a model to calculate the behaviour of the fixture. It was assumed that the variation of the radius of the lower half of the workpiece was the dominant factor for the behaviour of the fixture.

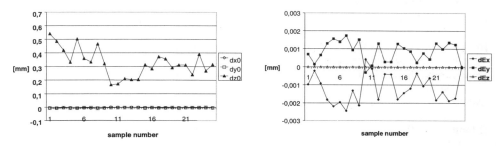

Figure 7; *Location deviation vector δP_0*   **Figure 8**; *Orientation deviation vector δE*

It can be seen that the deviation of the radius has only a limited influence on the orientation and the location deviation vector. The analysis of the same workpiece in a nearly worn out tool showed, that the cylindricity deviation was relatively small in relation to the deviation of the increase of the diameter of the cylinders due to wear, blast cleaning and polishing of the die.

5.   CONCLUSIONS

A method to improve the clamping of forged parts in machining using Vectorial Tolerancing was shown. The method consists of the following parts:
1.   Analysis of forging process variations using VT
2.   Design of fixture based on the vectorial tolerances
3.   Analysis of fixture behaviour
4.   Design of measuring jig based on the design of fixture
An example using real workpieces from Raufoss was shown in order to illustrate the method.

REFERENCES

[Cecil et al. 1996] Cecil, J; Mayer, R.; Hari U.; An integrated methology for fixture design, In: *Journal of Intelligent Manufacturing (1996) 7,* pp. 95-106

[Hong et al.1996] Hong, M.; Payandeh, S.; Gruver, W.A.; Moddeling and Analysis of flexible Fixturing Systems for Agile Manufacturing, In: *Proceedings of the IEEE International conferance Systems, Man and Cybernetics. Part 2 oct. 14-17 1996,Beijing, China,* pp1231 - 1236

[Jeng et al.,1995] Jeng, S. ;Chen, L. Chieng, W.; Analysis of minimum clamping force, *International Journal of Machine Tool and Manufacture Vol 35 No 9*, pp1213-1224

[Juran 1974] Juran, J.M.; Gryna, F.M., *Juran's Quality Control Handbook*, MacGraw-Hill, Inc., New York, 1974

[Martinsen, 1993] Martinsen K.; Vectorial Tolerancing for all types of surfaces, In: *Advances in Design Automation*, vol. 2, p. 187-198. ASME

[Martinsen, 1995] Martinsen K.; Statistical Process Control Using Vectorial Tolerancing, In *Computer Aided Tolerancing, (*Ed. by F. Kimura), Chapmann & Hall

[Martinsen, 1995] Martinsen K.; Vectorial tolerancing on manufacturing Systems, *PhD thesis University of Trondheim, Norway.* ISBN 82-7119-820-3

[Pham et al. 1990] Pham, D.T.; Lazaro D.; Autofix – an expert cad system for jigs and fixtures, In: *International Journal of Machine Tools and Manufacure, v30 No3*, pp 403-411

[Rong et al. 1997] Rong, Y.; Bai Y.; Automated generation of fixture configuration design, In *Journal of Manufacturing Science and Engineering, Transactions of the ASME Vol 119 May 1997*, pp. 208-219 ISSN: 1087-1357

[Shirinsadeh, 1996] Shirinsadeh, B.; Shen, Y.; Lee, W.; Identification and optimization of design paprameters based on workpiece geometry for flexible automated fixutring, *Japan/USA Symposium on Flexible Automation Vol. 2 ASME 1996* pp. 847 - 854

[Tuffentsammer, 1981] Tuffenstsammer K.; Automatic loading of Machining systems and Automatic clamping of Workpieces, Keynote paper, *Annals of the CIRP Vol 30/2*, pp. 553 - 558

[Xiuwen et al., 1996] Xiuwen G.; Fuh , J.Y.H. ; Nee A.Y.C; Modelling if frictional elastic fixture-worpiece system for improving location accuracy; In: *IIE Transactions (1996) 28*, pp. 821-827

[Wirtz, 1988] Wirtz, A.; Vektorielle Tolerierung zur Qualitaetssteuerung in der mechanischen Fertigung, *Annals of the CIRP*, Vol. 37/1 pp. 493-498

[Wirtz, 1991] Wirtz, A.; Vectorial Tolerancing for Quality Control and Functional Analysis in Design, *CIRP International Working Seminar on Computer Aided Tolerancing*, PennState University May 1991

[Wirtz, 1993] Wirtz, A.; Vectorial Tolerancing a Tool for Functional Analysis, *Proceedings of the 3rd CIRP Seminar on Computer Aided Tolerancing, France, Editions Eyrolles ISBN:2-212-08779-9*, pp. 115-128

[Wirtz et al., 1993] Wirtz, A.; A. Gaechter, C.; Wipf, D.; From Unambigously Defined Geometry to the Perfect Quality Control Loop, *Annals of the CIRP, Vol 42/1* pp. 615-618

Root Cause Analysis for Fixtures and Locating Schemes Using Variation Data

Johan S. Carlson
Mathematical Statistics
Chalmers University of Technology
S-412 96 GOTHENBURG
SWEDEN
jsc@math.chalmers.se

Abstract: This paper proposes a locating scheme method for diagnosis. The method utilises measurements/sensor readings to estimate the variation in contact points between the fixture and the workpiece. Kinematic analysis is used to derive a linear sensitivity equation that relates position error in locators to sensor readings. By using a subspace estimation technique based on the sensitivity equation the sensor variation is separated into noise and locator variation. The root cause of fixture failure is identified by ranking the estimated locator variation. The approach is attractive because it can deal with multiple coupled locator failures and is not limited to a 3-2-1 locating scheme, but works for an arbitrary deterministic locating scheme.
Keywords: Fixture diagnosis, rigid body, multivariate statistical analysis.

1 INTRODUCTION

A fixture is a high precision device for locating and restraining workpieces during manufacturing, inspection and transportation. Complex processes such as auto-body manufacturing requires hundreds of fixtures during assembly and therefore improvements in fixture design and diagnosis of fixtures are essential for reducing dimensional variation in the final product. Furthermore, an investigation performed during the launch of an auto-body assembly process showed that 72 percent of all root causes where due to fixture failures ([Ceglarek and Shi, 1996]).

The theory of positioning of an object is founded on that a rigid body has 6 DOFs which should be uniquely determined when brought into contact with some contact points. These contact points define a locating scheme and whether the points are provided by mating surfaces or fixture tooling elements are irrelevant when concerning development of theory for diagnosis. Therefore, the term locators will be used instead of contact points and fixture failure and locating failure will be used throughout the paper meaning the inability to correctly position an object.

1.1 Related work

Methods and theory considering fixture diagnosis based on in-line measurements have been neglected in literature in comparison with fixture design, but recently progress has been achieved by [Hu and Wu, 1992, Ceglarek et al., 1994, Khan et al., 1995, Ceglarek and Shi, 1996]. All papers known to the author apart from [Carlson and Ahlmark, 1997] use Principal Component Analysis (PCA) to identify the root cause of fixture failure. The reason for this is that dimensional variation in measurement points due to fixture failure are strongly correlated since there are only six underlying factors that can cause the variation .e. variation in locators.

The paper by [Ceglarek and Shi, 1996] pinpoints the dominant root cause for a 3–2–1 locating scheme by comparing the PCA eigenvector corresponding to the largest eigenvalue with a set of predetermined fault vectors each representing a single fault error in one out of six locators (pattern recognition). The power of the pattern recognition technique depends on the sensor location scheme and in [Khan et al., 1995] methods for choosing optimal sensor location is developed.

One essential part of both fixture design and diagnosis is the sensitivity equation relating locator errors to the resulting rigid movement of the workpiece. Different versions of the sensitivity equation, most of them based on the geometric perturbation technique presented in [Asada and By, 1985], are derived in literature. In [Weill et al, 1991, Cai et al., 1997, Wang, 1999] the sensitivity equation is derived and used for robust fixture design purpose, while in [Rong et al., 1995] the locating error effect on geometric errors produced in machining processes is analysed.

1.2 Outline

The method proposed in this paper for fixture diagnosis is novel in its approach dealing with multiple coupled locator failures for an arbitrary deterministic locating scheme. By estimating the locator variation causing the workpiece positioning error the root causes for fixture failure are pinpointed. Also contributions from locators to the total inspection point variation are ranked. These results constitute the fixture diagnosis.

The paper consists of three main parts. In the first part a sensitivity equation relating locator errors with sensor readings is derived (Section 2) The root cause analysis is accomplished in the second part, where both the locator estimating technique and corresponding interpretations are presented (Section 3). The last part attends to a numerical experiment testing the ability of the proposed method (Section 4).

2 DERIVATION OF THE SENSITIVITY EQUATION

The purpose of this section is to derive a linear sensitivity equation that relates errors in locators to the resulting workpiece movement. The sensitivity equation is derived with a small locator error assumption and with the geometric assumption that locators are placed on planar parts of the workpiece.

A deterministic locating scheme introduces a set of geometric constraints that uniquely defines the position of a rigid workpiece when brought into contact with the locators. If the locators are affected by source errors ϵ then the workpiece will be rigidly displaced from its desired location represented by a rotation matrix R and a translation vector t. Source errors can be due to fixture set-up errors, workpiece surface errors and locator wear. The displacement appearance depends on the position $\{p_i\}_{i=1}^6$ of the locators and on the workpiece normal direction $\{n_i\}_{i=1}^6$ at corresponding contact points.

We begin by formulating the geometric constraint equations corresponding to the perturbed locators being in contact with the displaced surfaces of the workpiece i.e.

$$n_i{}^T(R^{-1}(p_i + \epsilon_i - t) - p_i) = 0, \text{ for } k = 1, 2, \ldots, 6.$$

Next, the rotation operator is linearised into

$$R_\omega^{-1}\alpha \approx \alpha - \omega \times \alpha,$$

where the axis of rotation is parallel to ω and the angle of rotation is $\|\omega\|$. This relation is used to establish the linear version of the geometric constraints

$$n_i{}^T(\epsilon_i - \omega \times p_i - t) = 0, \text{ for } i = 1, 2, \ldots, 6.$$

Straight forward manipulation gives

$$
\begin{aligned}
n_i{}^T\epsilon_i &= n_i{}^T(\omega \times p_i + t) \\
&= (p_i \times n_i)^T\omega + n_i{}^T t.
\end{aligned}
$$

Or in a more compact form, by introducing a resultant vector containing the 6 DOFs $r = \begin{bmatrix} \omega^T & t^T \end{bmatrix}^T$,

$$D\epsilon = Jr,$$

where the normal directions are collected in $D = diag(n_1^T, n_2^T, \ldots, n_6^T)$ while

$$
J = \begin{bmatrix}
(p_1 \times n_1)^T & n_1^T \\
(p_2 \times n_2)^T & n_2^T \\
\vdots & \vdots \\
(p_6 \times n_6)^T & n_6^T
\end{bmatrix}.
$$

By assuming the above matrix J to be non-singular we can establish a sensitivity equation relating the locator perturbations to parameters describing the resulting rigid body movement, i.e.

$$r(\epsilon) = J^{-1}D\epsilon.$$

2.1 Source Errors Propagation into Sensors

In this section the linearised movement at each sensor, generated by source errors, is calculated using the sensitivity equation derived above. When making diagnosis the inverse problem, i.e. to determine the locator errors given the sensor movements, is the interesting one and therefore conditions on sensor positions and locator errors are specified to make the inverse problem solvable.

The movement in direction m registered by a single sensor, with position s, caused by a locator perturbation ϵ is

$$
\begin{aligned}
d_s &= m^{\mathrm{T}}(Rs + t - s) \\
&\approx m^{\mathrm{T}}(s + \omega \times s + t - s) \\
&= \begin{bmatrix} (s \times m)^{\mathrm{T}} & m^{\mathrm{T}} \end{bmatrix} r(\epsilon) \\
&= \begin{bmatrix} (s \times m)^{\mathrm{T}} & m^{\mathrm{T}} \end{bmatrix} J^{-1} D\epsilon \\
&= a^{\mathrm{T}} D\epsilon.
\end{aligned}
$$

And with p sensors

$$
d = AD\epsilon, \quad \text{where} \quad A = \begin{bmatrix} a_1^{\mathrm{T}} \\ a_2^{\mathrm{T}} \\ \vdots \\ a_p^{\mathrm{T}} \end{bmatrix}. \tag{1}
$$

The above problem is called diagnosable if each non-zero perturbation ϵ results in one and only one non-zero movement d at the sensors. A diagnosable problem is achieved if (i) the sensor positions generate a full rank matrix A and (ii) the direction of each locator error, a_i, is known and non-orthogonal to the corresponding workpiece normal direction n_i. The sufficiency and necessity of the above conditions are revealed using

$$
\epsilon = \begin{bmatrix} \delta_1 c_1 & \delta_2 c_2 & \cdots & \delta_6 c_6 \end{bmatrix}
$$

to reformulate the matrix vector multiplication, in equation (1), into

$$
D\epsilon = W\delta, \quad \text{where} \quad W = \begin{bmatrix} n_1^{\mathrm{T}} c_1 & 0 & \cdots & 0 \\ 0 & n_2^{\mathrm{T}} c_2 & \cdots & 0 \\ \vdots & \vdots & \ddots & \vdots \\ 0 & 0 & 0 & n_6^{\mathrm{T}} c_6 \end{bmatrix}.
$$

Thus we can see that condition (ii) implies a non-singular weight matrix W. Consequently, if condition (i) and (ii) are fulfilled then AW is full rank matrix which guarantees the problem to be diagnosable.

In practice the control directions are almost always specified to coincide with corresponding normal directions giving an identity weight matrix W. From now on we will only consider the diagnosable problem

$$
d = A\delta. \tag{2}
$$

3 LOCATING SCHEME FAILURE DIAGNOSIS

The diagnosis is realised in two steps. First, locator variation is estimated by a subspace technique described in the next section and then this variation is used to pinpoint the cause for locating scheme failure. With failure, we mean either the inability to locate a set of predefined inspection points on the object (Section 3.3) or the inability to locate the whole object at nominal position (Section 3.2). Whether the impact of locator variation is evaluated using only a few inspection points or the workpiece as a whole will often alter the diagnosis.

3.1 Locator Variation Estimates

Variation in data is due to both variation in locators and measurement errors. The purpose of locating scheme root cause analysis is to separate the variation and rank the contribution from each locator. The method decomposes the space of measurement observations into two orthogonal subspaces. A signal subspace defined by the linear sensitivity relation (2), where variation from the two sources are mixed and a noise subspace where only measurement variation exists. The estimation of locator variation is performed in two steps, first the measurement variation is estimated in the noise subspace, then this estimate is used to separate locator variation from measurement variation in the signal subspace.

The locator variation is estimated from n independent observations $\{x_j\}_{j=1}^{n}$ with p different sensors. Each observation x is assumed to follow from equation (2) with independent Gaussian noise added, i.e

$$x = A\delta + z,$$

where $z \sim N(0, \sigma^2 I)$ models the measurement uncertainty.

Estimating the locator perturbation mean μ_δ is straight forward by replacing the sensor mean μ_x in relation

$$\mu_x = A\mu_\delta$$

by its sample counterpart

$$\hat{\mu}_x = \bar{x} = \frac{1}{n}\sum_{j=1}^{n} x_j$$

and then using that A is a full rank matrix

$$\hat{\mu}_\delta = A^\dagger \hat{\mu}_x, \quad \text{where} \quad A^\dagger = (A^T A)^{-1} A^T.$$

The sensor variation is related to locator variation and measurement noise by their covariance matrices

$$\begin{aligned}
\Sigma_x &= A\Sigma_\delta A^T + \sigma^2 I \\
&= \Sigma_d + \sigma^2 I,
\end{aligned} \tag{3}$$

where $\Sigma_d = A\Sigma_\delta A^\mathsf{T}$ is introduced for convenience.

If the sensor covariance matrix is known then the measurement noise can be determined by solving an eigenvalue problem. This can be realised by the following argument. All vectors mapped by Σ_d into the null vector are eigenvectors to Σ_x with corresponding eigenvalue σ^2, i.e.

$$\Sigma_x v = \sigma^2 v, \text{ for all } v \in Nul(\Sigma_d).$$

Since the the rank r of A is 6, the multiplicity of the eigenvalue σ^2 is $p - r$.

In practice we have to estimate the measurement noise σ^2 by solving the eigenvalue problem with the population covariance matrix Σ_x replaced by its sample counterpart

$$\hat{\Sigma}_x = S_x = \sum_{j=1}^{n} (x_j - \bar{x})(x_j - \bar{x})^\mathsf{T}.$$

These $p - r$ eigenvalues are from the underlying noise sample covariance matrix S_z and thus, concerning estimation of the noise σ^2, we are in the same position as if we had observed the measurement noise directly, but with a loss of r dimensions.

To prove this, the above eigenvalue problem is circumscribed to

$$P_N S_x v = \lambda v, \text{ where } \lambda \neq 0,$$

where $P_N = I_n - AA^\dagger$ is the projection matrix onto the null space of A^T. By expanding the sample covariance matrix,

$$S_x = A S_\epsilon A^\mathsf{T} + S_z + A(\delta - \bar{\delta})(z - \bar{z}) + (z - \bar{z})(\delta - \bar{\delta})^\mathsf{T} A^\mathsf{T}$$

and using that the eigenvectors are contained in the Null space of A^T the above eigenvalue problem is proved equivalent with

$$P_N S_z v = \lambda v, \text{ where } \lambda \neq 0.$$

The $p - r$ eigenvalues $\tilde{\lambda}_k$ are estimates of the measurement noise σ^2 and we enhance the estimate precision by using the mean of the $p - r$ eigenvalues,

$$\hat{\sigma}^2 = \frac{1}{p - r} \sum_{k=1}^{p-r} \tilde{\lambda}_k,$$

to estimate σ^2. The sensor variation Σ_d is with relation (3) in mind estimated by

$$\hat{\Sigma}_d = \hat{\Sigma}_x - \hat{\sigma}^2 I_p$$

and the locator covariance matrix is consequently estimated by

$$\hat{\Sigma}_\delta = A^\dagger \hat{\Sigma}_d A^{\dagger\mathsf{T}}.$$

The diagonal elements of $\hat{\Sigma}_\delta$ are the variation at each locator, while the off diagonal elements are the couplings between different locators.

In the next sections we use the estimated locator covariance matrix to diagnose the locating scheme failure.

3.2 Diagnosis Method 1

Method 1 pinpoints the reason for locating scheme failure by ranking the locator variation. With this approach we ignore if the positioning of the object is more sensitive to errors in some locators than others. Also, possible couplings between locators are ignored.

3.3 Diagnosis Method 2

In method 2 the locating failure diagnosis is performed by ranking the locators with respect to the total variation contribution in predefined inspection points. The inspection points represent important workpiece parts where machining or measuring will be carried out. With this approach both the sensor variation at each locator and the correlation with the other locators affect the diagnosis.

If the sensor positions and corresponding measurement directions in equation (2) are replaced with inspection point positions and evaluation directions then

$$d = B\delta$$

gives the resulting movement in inspection points caused by locator errors, where the new matrix Λ has been renamed to B.

A measure of variability for a multivariate population is the total variance, which is calculated by taking the trace of the covariance matrix. The total variation in the inspection points is

$$
\begin{aligned}
\mathrm{Trace}(\Sigma_d) &= \mathrm{Trace}(B\Sigma_\delta B^{\mathrm{T}}) \\
&= \sum_{i-1}^{6} e_i{}^{\mathrm{T}}\Sigma_\delta B^{\mathrm{T}}Be_i \\
&= \sum_{i=1}^{6}\sum_{j-1}^{6} \sigma_{ij}e_j^{\mathrm{T}}(B^{\mathrm{T}}B)e_i \\
&= \sum_{i-1}^{6} \sigma_i,
\end{aligned}
$$

where $\sigma_i = \sum_{j} \sigma_{ij}e_j^{\mathrm{T}}(B^{\mathrm{T}}B)e_i$ is the contribution to the total variation from locator i. Thus, the coefficient $e_j^{\mathrm{T}}(B^{\mathrm{T}}B)e_i$ determines the exchange degree of σ_{ij} on the total inspection point variation.

4 NUMERICAL EXPERIMENTS

A numerical experiment is performed to test the proposed diagnosis method for locating scheme failure. The geometry for this experiment is given by a box positioned by a 3–2–1 locating scheme where 3 sensors *SP1–SP3* register movement in x–y–z-directions, see Figure 1. On top of the box an inspection point *IP1* is added, which is utilised in

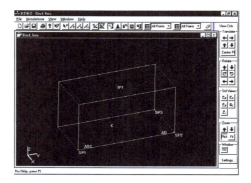

Figure 1: *Numerical experiment geometry*

diagnosis method 2. Variation data for locating scheme failure are generated by using the software *RD& T*, which is based on a Monte Carlo method. More specific, sensor readings are calculated by applying independent variation of given size to each locator and then Gaussian noise is added to simulate measurement noise.

Parameters for the experiment are given in Table I.

Locator positions	Control directions	Std of locator error	Sensor positions
$p_1 = (40, 10, 0)$	$n_1 = (1, 0, 0)$	0.012	$s_1 = s_2 = s_3 = (50, 0, 0)$
$p_2 = (40, 10, 0)$	$n_2 = (0, 1, 0)$	0.008	$s_4 = s_5 = s_6 = (50, 100, 0)$
$p_3 = (40, 10, 0)$	$n_3 = (0, 0, 1)$	0.003	$s_7 = s_8 = s_9 = (0, 100, 0)$
$p_4 = (40, 90, 0)$	$n_4 = (1, 0, 0)$	0.019	
$p_5 = (40, 90, 0)$	$n_5 = (0, 0, 1)$	0.016	
$p_6 = (10, 50, 0)$	$n_6 = (0, 0, 1)$	0.001	

Table I: *Experiment parameters*

4.1 Results

Given variation data based on different sample sizes and measurement noise levels the locator variations are estimated. The results for sample sizes $n = 10$, $n = 100$ and $n = 1000$ with measurement noise levels $\sigma = 0.001$ and $\sigma = 0.01$ are presented in Figure 2. Comparing the estimated variation to the true variation, we can see how the precision decreases with increased measurement noise level and how the precision increases with increased sample size.

Diagnosis method 1 pinpoints the locating failure by ranking the estimated locator variations. In all the simulation cases the main cause, i.e. variation in locator 4, is correctly identified. In fact the three main causes, variation in locator 4, 5 and 1 are determined in all cases. Problem with misjudgement in the ranking procedure affects only cases with a low signal-to-noise ratio combined with small sample size.

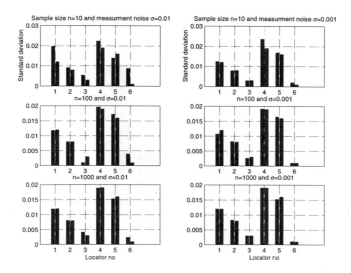

Figure 2: *Diagnosis method 1, estimated locator variance compared with correct locator variance.*

Diagnosis method 2 pinpoints the root causes for locating failure by calculating the contribution from each locator with respect to total variation in some inspection points. Contribution to total variation in the inspection point *IP1* from each locator is presented in Figure 3, which shows that variation in the inspection point is mainly due to variation in locator 5, and to some extent variation in locator 4 and 2. This conclusion differs from the one reached with method 1, which illustrates the importance of performing diagnosis with the appropriate measure of locating failure.

5 CONCLUSIONS

Fixture diagnosis is an important activity in automotive industry and this paper proposes a method which utilises workpiece measurements to pinpoint the root cause of fixture failure. The technique is novel since it handles multiple coupled locator faults for a deterministic locating scheme. A numerical experiment, where locator variation is estimated with high precision even at relatively small sample sizes and the root causes are pinpointed and ranked in an efficient manner, supports the practical relevance of the method.

On-going and future research includes (i) a formula relating the quality of locator variation estimates with parameters such as number of sensors, sample size and position of sensors, (ii) on-line surveillance of locator variation with a statistical process control approach and (iii) a test of the method on sheet metal assembly diagnosis following the stream-of-variation approach see [Hu, 1997].

120

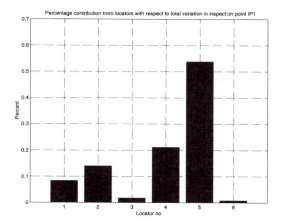

Figure 3: *Experiment result with diagnosis method 2.*

REFERENCES

[Asada and By, 1985] Asada, H.; By, A.; "Kinematic Analysis of Workpart Fixturing for Flexible Assembly with Automatically Reconfigurable Fixtures"; In: *IEEE Transaction on Robotics and Automation*, Vol. RA-1, pp. 86–94, 1985.

[Cai et al., 1997] Cai, W.; Hu, S.J.; Yuan, J.X.; "variational method of robust fixture configuration design for 3-D workpieces"; In: *J. Manuf. Sci. Eng.*, Vol. 119, pp. 593–602, 1997.

[Carlson and Ahlmark, 1997] Carlson, J.; Ahlmark, T.; "Production Quality Improvements Using Statistical and Geometrical Analysis of Car Body Measurements"; In: *Master Thesis*; Gothenburg; NO 1997-34/ISSN 0347-2809, 1997.

[Ceglarek et al., 1994] Ceglarek, D.; Shi, J.; Wu, S.M.; "A Knowledge-Based Diagnostic Approach for the Launch of Auto-Body Assembly Process"; In: *Journal of Engineering for Industry*, Vol. 116, pp. 491–499, 1994.

[Ceglarek and Shi, 1996] Ceglarek, D.; Shi, J.; "Fixture Failure Diagnosis for Autobody Assembly Using Patter Recognition"; In: *Journal of Engineering for Industry*, Vol. 118, pp. 55–518, 1996.

[Hu, 1997] Hu, S.J.; "Stream-of-Variation Theory for Automotive Body Assembly"; In: *Annals of CIRP*, Vol. 46, pp. 1–6, 1997.

[Hu and Wu, 1992] Hu, S.; Wu, S.M.; "Identifying Root Causes of Variation in Automobile Body Assembly Using principal Component Analysis"; In: *Transaction of NAMRI*, Vol. XX, pp. 311–316, 1992.

[Khan et al., 1995] Khan, A.; Ceglarek, D.; Shi, J.; Ni, J.; "An Optimal Sensor Location Methodology for Fixture Fault Diagnosis"; In: *Manufacturing Science and Engineering ASME*, MED-Vol, 2-2/MH-Vol.3-2, pp. 1165–1176, 1995.

[Rong et al., 1995] Rong, Y.; Li, W.; Bai, Y; "Locating error analysis for computer-aided fixture design and verification"; In: *Proceedings of the Computers in Engineering Conference and the Engineering Database Symposium*, pp. 825–832, 1995.

[Wang, 1999] Wang, M.; "An Optimum Design Approach to Fixture Synthesis for 3D Workpieces"; To appear in: *IEEE International Conference on Robotics and Automation*, 1999.

[Weill et al., 1991] Weill, R.; Darel, I.; Laloum, M.; "The Influence of Fixture Positioning Errors on the Geometric Accuracy of Mechanical Parts"; In: *Proceedings of CIRP conference on PE and Ms*, pp. 215–225, 1991.

On-machine quality control system for TRUE-CNC manufacturing system

Jiancheng Liu, Kazuo Yamazaki, Eric Kong
IMS- Mechatronics Laboratory
Dept. of Mechanical & Aeronautical Engineering
University of California, Davis
Davis, CA 95616
Phone: (530) 754-7687
Fax: (530) 752-8253
E-Mail: jliu@ucdavis.edu

Abstract: This paper deals with the design and development of a new quality control system for the CNC machine tool manufacturing system. The proposed quality control system can monitor product or part quality in a machining process, diagnose the system cause that relates to the unsatisfactory result in both autonomously and interactive mode, generate recommendation to take immediate corrective action through CNC controller, and automatically record all the results of monitoring diagnosis and recommendation in the database for the future diagnosis. The system consists of several functional modules. The functionality and structure of each module is discussed in detail in this paper.
Keywords: quality control, manufacturing system, CNC controller, machining process.

1. INTRODUCTION

The common industrial practice today is to incorporate factory automation to increase quality and productivity. Quality control is an activity used to evaluate a product quality with the designed specification or requirement and take remedial action if a difference between the actual and the standard is too great or unacceptable. In manufacturing, quality control consists of inspection, measurement, statistical process control (SPC), monitoring, error detection, analysis, and corrective action. Despite its importance, quality control related activity is still not productive in the manufacturing process [Hung et al., 1998].

Some quality control systems with hardware and software combinations are made available and are in common use. However, these systems are not able to well meet the requirement of the ever-increasing product quality and especially new manufacturing environment. Using an existing approach could provide most of the inspection results, but the decision is made mainly based on the measurement of the part instead of the overall machining performance and processes. Its emphasis is only in inspection of finished products or the part itself. Based on the inspection results, an operator performs labor-intensive manual diagnosis and then corrective action. Diagnosis knowledge not effectively retained and shared is another problem with quality control in manufacturing due to the complexity of its diagnosis operation. There are many decisions that the operator must make during the diagnosis procedure; so the operator would require a specialized skill that is acquired through hands-on experience. Expert systems for quality control are not robust in situations beyond the limits of their static knowledge bases [Kline et al, 1990]. This implies that to define cause-and-effect relations seems very difficult because each relation is varied from machine to machine and from machining process to machining process and also dynamically changes. The machining error cause and effect relationship in machining schematic has been well organized by T. Moriwaki [Moriwaki, 1984]. But, there is no research that shows how to find the relationships between any entity or how to make a detail relational mapping between them.

However, the authors recently proposed an autonomously proficient and autonomously intelligent CNC control manufacturing system, called the TRUE-CNC manufacturing system [Yamazaki et al., 1997]. This system is able to implement rapid inspection process through allowing the machine system to perform autonomous coordinate measurement planning for CMM (Coordinate Measuring Machine) and in-line surface roughness measurement after one machining operation is completed without or with minimum human intervention. By use of intelligent sensor systems developed, this system is able to diagnose and judge the machining process performance and perform the environmental inspection and structure deformation compensation. Hence, the behavior of the machine tool and machining process is able to be predicted, monitored on-line, and timely sent to the control system. Therefore, the TRUE-CNC system can provide a solution to overcome the above-mentioned obstacles existing in quality control.

This research presents an on-machine quality control system as one of many modules of TRUE-CNC manufacturing system to realize product quality control in an active, rapid, and autonomous manner through dynamically relating product quality with machine tool performance as well as machining process. The concept, design, functionality and implementation of the system will be addressed in the following sections.

2. SYSTEM REQUIRMENTS

The true meaning of quality control is checking the part quality, watching the machining behavior to diagnose the cause of the error in manufacturing system, and autonomously recommending the correct actions. Thus, the newly developed quality control system should have the following features: 1) Automatically monitoring both system performance and product quality in machining process through accessing TRUE-CNC Database [Ben et al., 1997]. 2) Autonomously detecting unsatisfactory results and self-diagnosing the system cause that relates to the unsatisfactory results in real time. 3) Generating recommendation to take immediate corrective action. 4) Recording all the results of monitoring diagnosis and recommendation into database for future diagnosis through CNC controller. 5) Directly interacting with as well as easily being integrated into TRUE-CNC manufacturing system to act as one functional module.

3.   SYTEM CONCEPTUAL DESIGN

In order to realize the aforementioned requirements to the on-machine system, the proposed system is made up of five functional modules: 1) Data Acquirer, 2) Event Observer, 3) Diagnoser, 4) Recommender, and 5) Maintenance Planner. In addition, a diagnosis database is designed to store all the diagnosis and recommendation knowledge.   Its overall component-wise system conceptual design with inputs and outputs to the TRUE-CNC system is shown in Figure 1. The main input of the system is part and measurement data of the machined or finished product, machine tool system

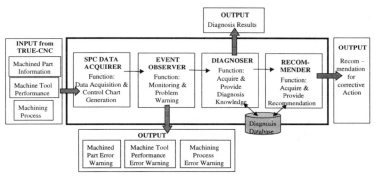

Figure 1 *System structure diagram*

performance and machining process related information. The main outputs are the autonomous diagnosis results and recommendation for corrective action. When the cause and effect relationships among product quality, machine tool performance, and machining process are established, a maintenance schedule will be an additional output.

3.1 Data Acquirer

The Data Acquirer shown in Figure 2 performs statistical process control data collection and control chart generation with a user interface for data input manually. It also allows product quality related data exchange with other TRUE-CNC modules automatically. TRUE-CNC database provides the Data Acquirer with machine tool data such as thermal deformation, spindle temperature, motion error, etc and machining operation data such as machining conditions (feed rate, depth of cut, spindle speed, coolant utilization etc..), cutting tool, material removal, tool path and so on. The geometric information of the machined part are obtained from the Autonomous Coordinate Measurement Planning (ACMP) output, one of the modules in TRUE-CNC system, which is able to automatically conduct the part geometry measurement. The acquirer

124

offers functions to include acquiring product inspection related to input measurement data, machine tool performance data, and machining process data. Also, Data Acquirer generates control charts for output, which is used for machine tool performance control.

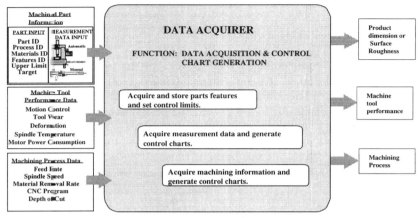

Figure 2 Data Acquirer

3.2 Event observer

The Event Observer gets the input from Data Acquirer as illustrated in Figure 3. The Event Observer serves for monitoring and error detection. The observer provides displays of control charts for monitoring and warning when any error is detected through the GUI. One of the inputs for observer is a control chart with measurement data from the Data Acquirer. The other two inputs are control charts of machine tool performance and machining process. The Event Observer monitors machine tool performance and machining process by accessing the TRUE-CNC Database in real-

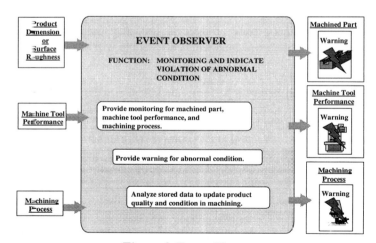

Figure 3 Event Observer

time.   The observer will trigger the Diagnoser module if any error occurs in product quality, machine tool performance, or machining process.   Outputs of the Event Observer include error warnings for product quality, machine tool performance, and machining process.

3.3 Diagnoser

The Diagnoser shown in Figure 4 provides users interfaces to acquire and provide diagnosis knowledge. An error warning of product quality, machine tool performance, and machining process from the Observer automatically triggers the Diagnoser. Diagnoser runs in either autonomous mode or interactive mode. In the interactive mode, the Diagnoser provides a list of selection choices from previously stored knowledge or blank space for new input from the GUI. Diagnoser organizes acquired knowledge and updates the listing with new verified knowledge that will be transferred into the Diagnosis Database. This retained diagnosis knowledge is used for future autonomous diagnosis. The interactive sequence in the Diagnoser is essential in organizing information for future autonomous diagnosis. If the Diagnoser runs in autonomous mode, all the judgement will be automatically performed.

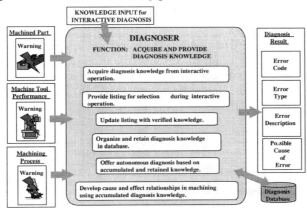

Figure 4 Diagnoser

Figure 5 illustrates an example of the operation mechanism of the Diagnoser in the case of that the surface finish machined is not acceptable. The Diagnoser organizes information into three levels. They are category, symptom, and possible cause in hierarchical order. The recommendation in the fourth level is included in the Recommender discussed in the next section. Each level has predefined selections and blank spaces or objects for adding new information. The operator could either select the available choices or input new information. Since the three levels are in hierarchical order, the selection of the previous level determines the direction of the next level. The hierarchical level setting organizes essential information to ease the diagnostic process. Therefore, information can easily be selected or entered into the Diagnoser. In the autonomous mode, the diagnostic result is automatically generated based on the

126

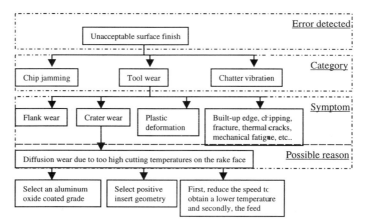

Figure 5 Diagnoser operation mechanism

provided key information matching with historical diagnostic data from the Diagnosis Database.

3.4 Recommender

The Recommender provides suggestions for corrective action based upon the diagnostic result for finding the cause of error. Its structure is shown in Figure 6. Similar to the Diagnoser, two operational modes are available in the Recommender; one is interactive and the other is autonomous. In the interactive mode, the Recommender has predefined

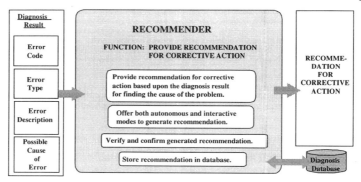

Figure 6 Recommender

selection and blank spaces or objects for adding new information. The operator could either select the available choices or input new information. In the autonomous mode, recommendation is conducted based on the historical diagnostic result and recommendation stored in the Diagnosis Database. This historical data is acquired and accumulated through the interactive mode. In some cases several recommendations could be suggested. The Operator performs step by step verification to check each of the possible recommendations until the problem is solved. In the beginning, interactive mode is frequently used to accumulate knowledge. As knowledge continues to accumulate, autonomous mode gradually takes place in the operation

3.5 Maintenance Planner

The Maintenance Planner shown in Figure 7 makes proper maintenance schedules based on the cause and effect relationships and the rules established in the Diagnosis Database. As diagnostic knowledge is accumulated and organized to map the cause and effect relationships among product quality, machine tool performance, and machining process, the influence degree of these factors can gradually be established as well as the occurrence of each error. The maintenance schedule is governed by two factors, the influence degree and occurrence frequency.

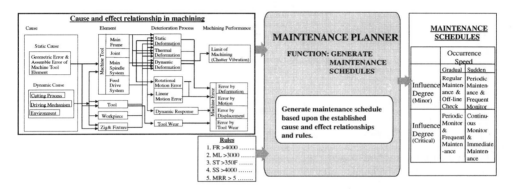

Figure 7 Maintenance Planner

4. SYSTEM DESIGN

The system design consists of three parts: database structure design, function algorithm design for each module, and Graphic User Interface design. Based on the conceptual design, the on-machine quality control system was designed by use of Microsoft Access Database Management Program (Version 97). The operation interface portion was made as a windows-based application for the Microsoft Windows NT or Windows 95 operating system. The hardware environment is a personal computer system equipped with 200MHz Intel Pentium Pro CPU with 128 MB of RAM. The database structure is organized by use of database elements (Tables, Query, Form, Report, Macro, and Module) in the Microsoft Access Database Management Program. The required database and table for the system are shown in Figure 8.

In the whole system, each module has specific operation rules. Figure 9 shows the designed Diagnoser function algorithm. To begin system operation, information consisting of the machined part, machine tool performance, or the machining process is input into Data Acquirer to make a control chart that is monitored by the Event Observer. Depending on the defined criteria, recognition of unacceptable data triggers the Diagnoser. If the historical case exists, the autonomous diagnosis will proceed to provide possible cause of error. A machining error occurs, then the Recommender will provide recommendation for corrective action. Otherwise, interactive diagnosis will take place to acquire diagnostic knowledge from the operator.

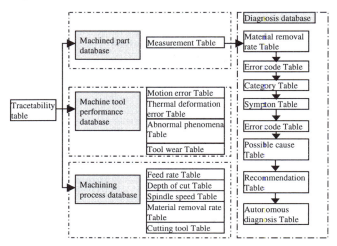

Figure 8 Data structure of Diagnoser Database

When in interactive mode, because the Event Observer monitors product quality, machine tool performance, and the machining process simultaneously, an operator or quality assurance personnel can check these three entities to help make diagnostic decisions. For example, the operator can check motion error, tool wear, or spindle speed for any possible problem. In addition to having machining information available, a user interface from the Diagnoser displays is ready to request input from the operator. This interface consists of three input level, category, symptom, and possible cause in hierarchical order. The Recommender contains user interfaces that acquire recommendation knowledge based on diagnostic results from the Diagnoser. After a recommendation is selected, verification is conducted by taking action from the recommendation for the machining process. The operator takes this recommendation, makes the necessary corrections and performs the machining process. Measurement is then conducted for this verification machining process.

After the verification step approves the corrective action, the last diagnosis and recommendation information is then permanently stored into the Diagnosis Database for future autonomous diagnosis. Each record is assigned a frequency of hits to illustrate how often a particular event has occurred. Any new information is added to the selection list at each request menu. The diagnosis and recommendation information continues to expand as more information is accumulated. Once the Diagnosis Database reaches a certain size, the autonomous mode can be activated more frequently, utilizing the Diagnosis Database more appropriately and effectively. It will gradually function in its desired purpose – direct matching. Each time, when the autonomous modes of the

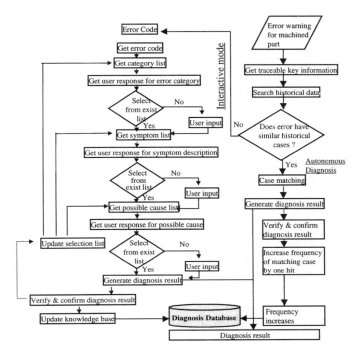

Figure 9 Diagnoser operation algorithm

Diagnoser and Recommender are running, a frequency counter is assigned to each record that has been called to keep track of the frequency of occurrence. The complete operator interface portion of the system is made as a window based application. The interface is developed to accept an operator's input to collect inspected part information and cause-result relation knowledge.

5. SYSTEM OPERATION STRATEGY

In the beginning of system operation, the major focus is the data acquisition for inspection measurement of a finished part, machine tool performance, or machining process to accumulate knowledge for the Diagnosis Database. Up to this point, the system mainly runs in interactive mode. In each system operation, information of the machined part, machine tool performance, or machining process is input into the Data Acquirer to make a control chart that is monitored by the Event Observer, if any abnormal situation (user defined) occurs, an error signal generated from the Event Observer will trigger the Diagnoser to autonomously find the cause of error. However, because knowledge continuously accumulates into the system, this knowledge will eventually help to establish the cause and effect relationships among the product quality, machine tool performance, and machining process. After having collected enough information and the cause and effect relationships have been established, monitoring the machine tool performance and the machining process can help predict whether the machining operation will run properly or not. Once this is done, any indication of possible failure can be stopped immediately upon detection. In addition to active monitoring, the organized information helps to develop better action plans for machine tool maintenance or repair.

6. CONCLUDING REMARKS

As one function module of TRUE-CNC manufacturing system, an on-machine quality control system was developed in this research. This system has the functionality to collect measurement data input with the data acquirer, recognize unacceptable product quality, and dynamically capture real-time machine tool performance and machining process information through interacting with TRUE-CNC database both manually and automatically. This system is able to perform autonomous and interactive diagnosis for possible cause and corrective action, provide corrective action to prevent unacceptable part production, and assist in maintenance schedule planning.

ACKNOWLEDGE

The authors express their sincere appreciation for the generous support of the TRUE-CNC consortium of sponsoring companies to make this research possible.

REFERENCES

[Hung et al., 1998] Hung Ng, et al.; "Autonomous coordinate measurement for TRUE-CNC," *CIRP General Assembly*, Vol.47/1, pp. 455-458.

[Kline et al., 1990] Kline, W.A.; Devor, R.E. ed.; Kapoor, S.G. ed.; "On-line Tool Condition Monitoring in Untended Manufacturing systems," *Quality, Design, Planning, and Control*, ASME PED-Vol.27: pp.1-3.

[Lam et al., 1997] Lam, B.; Liu, J.; Yamazaki, K.; "Dynamically Augmenting Database-Based Tool Data Management for TRUE-CNC," *CIRP International Design Seminar Proceedings: Multimedia Technologies for Collaborate Design and Manufacturing*, pp.110-116, Los Angeles.

[Moriwaki, 1994] Moriwaki, T.; " Intelligent Machine Tool: Perspective and themes for future development," *Manufacturing Science and Engineering*, Vol. 2, ASME, PED-Vol. 68-2, pp.841-849, New York.

[Yamazaki et al., 1997] Yamazaki, K.; Hanaki, Y.; Mori, M.; Tezuka, K.; "Autonomously Proficient CNC Controller for High-Performance Machine Tools Based on an Open Architecture Concept," *Annals of CIRP*, vol. 46/1: pp. 275-78.

Tolerance Representation Scheme
for Integrated Cutting Process & Inspection Planning

Sungdo Ha, Inshik Hwang, Kwanbok Lee, and Hyung-Min Rho
CAD/CAM Research Center, Korea Institute of Science and Technology
P.O.Box 131, Cheongryang, Seoul, 130-650, Korea
sungdo@kist.re.kr

Abstract: In order to improve machining quality of parts, tolerance information should be easily exchanged among product designers, process planners, operators and inspectors. The tolerance information exchange is not yet accomplished since common formats for tolerance information management are being under development. So far the tolerance is recorded as text information in the drawings of commercial 3D CAD systems, which requires human interpretation for its use in cutting process planning and inspection planning processes.

In this work a tolerance-input system is developed, which enables the assignment of geometric tolerances to 3D geometry. The tolerance information is to be used in the cutting process planning process such that more appropriate cutting processes can be selected for satisfying the assigned geometric tolerance. It is also to be used in the inspection planning process for the determination of measurement points and measurement sequences in 3D coordinate measuring machines.

Keywords: Tolerance, Tolerance Database, Process Planning, Inspection Planning

1. INTRODUCTION

Dimensional accuracy of machined parts is critical for product quality since it determines the assembly conditions that have an effect on the product performance and reliability. Designers specify the part dimensional accuracy with tolerances, and the machining and inspection processes are performed according to the assigned tolerances. Hence the tolerance information should be easily exchanged among product designers, process planners, operators and inspectors in order to improve the product quality.

It was found that bridging the gaps among design, machining, and inspection has difficulties due to the lack of a proper handling method of the non-geometric information such as tolerance information. In this work a tolerance-input system is developed such that the tolerance information is linked to the geometric information from design processes. And then the tolerance information in a common database is used in process planning and inspection planning processes. The process planning is based on the machining feature information and the tolerance information related to the feature is useful in deciding the proper processes. The inspection planning uses the tolerance information for the determination of measurement points of 3D coordinate measuring machines.

2. TOLERANCE INFORMATION MANAGEMENT - BACKGROUND

3D CAD systems not only support solid geometric modeling but also enable to provide necessary information for evaluation of product physical characteristics. They also have such data structure that geometric entity information and topology information are stored together, which helps CAD/CAM interface through geometric reasoning.

However wide use of 3D CAD systems recently reveals the necessity of tolerance representation.[Quilford et al., 1993][Chao et al., 1994][Shah et al., 1995] Since current 3D CAD systems are based only on geometric and topological relations, important non-geometric information such as dimensional tolerance, geometric tolerance, and surface finish is not handled in a proper way. The non-geometric information is necessary for both selecting processes, equipment, and fixtures, and determining process sequences, operations, tools and process conditions. NC programming and inspection planning also require the non-geometric information.

Current 3D CAD systems have the capability of generating 2D drawings that are still necessary in manufacturing preparation processes. The non-geometric information is recorded as text information in the 2D drawings. Human intervention is needed for the interpretation of this text information in the following processes. Hence it is necessary to develop a method for managing non-geometric information along with geometric information in CAD models.

Data structures for both geometric information and tolerance information are necessary. In order to develop such data structures, systematic ways of representing tolerance information need to be prepared. It was not until ANSI standard Y14.5 was published in 1982 that the development of tolerance representation in CAD systems began.[ASME, 1983] Shah and Miler discussed the representation of Y14.5 tolerances within the frame work of a general CIM system.[Shah et al., 1989][Shah et al., 1990] They proposed an object-oriented system for encoding tolerances so that new tolerances can be added without major revisions. Roy and Liu described a system where tolerances are expressed as attributes in the model [Roy et al., 1988] and Ranyak and Fidshal proposed a hierarchical model where tolerances are represented with general classes augmented by qualifiers.[Ranyak et al., 1988]

Allied Signal Inc. announced FBTol that is designed for the tolerance information management.[Brown et al., 1994] It extended the foundation provided by CAM-I D&T model [CAM-I, 1986] and uses Booch's object-oriented methodology.[Booch, 1994] The modeler is based on the interrelationship among five class objects.

3. TOLERANCE-INPUT SYSTEM

The tolerance-input system developed in this work accepts geometric information in ACIS format. The system incorporates the functions of displaying geometry, identifying geometric entity, assigning datum surfaces and tolerances, and storing tolerance information into a common database. The system is developed with MicroSoft Visual C++ and ACIS Geometric Kernel 4.0 in a Windows 95 environment.

The system consists of 6 modules as shown in Figure 1. The ACIS file input module opens ACIS files and obtains geometric information. Geometric elements are given persistent ID's for further manipulation in the persistent ID definition module. The datum/tolerance selection and tolerance assignment module guides users to input datum and tolerance information by selecting target surfaces through a graphic user interface. The ACIS file output module stores the geometric elements with their persistent ID's and the ACIS model output module displays the geometry as well as the tolerance information on the screen. The DB interface module stores a tolerance table into a common database.

Table I is a tolerance data table in the common database. The database is in the form of a relational database. The stored tolerance information is used in the cutting process planning and inspection planning processes, where a feature is used as an information unit. The 'TOL_ID' is the geometric element ID to which tolerance is assigned. The geometric element needs to be related to the feature. The mapping information between features and geometric elements is also stored in the common database.

Figure 2 is an example screen display when a perpendicularity tolerance is assigned to the floor surface of the top pocket with the front side as a datum.

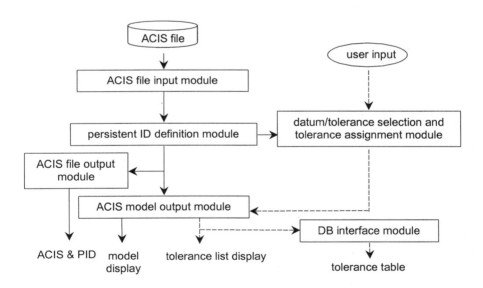

Figure 1; *Tolerance-input System Structure*

Column Name	Description	Column Name	Description
TOL_ID	tolerance ID	TOL_MINUS	- value of tolerance
PRODUCT_ID	product number	TOL_CONSTRNT	Tolerance constraint
ITEM_ID	item number	DTM_1	1st datum
TOL_TARGET	tolerance assigned geometric entity	DTM_1_CONSTRNT	Constraint of 1st datum
TOL_TYPE	tolerance type	DTM_2	2nd datum
TOL_VALUE	tolerance value	DTM_2_CONSTRNT	Constraint of 2nd datum
DIM_VALUE	dimension value	DTM_3	3rd datum
TOL_PLUS	+ value of tolerance	DTM_3_CONSTRNT	Constraint of 3rd datum

Table I; *Tolerance Data Table*

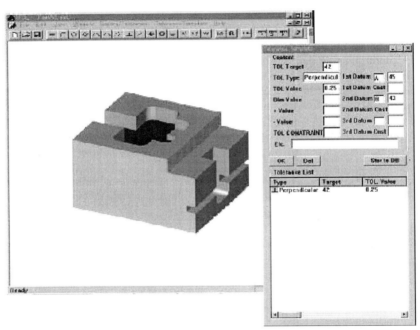

Figure 2; *Tolerance-input System Screen Display*

4. USE OF TOLERANCE INFORMATION

Tolerance information stored in a common database as explained in the previous section is to be used in the cutting process planning and inspection planning processes. Since tolerance information is linked to the geometric information, human interpretation is not necessary, which allows the close integration among design, process planning and inspection planning processes. Figure 3 shows the use of tolerance information among a tolerance-input system, a process planning system, and an inspection planning system.

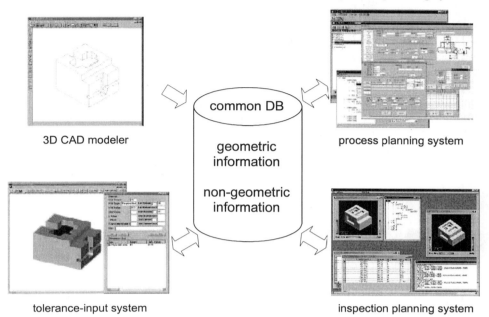

3D CAD modeler

common DB

geometric
information

non-geometric
information

process planning system

tolerance-input system

inspection planning system

Figure 3; *Use of Tolerance Information*

4.1. Cutting process planning

Process planning determines the process sequences, equipment, tools, and operation conditions based on both geometric and non-geometric information from the design process. The process planning in this work is based on the manufacturing features that are recognized from 3D CAD file.[Rho et al., 1998] The manufacturing features consist of pre-defined atomic features for milling and turning operations as well as compound features.[Rho et al., 1996] The tolerance information is linked to the features through the mapping information between geometric elements and features.

The tolerance information determines the process sequence because datum surfaces should be machined first. It also selects the types of equipment and tools since tighter tolerances require more accurate equipment or additional processes.

4.2. Inspection planning

Automation of the inspection planning has been hardly realized because it is difficult to replace human inspection planners with an intelligent program. The inspection planners recognize geometry, dimensions and tolerances from drawings where all the information is contained as figures, texts, symbols, etc. Since the tolerance information was not related to the geometric data even in recent CAD software, it was impossible to develop an automated inspection planning system.[Hwang et al., 1997]

The inspection planning system aims to generate inspection plans for 3D coordinate measuring machines. The 3D coordinate measuring machines are widely used in CIM environments because of their accuracy and ease of automation. The inspection plan consists of inspection sequences, positions of inspection points and trajectories of inspection probes. The system requires geometric information, feature information as well as tolerance information from the common database and generates inspection plans in DMIS format as shown in Figure 4. The DMIS format is a communication standard between measurement devices and computers and is used to operate CNC-controlled 3D coordinate measuring machines.[CAM-I, 1995]

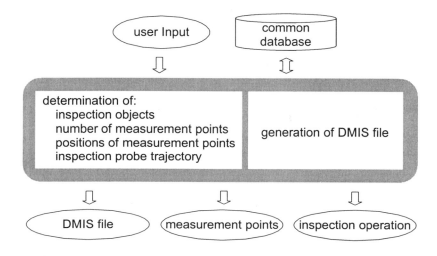

Figure 4; Inspection Planning System Structure

Inspection planning consists of the determination of inspection objects, number of measurement points, positions of measurement points, and inspection probe trajectories, and the generation of DMIS file. The inspection object is basically the part of measurable surfaces where tolerances are assigned. The measurability of a surface is determined by set-up directions, probe installation directions, sizes of the probe ball tip and lengths of the probe styli. The inspection object also includes datum surfaces.

It is necessary to select the number of measurement points that guarantee the measurement accuracy in consideration of the assigned tolerances. The number also depends on the size and shape of the inspection objects. The positions are determined with the Hammersley sequence. The Hammersley sequence shows a nearly quadratic reduction in the number of points needed by the uniform sampling method for the same level of accuracy.[Woo et al., 1993][Lee et al., 1997] All the inspection objects are classified as rectangles, circles and conical surfaces, and the Hammersley sequences for each object are developed.

The inspection probe trajectory is based on the hierarchical structure of features. The possible trajectories are all included in the machined volume. Hence the trajectory of the inspection probe follows the hierarchy of the feature structure from high level features to lower level features. The probe trajectories for each feature measurement are defined according to the assigned tolerances.

A DMIS file consists of header, geometry-oriented commands, and process-oriented commands. The header defines the general specification of measurement devices and selective options. The geometry-oriented commands are similar to the APT language and the process-oriented commands have similar structures as the G Code in NC programs. The user determines the header through graphic user interfaces. The system generates the geometry-oriented commands and the process-oriented commands based on the geometric information and tolerance information.

5. CONCLUSIONS

This work is based on the notion that the current 3D CAD systems do not have the capabilities of managing non-geometric data such as tolerance information. The exchange of tolerance information among the design, process planning and inspection planning processes is very important for the improvement of product quality.

The tolerance-input system was developed such that tolerances are assigned to the geometric entity designed in ACIS format. The tolerance information is stored in the common database that also has the mapping information between the surfaces and machining features. Both machining feature information and tolerance information are used in the process planning and inspection planning processes. The inspection planning system determines the inspection objects, number of measurement points, positions of measurement points and inspection probe trajectory. The inspection plan is generated in DMIS format such that it is directly applied to 3D coordinate measuring machines.

The functions such as suggesting tolerances according to the characteristics of the product, and checking tolerance consistency for assembled parts need to be developed further in the tolerance-input system. Also the inspection planning system should be studied further such that it can determine inspection setups and inspection sequences based on tolerance information.

REFERENCES

[ASME, 1983] American Society of Mechanical Engineers; *Dimensioning and Tolerancing*; ANSI Y14.5M-1982; U.S.A.; 1983.

[Booch, 1994] Booch, G.; *Object-Oriented Analysis and Design*; The Benjamin/Cummings Publishing Company, Inc.; 1994.

[Brown et al., 1994] Brown, C.W.; Kirk, W.J.III; Simons, W.R.; Ward, R.C.; Brools,

S.L.; "Feature-based tolerancing for advanced manufacturing applications" In: *Applications in Manufacturing and Services Industries*; MACIS; Kansas City, MO, U.S.A.; Oct. 1994.

[CAM-I, 1986] Computer Aided Manufacturing International, Inc.; *CAM-I D&T Modeler Version 1.0 - Dimensioning and Tolerancing Feasibility Demonstration Final Report*; Arlington, TX.; PS-86-ANC/GM-01; Nov. 1986.

[CAM-I, 1995] Consortium for Advanced Manufacturing – International (CAM-I); *Dimensional Measuring Interface Standards (DMIS) Version 3.0*; 1995.

[Chao et al., 1994] Chao, H.; Alting, L.; *Computerized Manufacturing Process Planning Systems*; Chapman & Hall; pp.230-255; 1994.

[Hwang et al., 1997] Hwang, I.; Lee, K.; Ha, S.; "A Study on 3D CAD Tolerance Information Handling for Inspection Planning" In: *Proceedings of '97 KSPE Fall Conference*; Korean Society of Precision Engineering; Suwon, Korea; Nov. 1997 (in Korean).

[Lee et al., 1997] Lee, G.; Mou, J.; Shen, Y.; "Sampling strategy design for Dimensional Measurement of Geometric Features Using Coordinate Measuring Machine" In: *International Journal of Machine Tools and Manufacturing*; Vol.37; No.7; 1997.

[Ranyak et al., 1988] Ranyak, P.S.; Fridshal, R.; "Features for tolerancing a solid model" In: *Proceedings of 1988 ASME Computers in Engineering Conference*; pp.275-280; 1988.

[Quilford et al., 1993] Guilford, J.; Turner, J.; "Representational primitives for geometric tolerancing" In: *Computer- Aided Design*, Vol.25, No.9, 1993.

[Rho et al., 1996] Rho, H.M.; Lee, C.S.; "Manufacturing Features Applied to the Milling Operation Planning" In:*3rd CIRP Workshop on Design and Implementation of Intelligent Manufacturing Systems(IMS)*; Tokyo; 1996.

[Rho et al., 1998] Rho, H.M. et al.; "Process Planning and Quality Management Technology Development" In: *Research Report*, Korea Institute of Science and Technology; BSM0753-6225-2; 1998 (in Korean).

[Roy et al., 1988] Roy, U.; Liu, C.R.; "Feature-based representational scheme of solid modeler for providing dimensioning and tolerancing information" In: *Robot & Computer Integrated Manufacturing*; Vol.4; No.3; pp.333-345; 1988.

[Shah et al., 1989] Shah, J.J.; Miller, D.W.; "A Structure for Supporting Geometric Tolerances in Product Definition Systems for CIM" In: *Manufacturing Review*; Vol.3; No.1; pp.23-31; 1990.

[Shah et al., 1989] Shah, J.J.; Miller, D.W.; "A Structure for integrating geometric tolerances with form feature and geometric models" In: *Proceedings of 1989 ASME Computers in Engineering Conference*; 1989.

[Shah et al., 1995] Shah, J.J.; Mantyla, M.; *Parametric and Feature-Based CAD/CAM*; JOHN WILEY & SONS; pp.9-18; 1995.

[Woo et al., 1993] Woo, T.C.; Liang, R.; "Dimensional measurement of surfaces and their sampling" In: *Computer-Aided Design*; Vol.25; No.4; 1993.

Towards easier and more functional semantics for geometrical tolerancing

E. Pairel, M. Giordano, S. Samper

LMécA/CESALP (Laboratoire de Mécanique Appliquée)
Ecole Supérieure d'Ingénieurs d'Annecy - Université de Savoie
BP 806, 74016 ANNECY Cedex - FRANCE
pairel@esia.univ-savoie.fr

Abstract: Geometrical defects of a manufactured feature are limited by form, size and location tolerances. The standardized semantics of those tolerances, i.e. their interpretation on the manufactured feature, presents a lot of drawbacks. We propose new semantics, based on the fitting of a unique theoretical feature on the manufactured feature. The fitting criterion we propose, named minimum volume criterion, leads to the actual mating envelope of the manufactured feature. For some functional requirements the tolerancing shown to become simpler and more functional.
Keywords: tolerancing, semantics, fitting criterion, minimum volume criterion

1   THE THREE DEFECTS OF A FEATURE: FORM, SIZE, LOCATION

The tolerancing principle consists in considering that the actual feature presents three kinds of defect: form, size and location defects.

Consequently, we find three types of tolerances for each feature (figure 1.a).

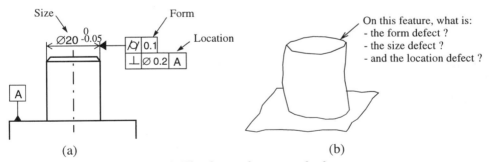

(a)   (b)

Figure 1; *The three tolerances of a feature*

Note: the word «location» will be used to refer indifferently to location or orientation tolerance.

139

Let us note that this division in three defects is arbitrary because the actual feature has a shape on which form, size and location defects are not intrinsically defined (figure 1.b). So defining those different defects regardless of the actual feature is necessary.

The main study on this problem is due to A.A.G. Requicha [Requicha 1983]. But the semantics that he has proposed is not really functional and does not allow to define the values of form, size and location defects of an manufactured feature.

The other main approach consists in defining these three defects with a theoretical feature fitted to the manufactured feature [Hillyard 1978] [Wirtz 1989].

Let us note that the standardized semantics noes not correspond to any of these two approaches.

2   PROPOSED SEMANTICS

We propose to define the form, size and location defects of the actual feature by comparing it to a unique theoretical feature fitted to the actual feature :

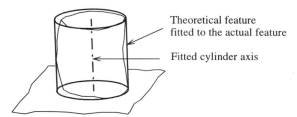

Theoretical feature
fitted to the actual feature

Fitted cylinder axis

Figure 2; *Theoretical feature fitted to define the defects of a surface*

The fitting criterion that we choose, presented in section 4, leads to obtain the actual mating envelope of the manufactured feature. For a shaft, it is the smallest theoretical cylinder fitting the outside of the shaft. Nevertheless, the designer should be able to choose another criterion.

2.1 Size tolerance semantics
The size tolerance concerns the size of the fitted feature. Its value has to be between the minimum and the maximum sizes authorized by the tolerance.

The size defect is the deviation between the size of the actual matting envelope and the nominal size defined by the designer.

Case of size tolerance applied to angle :
The standards allow the size tolerancing of the angle between two planes. However the standardized interpretation of such a tolerance is ambiguous because of its incomplete definition. With the proposed semantics, there is no ambiguity: the angle tolerance concerns the angle between the two planes fitted to the manufactured features.

2.2 Form tolerance semantics

The form defect is the maximum distance between the theoretical feature and the actual feature along a perpendicular direction to the theoretical feature. It has to be inferior to the form tolerance. This semantics is very close to standardized semantics.

2.3 Location tolerance semantics

The location tolerance concerns the fitted feature axis along the nominal height. This axis, limited by two extreme points, has to be within the tolerance zone defined by the location tolerance:

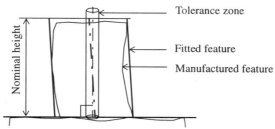

Figure 3; *New semantics for a location/orientation tolerance*

The location defect can be defined as the size of the minimum tolerance zone.

3   STANDARDIZED SEMANTICS

3.1 Standardized interpretation of the size tolerance

International standards [ISO 8015] indicate that the size tolerance only limits the «actual local sizes» of the actual feature :

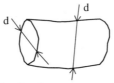

Figure 4; *Standardized semantics of a size tolerance*

This semantics is based on the shop floor practices using micrometers and callipers which enable to measure the distance between two points.

Different studies were conducted to give a definition of the actual local size [Srinivasan 93]. The proposed definitions are complex and lead to more difficulties in metrology.

Whatever its definition, the actual local size concept involves problems for the geometrical specification and verification of products because local sizes do not always exist on parts.

For instance, in many cases, the tolerances cannot always be verified on every point of the feature:

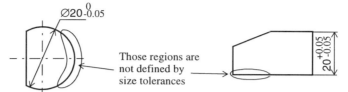

Figure 5; *Features incompletely defined by the size tolerance*

A lot of specifications are irrelevant with the standardized interpretation :

Figure 6; *Irrelevant specifications with the standardized semantics*

With the proposed semantics:

With the proposed semantics, each one of the above specifications is relevant.

For cylindrical features, it is always possible to fit theoretical cylinder on them even if they are not complete.

For a size tolerance between two planes, the theoretical entity to be fitted is composed of two parallel planes with a variable distance :

Figure 7; *Two parallel plane entity*

Furthermore, interpretation in terms of actual local sizes has not any real functional justification. So, in our opinion, this concept should be abandoned for a interpretation based on feature fitting.

3.2 Standardized interpretation of a form tolerance

The manufactured feature has to be between two theoretical features obtained by offsetting of the nominal feature to the exterior and interior sides.

This condition can be verified by fitting the exterior offsetted feature on the actual feature, with the minimum form defect criterion. So, the only difference between the standardized interpretation and the proposed interpretation is the fitting criterion.

3.3 Standardized interpretation of a location tolerance

According to the standardized interpretation, a location tolerance concerns an «actual» feature of the part.

For the cylinder, the concerned feature is the «actual axis» of the feature :

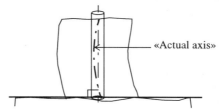

Figure 8; Standardized interpretation of a location tolerance

But the «actual axis» is a virtual feature on the manufactured part. So it is necessary to define how this feature is built from the actual feature. In our knowledge, standards on tolerancing do not give this definition. This gap leads to several metrology practices:

• The first method consists in building approximate centers of some sections of the shaft.

• The second method consists in verifying the location of a perfect feature fitted to the manufactured feature.

Although this method is not consistent with standardized interpretation, it is more and more commonly used because of the growing use of CMM.

Besides, it is generally more adapted to the designer's intention.

The semantics proposed here corresponds to this verification method.

Projected tolerance zone:

A location tolerance with a projected tolerance zone enables to specify that the continuation of the actual feature has to respect the location tolerance (figure 9.a).

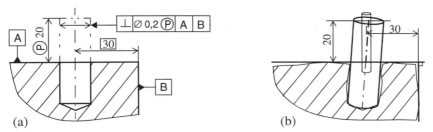

Figure 9; Projected tolerance zone specification and interpretation

In this case, it is impossible to consider the «actual axis» of the hole. The axis concerned by the specification is necessarily the axis of a theoretical feature fitted to the hole (figure 9.b). So, the standardized interpretation is not coherent with this kind of specification. On the other hand, the proposed semantics is relevant in this case.

4 DEFINITION OF THE FITTED FEATURE - MINIMUM VOLUME CRITERION

The fitted feature has to represent the manufactured feature as functionally as possible. In our opinion, it has to be the actual mating envelope of the manufactured feature so as to allow to control the mating requirement directly.

Thus an original and general fitting criterion is proposed, named the minimum volume criterion.

It consists in defining the theoretical feature, outside of the material, which minimizes the volume of the space between the theoretical and the manufactured features. Figures below illustrate fittings of a plane and of a cylinder according to this criterion.

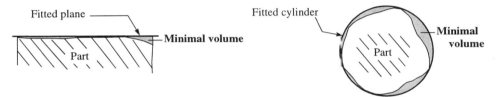

Figure 10; *Fittings according to the minimum volume criterion*

In the case of the cylinder, the minimum volume criterion leads to the smallest size theoretical cylinder fitting the outside of the shaft.

In some cases, the minimum volume criterion can give several solutions, i.e. several theoretical features minimizing the volume. The chosen feature will be the one which also minimizes the maximum distance between the theoretical and the manufactured features.

The theoretical features are better fitted to actual features with the minimum volume criterion than by other criteria. In particular, the minimum form defect criterion, which consists in minimizing the maximum distance between the theoretical feature and the actual feature, can leads to obtain fitted features more distant to actual features. The actual features in the above figure are in this case:

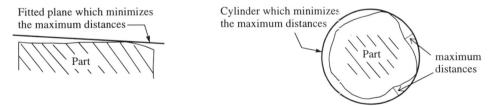

Figure 11; *Exterior theoretical features obtained by the minimum form defect criterion*

Another advantage of the minimum volume criterion is that it formalizes and generalizes standardized fittings of datums to actual features.

Let us note that, with the proposed semantics, when an actual feature is a first datum feature of the part, the same theoretical fitted feature defines the datum and the geometrical defects of the actual feature.

The minimum volume criterion can be computerized [Pairel 1995]. Nevertheless, even if the least square criterion is used, the verifications done now on CMM are closer to the proposed semantics than to the standardized semantics.

5 SOME CASES OF FUNCTIONAL TOLERANCING

5.1 Mating requirement

The proposed size tolerance semantics enables to control the maximum and minimum clearances between two mated features very simply. It is more functional than the envelope requirement [ISO 8015] because it does not limit actual local sizes which are not functional for the assembly.

On the example below, the radial clearance is maximum when the actual mating envelope of the shaft has the minimum diameter allowed by the tolerance. The fact that some actual local sizes have smaller values does not increase the maximum clearance.

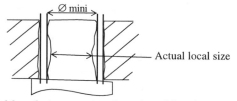

Figure 12; Actual local sizes are not functional for the assembly

With the proposed semantics, the actual local sizes result from the form defects of the actual feature. So they are limited by the form tolerance of the feature.

5.2 Guiding precision requirement for a shaft

For a long shaft, for example a guiding shaft for a collar, the size tolerance can be applied along part of the length of the shaft (the collar length):

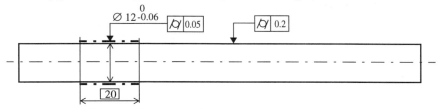

Figure 13; Proposed tolerancing for a long shaft

According to the proposed semantics, anywhere along the shaft, the diameter of the actual mating envelope, of a length of about 20 mm, has to be between the minimum and the maximum sizes allowed by the tolerance.

The form tolerance concerns the actual mating envelope too: the distance between the shaft and any 20 mm-long mating envelope has to be inferior to the form tolerance value.

In order to limit the straightness defect of the shaft, a cylindricity tolerance for the whole shaft i specified (figure 13). This is almost equivalent to a straightness tolerance applied to the generators of the shaft. But the semantics of this last kind of tolerance is more difficul to define.

6   CONCLUSION

The semantics proposed in this paper for form, size and position tolerances is greatly inspired from verifications on CMM: What is check is the size and the position of the fitted feature.

The idea to define the form, size and position defects regardless of a theoretical feature, fitted to actual feature, was already proposed by A. Wirtz [Wirtz 1989]. But the «vectorial tolerancing» that he propose, uses a very different principle of the tolerancing by tolerance zones. It is more difficult to use and less functional than the geometrical tolerancing as it is shown by S. Bialas [Bialas 1997].

The originality of the proposed semantics is the fitting criterion which leads to obtain the mating envelope of the manufactured feature. Thus, the size of the fitted feature has a functional interpretation.

The proposed semantics allows to define the form, size and position/orientation defects. Thus the geometrical verification according to this semantics enables to give the values of these defects. On the other hand, the geometrical verification according to standardized semantics can give only a binary response to the conformance question of the feature.

Furthermore, let us note that form, size and position tolerances are completely independent with the proposed semantics. Thus, the form tolerance can be larger than the position tolerance. It is not true for the standardized semantics: the position tolerance limits also the form defect of the feature. So the principle of independency between the tolerances is better respected.

Finally, the proposed semantics corresponds to the ones which are generally considered in models developed for the computer aided tolerancing systems.

REFERENCES

[Bialas 1997] Bialas S.; Humienny Z.; Kiszka K.; «Relations between ISO 1101 Geometrical Tolerances and Vectorial Tolerances - Conversion Problems»; Proceedings of 5th CIRP International Seminar on CAT; April 1997 Canada; pp. 37-48

[Hillyard 1978] Hillyard R.C.; Braid I.C.; «Characterizing non-ideal shapes in term of dimensions and tolerances»; Computer Graphics, Vol. 12, N°3, pp. 234-238

[ISO 8015:1985] ISO 8015:1985, Technical drawings - Fundamental tolerancing principle.

[Pairel 1995] E. Pairel; «Métrologie fonctionnelle par calibre virtuel sur machine à mesurer tridimensionnelle"; Thèse de Doctorat de l'Université de Savoie; 1995

[Requicha 1983] Requicha A.A.G.; «Toward a theory of geometric tolerancing», The International Journal of Robotics Research, Vol. 2, No. 4, Winter 1983.

[Srinivasan, 1993] V. Srinivasan, «Recent Effort in Mathematization of ASME/ANSI Y14.5M»; Proceedings of the 3rd CIRP Seminars on Computer Aided Tolerancing; April 1993 France; pp223-232

[Wirtz 1989] A. Wirtz, «Vectorial tolerancing»; Proceedings of the CIRP International Conference on CAD/CAM and AMT; Dec. 1989 Israel.

A tolerance system to interface design and manufacturing

Helmut Bley, Ralf Oltermann, Oliver Thome, Christian Weber
University of the Saarland
Institute of Production Engineering/CAM
P.O. Box 151150, D-66041 Saarbruecken, Germany
phone: +49 681 / 302-3210, fax: +49 681 / 302-4372, E-mail: bley@cam.uni-sb.de

Abstract: In this article the concept and a prototype of a feature based integrated tolerance support system is described. The system consists of a "functional view" and a "manufacturing view" onto the tolerances. The "functional view" supports the designer finding appropriate solutions concerning dimensions and tolerances for certain functions like "bearing seat" or "transfer torque". The "manufacturing view" contains information about the feasibility, production cost and time of a tolerance. This knowledge based on experiments is stored in a SQL database. The manufacturing information is given to the designer and he/she can alter his/her decisions in a very early phase of the design process if problems arise. Implementation platform of the fully CAD integrated support system is CoCreate's SolidDesigner.
Keywords: CAT, NC-programming, Feature technology

1.   INTRODUCTION

Tolerances in engineering design are usually assigned in accordance to technical and functional aspects. Thereby, the designer only has very crude means of estimating the effects of the tolerance scheme (to be) applied. The absence of appropriate tools for the evaluation of given tolerance combinations affects two fields of interest: Firstly, the influence of the tolerances on manufacturing expenditure and secondly the cost caused by machine scheduling problems during the realisation of inadequate tolerances. Basically these problems occur because there is not enough information from the shop floor's viewpoint available during design. Furthermore today's design support systems are often badly integrated into the overall work flow.

To overcome this situation it is necessary to integrate tolerance information into CAD systems. This information includes the feasibility and cost of a specific tolerance. The designer retrieves this additional information from downstream processes (e.g. manufacturing) during an early stage of the design process. Thus, the designer develops a deeper understanding of the influence which assigned tolerances have on production cost and time.

2. CONCEPT FOR A FEATURE-BASED INTEGRATED TOLERANCE SYSTEM (FIT)

2.1 Feature technology

According to the FEature Modelling EXperts group "FEMEX" a feature is defined as "an information unit (element) representing a region of interest within a product" [Weber, 1995]. Thus, a feature is described by an aggregation of properties of a product. It has to be seen in the scope of a specific view onto the product description with respect to the classes of properties and to the phases of the product life cycle. A feature can be described in a matrix as shown in Figure 1.

As a main topic of this contribution a concept for fulfilling one of the main objectives of feature technology is shown: to improve the communication throughout the product Life-cycle, especially between different CAx-systems, and by this to reduce development time.

Figure 1; Matrix representation of the different stages of the product life-cycle, the different classes of properties describing the target tolerance feature.

In [Weber et al., 1998] three different classes of CAD integrated feature based tolerancing moduls are introduced:

The Feature based Modelling, Dimensioning and Tolerancing module (FbMDT) comprises the automatic generation of geometry as well as dimensioning and tolerancing and thereby covers almost the whole design process. The designer starts with functions to be fulfilled, e.g. "transfer a torque of 20 Nm". Then he/she has to find appropriate solutions for these functions. E.g. he/she can choose a certain solution principle from a predefined, structured table containing different alternatives and can thus generate the detailed design (semi-) automatically. The principles and solutions can be listed by cost to enable an easy way for selecting problem solutions as cheap as possible.

In contrast to this, the Feature based Dimensioning and Tolerancing (FbDT) module is restricted to the detail phase of the design process. The embodiment of a solution which has to be toleranced is already existing. Only the details concerning distinct functional requirements (e.g. bearing seat dimensions, sealing seat diameter) have to be completed.

As an enhancement of the function oriented scope of these two kinds of features the FbMDT&DfX module additionally includes further Design for X information. The combination of the designer's functional view and the manufacturing view represents the target feature of this contribution as shown in Figure 1, classed into the matrix representation introduced by FEMEX. Potentially, this matrix representation of the different stages of the product life-cycle, the different classes of properties and the views of special features will be part of a draft proposal of a feature guideline devised by the VDI (the German society of engineers).

2.2 Determination of technological data

An inexperienced CAD designer has only little insight into the manufacturing effects of tolerances assigned by him. A simple decision, whether a specific tolerance can be produced within the enterprise or not, often requires the consultation with the production department. Also the cost resulting in this case remain concealed. Up to now the designer only finds inadequate support in the form of factory standards and internal design guidelines. The essential disadvantage of this is bad integration into the work flow and rigidity of this concept. An ideal approach integrates manufacturing information directly into the work flow, in this case into the CAD system of the designer. Decision criteria that are important for the designer during allocation of a tolerance (from the point of view of manufacturing) are:

- feasibility (internal/external)
- cost (high/medium/low)
- machine availability
- production time.

These criteria have partially a very high complexity. For example, the question of feasibility can not be seen independent of the production cost. It can often be more reasonable for an enterprise to order highly precise parts externally instead of producing them itself at high expenses. The same can be true for simple parts with small accuracy requirements. For this reason the designer needs a system based on rules and heuristical knowledge. This knowledge for logical reasons is gained by the evaluation of

152

manufacturing processes. The system of rules should be based on experience from production planning.

A systematic for collecting and storing tolerance relevant manufacturing information was presented in [Bley et al., 1998].

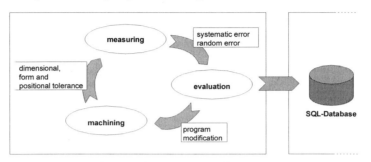

***Figure 2**; Experimental circuit.*

In order to gain information about the interdependencies between production parameters and attainable accuracy a package of experiments should be carried out. In an experimental circuit from machining – measuring – evaluation (Figure 2) technological and economical knowledge can be collected and prepared. Cost and data structures of the manufacturing process can correspondingly be stored in a database. It is important to separate between random errors and systematic errors in the step of evaluation. The modification of the NC programs based on the obtained results closes the experimental circuit. Finally, the experimental results in the SQL database are evaluated and summarised.

This data can then be integrated into the working environment of the CAD designer, using an SQL-CAD Interface. There are some special requirements for this interface to prepare the information.

2.3 NC-Programming

The semi-automatic generation of NC programs as part of the production process is nowadays state of the art. A number of different systems is available on the market. They can be distinguished into universal programming systems, integrated CAD systems and workshop-systems depending on the degree of integration [Kief, 1998]. However, they still require more or less user interaction to enter technological data as feed, cutting speed, way of clamping, etc. Determination of the process parameters is up to now largely based on empirical knowledge. Attaining specific shape and position tolerances is mostly not predictable. In most cases the selection of cutting parameters will be based on recommendations of the cutting tool manufacturers [Pryzybylski et al., 1997].

However, this procedure has only a small benefit in the sense of an automation of the production process. To estimate the cost resulting from a specific tolerance, it is necessary to determine the required processing parameters. Consequently, these parameters can be used for the NC programming. In this way, the NC program will be generated simultaneously to the product design.

A vectorial approach for the description of tolerances in the CAD system seems to be very suitable for the conversion into an NC programming system.

3.   PROTOTYPE

A prototype of the feature-based integrated tolerance support system (FIT, see Figure 3) is being implemented within the CAD-system CoCreate SolidDesigner using the powerful LISP integration kit in combination with an interface to an SQL database. The database interface is realised using a intranet server (Apache HTTP server) and a server-side, HTML embedded scripting language (PHP: Hypertext Preprocessor). The usage of TCP/IP based intranet techniques enables the development of system-independent solutions.

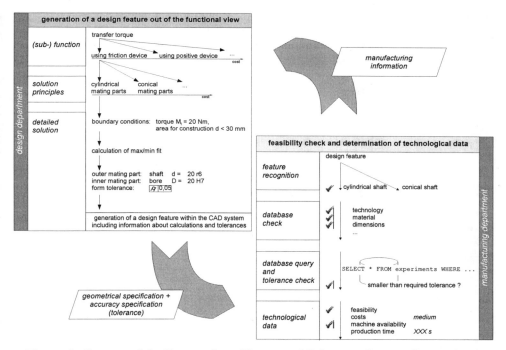

Figure 3; Concept of the Feature-based Integrated Tolerance Support System (FIT).

3.1 Feature-based tolerancing out of the functional view

As already mentioned before, the process of FbMDT covers the whole design process except the clarification of the task. The designer starts with a function to be fulfilled,

154

e.g. "transfer a torque of 20 Nm from shaft to hub", and has to find an appropriate solution to this function.

With tables structured as shown in Figure 3 (from functions to solution principles) he/she picks a certain way to get the detail design. The principles and solutions are listed by cost. Hereby, an easy possibility for the selection of the most cost advantageous problem solution is given. In Figure 3 the principle "transmission using friction between cylindrical mating parts" has been selected. After this decision is made, a calculation starts to verify the solution with respect to constrains included in the design task (e.g. torque 20 Nm, d < 30 mm). If the calculated result is dissatisfactory, this solution is abandoned. The designer has to step backwards and has to select another principle (e.g. "transmission using friction between conical mating parts") or another possibility of transfer (e.g. "transfer using form closure / positive device").

After a sensible solution is found, an instance of the design feature will be generated consisting of model data, calculated dimensions and specific tolerance information. The feature data can be read by every engineer involved downstream in the production process. Thereby the attached tolerance data eases the communication between the departments. Since every engineer can retrieve the current status of the product data the decision making process is significantly accelerated and modifications, e.g. to meet additional requirements or constraints, are easier to be realised.

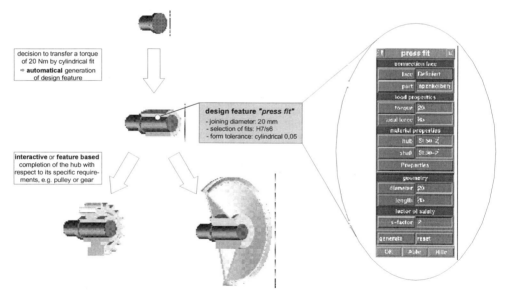

Figure 4*; Feature-based Modelling, Dimensioning and Tolerancing (FbMDT).*

After the selection of the cylindrical fit, shown in Figure 4, the geometry of the inner and outer mating parts are generated automatically. The thickness of the hub is calculated with reference to the stress imposed by the press fit on that ring. The designer has to complete the design of the hub with respect to its specific requirements and functions. For example, if the automatically generated ring will be used as basis of a gear wheel the missing gear geometry has to be generated around the ring to complete the hub.

3.2 Feature-based tolerancing out of the manufacturing view

The geometrical and tolerance specification from the CAD system is the starting point for a feasibility check in the manufacturing department (see Figure 3). The first step is to identify and select the appropriate manufacturing feature. The feature describes the parameters decisive for manufacturing this tolerance. The next step is to check the database for some boundary conditions. If the experimental database includes no information about the required technology, material or dimensions, it is not possible to feed back any manufacturing information. If the database includes enough information, in the next step the database query can be carried out. Based on the geometrical and tolerance specification an SQL query will be done. The result shows the feasibility of this tolerance and the required manufacturing parameters. Using these results, in the last step, production time, cost and other information can be calculated. Some of this information (e.g. concerning cost) will be reported to the design department, other pieces of information (e.g. manufacturing parameters) will be used for NC programming.

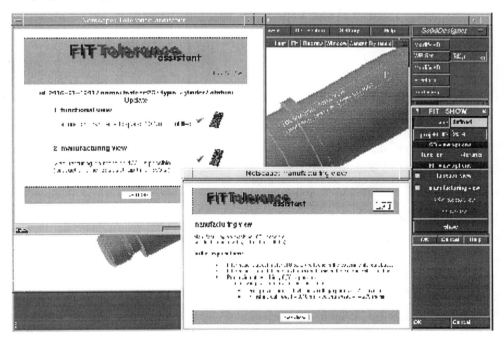

Figure 5; *The tolerance assistant and the result of the manufacturing view.*

The designer is able to see and control the results through an intranet-interface. With a click on the calculator inside the "tolerance assistent"-window he/she gets more and detailed information about the functional and/or the manufacturing view within another window (Figure 5).

The designer has now also the chance to change his tolerance specification if the manufacturing needs too much time, is too expensive or simply not possible. The manufacturing view can also make recommendations how to reduce cost and time.

4. OUTLOOK

This contribution has focused mainly on feature-based tolerancing from the designer's functional view and the manufacturing view. The next step is to capture further "design for X"–information. For this purpose the departments of design, manufacturing and inspection have to co-operate with the business administration division.

A precise calculation during the design process and a precise calculation of the expected process cost have to be realised to capture the cost of the tolerance schemes. Especially tolerances which are too tight and in consequence too expensive to realise should be detected and eliminated by the designer in an early stage.

REFERENCES

[Bley et al., 1998] Bley, Helmut; Oltermann, Ralf; "Determination of technological data during NC programming based on CAD-integrated tolerances"; In: *Proceedings of the 14th International Conference on Computer-Aided Production Engineering*, pp. 143-148; Tokyo/Japan

[Kief, 1998] Kief, H. B.; "NC/CNC Handbuch '97/98"; Carl Hanser Verlag München Wien; ISBN 3-446-18989-0

[Pryzybylski et al., 1997] Pryzybylski, L.; Slodki, B.; "The influence of Local Operating Features on the Cutting Parameters in Turning"; *The International Journal of Advanced Manufacturing Technology*, pp. 233-236, Springer-Verlag London

[Weber, 1996] Weber, C.; "What is a Feature and What is its Use? - Results of FEMEX Working Group I"; *Proceeding of the 29th International Symposium on Automotive Technology and Automation 1996*, pp. 109-116; Florence, ISBN 0-94771-978-4

[Weber et al., 1998] Weber, C.; Thome, O.; Britten, W.; "Improving Computer Aided Tolerancing by Using Feature Technology"; In: *Proceedings of Design '98 – 5th International Design Conference*, pp. 117-122; Dubrovnik; ISBN 9-53631-320-0

Contribution of a mathematical model of specifications of a part to their coherence analysis

Eric BALLOT*, Pierre BOURDET**, François THIEBAUT**
*Centre de Gestion Scientifique, Ecole nationale supérieure des Mines de Paris
60, bd Saint Michel - 75272 PARIS cedex
eric.ballot@cgs.ensmp.fr
**Laboratoire universitaire de recherche en Production Automatisée
Ecole Normale Supérieure de Cachan, 94235 Cachan Cedex
bourdet@lurpa.ens-cachan.fr, thiebaut@lurpa.ens-cachan.fr

Abstract : This paper presents a method to check the validity of the geometric specifications of a given part. We use a substitute part constituted with exact surfaces and we assemble this part on another which represents both the specified datum frame and the location of the tolerance zone. This model permits to know if some part features are over constrained or if the specification is not sufficient to define the location of the tolerance zone.

Keywords : tolerance modeling, tolerance analysis, tolerance coherence, deviation area

1. INTRODUCTION

The analysis of the tolerancing validity of the parts of a mechanism is a research field developed in numerous papers [Clément et al., 1991], [Requicha, 1983], [Turner, 1990]. The usually used methods suppose that the tolerancing of the parts of the mechanism is known and correct. In this paper, we propose a method which permits to check the coherence of the tolerancing of a part by a formal mathematical analysis of every geometric specification on the part. This analysis is done by the assembly of the substitute part of the actual part on a true part which represents the geometric specification. This analysis permits to know and to quantify the specification lacks and the over-constrained features on the part. This study completes a first study [Leveaux 92] which gives both a syntax analysis of the geometrical specification and a validation of the specification based on the pertinence of their inspection.

2. GEOMETRIC TOLERANCING PRINCIPLES

2.1 Without options: MMC (M), LMC (L), projected tolerance zone P:

ISO standard permits to point out either actual features or true features derived from the true geometric counterpart of features according to constraints.

So actual features are either the toleranced feature pointed out by an arrow or the datum features pointed out by a triangle. These features may be established by datum targets and be a combination of features, or be an area of the feature. The toleranced feature may also be a derived median plane of two actual parallel planes, or the derived axis of an actual cylinder or cone.

The true features are:

- Specified (single or common) datum features, and specified datum frames. These are mainly points, lines and planes which may be geometrically constrained in orientation or in position. Single datum or common datum will be derived from the actual features accordingly to the same association criteria. The specified datum of a datum frame will be associated one after the other, in a sequential and arranged way.
- Tolerance zones which may be constrained in orientation or in position relatively to the specified datum.

The toleranced features have to be included in the described tolerance zones. Each specification needs to be satisfied regardless of the other specifications.

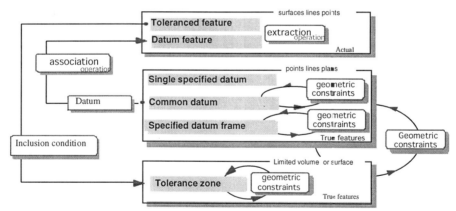

Figure 1 : ISO tolerance specification reading (without any option)

2.2 With options: MMC (M), LMC (L), projected tolerance zone P

In the case a least or a maximum material condition option is used, the signification of the specification is changed. ISO standard presents the virtual condition for the toleranced feature and the maximum (or minimum) material size for the datum feature simulator, these two states must not be crossed by the material (or being in the material). The virtual state is defined by the true envelope which dimension is given by the cumulated effect of the maximum material dimension (or least material dimension) and of the geometric tolerance.

2.3 Specification expression

In order to illustrate our approach, we will use the part presented in figure 2 as an example. This part, coming from the industry and toleranced by a group of expert, is reproduced here in a stylized way.

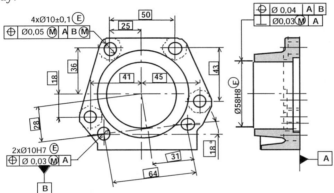

Figure 2 : *definition drawing of a cover*

Specification	Toleranced feature	Datum feature	Datum and maximum material condition datum	Tolerance zone or virtual condition size
⊥ Ø0,03Ⓜ A	Actual bore (Ø58)	Actual feature A	True plane A associated with feature A	Cylinder Ø57.97 normal to planeA
⊕ Ø0,05 Ⓜ A B Ⓜ	Four Actual bores (Ø10±0,1)	Actual feature A and two actual bores (Ø10H7)	True plane A associated with feature A and two cylinders Ø10, normal to A, in a theoretical position.	Four cylinders Ø9,85, normal to plane A, in a theoretical position.

Table I : *Studied tolerance description examples*

3. SPECIFICATION MODEL BY ASSEMBLY CALCULATION

In order to define a mathematical model for the geometrical specifications of a part, we will apply a calculation method of three-dimensional chain of dimension to a part and its specification. We will first describe why this transposition is possible and then give the principles and the modes of enforcement.

3.1 Identity between the parameters used in three-dimensional metrology, the three-dimensional chain of dimension and the specifications

The small displacement torsors permits a representation of the deviations between a nominal (or theoretical) geometry and a substitute geometry. This substitute geometry is already used in three-dimensional metrology to identify actual surfaces and build some specified reference frames.

On a complementary way, numerous model of three-dimensional chain of dimensions are based on a similar representation. The small displacement torsor is the tool which permits to express the deviation between the nominal representation and the substitute representation of the surfaces, the gap between two surfaces belonging to different parts and the position of the part relative to their nominal position.

The same tool can also be used to mathematically describe the tolerances. Then the components of the torsor provide a representation of the geometrical deviation of the substitute feature in a different space: the space of the small torsor displacement of the toleranced feature relative to the nominal feature cf. figure 1. The limits corresponding to the tolerance zone define then another expression of the specifications: the specified deviation area. A systematic methodology has then to be given to calculate the constraints on the small displacement torsor in order to mathematically define the specifications.

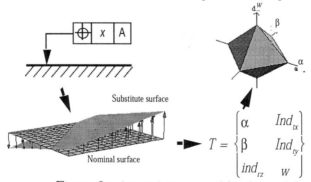

$$T = \left\{ \begin{matrix} \alpha & Ind_{tx} \\ \beta & Ind_{ty} \\ ind_{rz} & w \end{matrix} \right\}$$

Figure 3 : *changed space used for the model*

3.2 General principle of the calculation of the specified deviation area

An equivalence can be provided between a geometrical deviation between some features of a part and a functional condition relative to a mechanism. Indeed, the substitute part will verify the specification if it can be assembled in true parts which represent the tolerance zone of each specification. The specified datum frame of each specification defines the assembly conditions between the substitute and the true model of the part. The tolerance zone is then defined as a condition to be verified by the substitute feature relative to the nominal position.

This equivalence between the functional condition expressed on a mechanism and the specification relative to features of a part permits to use the results of the model study of the three-dimensional chain of dimension presented during the last CIRP seminar on Tolerancing [BALLOT 97]. The geometric laws of comportment expressed on the feature of a part give the relations to verify between the substitute and the nominal features. The deviations calculated with the behavior laws give the relation to be verified by the substitute surfaces to verify the specification.

The expression of these deviations on a sampling of the tolerance zone defines then the mathematical constraints which represent the specification in the space of the small displacement torsor parameters.

Table 2, presented below, resumes the correspondence between the standard geometric specification feature and the features used in the model that we will describe.

Standard feature	Proposed model
Toleranced feature	Substitute feature
Datum feature	Substitute feature
Specified datum	Theoretical feature or their assembly
Tolerance zone	Geometric zone where the deviation calculated with the behavior law is sampled
Options (E) (M) (L)	Always taken into account by the model. Like the relations are expressed between the features, they correspond to the use of the least or maximum material condition for lack.

Table II : *correspondence between the standard and the model features*

The systematic calculation of the deviation area is based on the use of the model of calculation of the laws of geometrical behavior presented during the last CIRP seminar on tolerancing. That is why we will describe above all his application in this new utilisation context: the assembly of an substitute part in a set of true parts which represents the specified datum frames and the tolerance zones.

3.3 The small displacement torsor

Like numerous other works in this field of research, we use the small displacement torsor to characterize the deviations between the substitued model and the nominal model [Bourdet et al., 1996]. Three kinds of deviation are represented by different torsor:

The deviation between a substitute feature and its nominal definition is characterized with a deviation torsor. The form of this torsor is deduced of the geometrical feature type (plane, cylinder, etc).

The gap between two features belonging to two different parts is characterized by a gap torsor. For the presented application, the gap deviation is defined between the substitute surface of the part and a true feature representing the tolerance zone and placed relatively to the datum frame. The form of this torsor is determined by the (same) type of the two surfaces.

The small displacement torsor which is necessary to place the part on each of the specified datum frame is characterized by a part torsor.

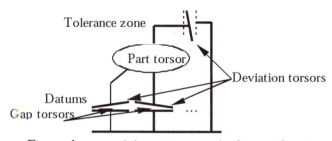

Figure 4 : *general description graph of a specification*

3.4 Specified datum frame: assembly between a substitute model and the true model

We will successively examine the case of a single datum and the case of a datum frame.

The positioning of a tolerance zone relatively to a simple reference represents an assembly between the substitute feature (which is the toleranced feature) and the specified datum. In this case the composition of the small displacement torsor of the substitute feature and of the datum are sufficient to give the tolerance zone position.

The positioning of a tolerance zone relative to a datum frame is equivalent to the aggregate some parallel assembly links. In a similar way than the one used in assembly, the algorithm which permits to calculate the position of the part is based on the undeterminated components of each gap (their degree of freedom) in function of the complementary gap components of the other links [Ballot, 1995].

In the case of a datum frame, standard indicate explicitly the hierarchy of the contacts. Then, the specification shown in the figure 5 is treated with the following method: positioning relative to the feature 2 (suppression of the mobility in rotation around x and y and in translation along z) then suppression of the supplementary degrees of freedom between the features 3 and 4 (by example $J[tx,3,13] = J[ty,3,13] = J[tx,4,14] = 0$ or one of the three other possibilities given by the compatibility system.

-J[ty,3,13] + J[ty,4,14] + J[tx,4,14] - J[tx,3,13] - dr[3,B].Sin[t[3,13]] + dr[4,B].Sin[t[4,14]] + w[3,B] - w[4,B] + (b[2,B] - b[3,B]).z[3,13] + (-b[2,B] + b[4,B]).z[4,14] + (61.(-(Cos[t[3,13]].dr[3,B]) + Cos[t[4,14]].dr[4,B] + v[3,B] - v[4,B] + (-c[2,B] + c[3,B]).z[3,13] + (c[2,B] - c[4,B]).z[4,14]))/20 = 0

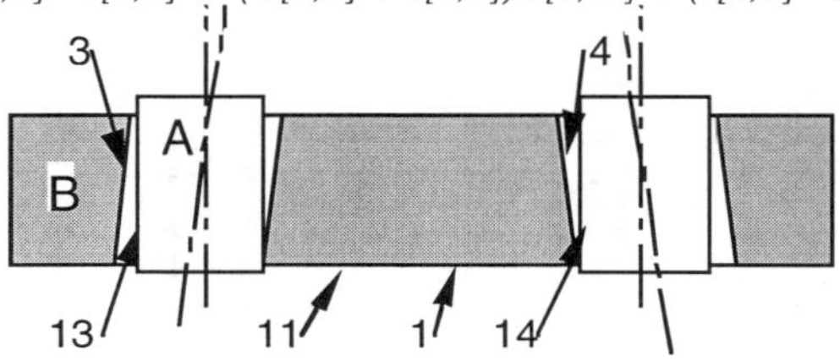

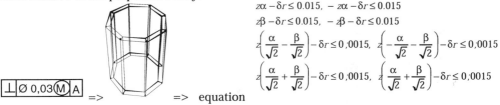

Figure 5 : *reference frame of a localization*

3.5 Tolerance zone : expression of a functional requirement

A geometric specification defines, by the mean of its symbols, the feature on which the specification is applied, and the orientation, the form and the size of the tolerance zone. We will use this information to calculate the deviation area which models the specification. The tolerance zone is determined, on each point of the zone, with the deviation calculated between the substitute feature and the nominal feature; Figure 6, created with the software [Wolfram, 1991] represents the set of deviation components which represent the tolerance zone relative to the perpendicularity.

$$z\alpha - \delta r \le 0.015, \ -z\alpha - \delta r \le 0.015$$
$$z\beta - \delta r \le 0.015, \ -z\beta - \delta r \le 0.015$$
$$z\left(\frac{\alpha}{\sqrt{2}} - \frac{\beta}{\sqrt{2}}\right) - \delta r \le 0.0015, \ z\left(-\frac{\alpha}{\sqrt{2}} - \frac{\beta}{\sqrt{2}}\right) - \delta r \le 0.0015$$
$$z\left(\frac{\alpha}{\sqrt{2}} + \frac{\beta}{\sqrt{2}}\right) - \delta r \le 0.0015, \ z\left(\frac{\alpha}{\sqrt{2}} + \frac{\beta}{\sqrt{2}}\right) - \delta r \le 0.0015$$

⊥ | Ø 0,03 (M) A
=>   => equation

Figure 6 : *model of a cylindrical tolerance zone (increment of $\pi/4$)*

4. DEMARCH APPLICATION ON THE EXAMPLE

In order to illustrate our approach and the type of our results, we have applied at the part of the figure 2 the algorithms used for the behavior law study.

4.1 Part model and notations

We are working here on the numerated surfaces shown on figure 7. At each surface is associated a deviation torsor, at each frame feature is associated a gap torsor and a part torsor is associated to the actual part to express the displacements of the part relative to the datum frame.

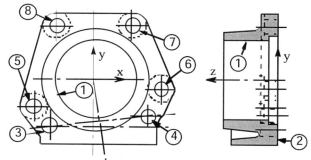

Figure 7 : *substitute feature locations*

4.2 Result examples

Specification	Generic inequality representative of the deviation
Ø58H8 Ⓔ	$\partial r[1, B] \leq 0.046$ $\partial r[1, B] \geq 0$
⊥ |Ø0,03 Ⓜ| A	$(z0-z1).(cost\ (c[1,B]- c[2,B]) - sint\ (b[1,B] - b[2,B])) - \partial r[1,B] \leq 0,015$
⊕ | Ø0,05 Ⓜ | A | B Ⓜ |	$(-122.z.c[2,B].Cos[t] + 122.z.c[5,B].Cos[t] + 76.Cos[t-t3].dr[3,B] -$ $76.Cos[t+t3].dr[3,B] + 46.Cos[t-t4].dr[4,B] + 76.Cos[t+t4].dr[4,B] -$ $122.dr[5,B] + 122.z.b[2,B].Sin[t] - 122.z.b[5,B].Sin[t] + dr[3,B].Sin[t-t3] -$ $dr[3,B].Sin[t+t3] - dr[4,B].Sin[t-t4] + dr[4, B].Sin[t+t4] - 122.Cos[t].v[4,B] +$ $122.Cos[t].v[5,B] + 2.Cos[t].w[3,B] - 152.Sin[t].w[3,B] - 2.Cos[t].w[4,B] +$ $30.Sin[t].w[4,B] + 122.Sin[t].w[5,B] + 2.b[2,B].Cos[t].z[3,13] -$ $2.b[3,B].Cos[t].z[3,13] - 152.b[2,B].Sin[t].z[3,13] + 152.b[3,B].Sin[t].z[3,13]$ $- 2.b[2,B].Cos[t].z[4, 14] + 2.b[4,B].Cos[t].z[4,14] +$ $122.c[2,B].Cos[t].z[4,14] -$ $122.c[4,B].Cos[t].z[4,14] + 30.b[2,B].Sin[t].z[4,14] -$ $30.b[4,B].Sin[t].z[4,14])/122 \leq 0,015$

Table III : *Geometrical behavior representative of the behavior laws examples*

Specification	Deviation area
⊥ |Ø0,03 Ⓜ| A	$30\big(\gamma(1,B)-\gamma(2,B)\big)-\delta r \le 0.015,\quad 30\left(-\dfrac{\beta(1,B)-\beta(2,B)}{\sqrt{2}}+\dfrac{\gamma(1,B)-\gamma(2,B)}{\sqrt{2}}\right)-\delta r \le 0.015$
	$30\big(-\gamma(1,B)+\gamma(2,B)\big)-\delta r \le 0.015,\quad 30\left(-\dfrac{\beta(1,B)-\beta(2,B)}{\sqrt{2}}-\dfrac{\gamma(1,B)-\gamma(2,B)}{\sqrt{2}}\right)-\delta r \le 0.015$
	$30\big(-\beta(1,B)+\beta(2,B)\big)-\delta r \le 0.015,\quad 30\left(\dfrac{\beta(1,B)-\beta(2,B)}{\sqrt{2}}-\dfrac{\gamma(1,B)-\gamma(2,B)}{\sqrt{2}}\right)-\delta r \le 0.015$
	$30\big(\beta(1,B)-\beta(2,B)\big)-\delta r \le 0.015,\quad 30\left(\dfrac{\beta(1,B)-\beta(2,B)}{\sqrt{2}}+\dfrac{\gamma(1,B)-\gamma(2,B)}{\sqrt{2}}\right)-\delta r \le 0.015$

Table IV : *Example of representative area of the specifications*

5. ANALYSYS OF THE SPECIFICATIONS

The definition of the deviation zone permits the integration of the already defined specifications to the chain of dimension. The calculation of the chains of dimension in which some specifications are already included constitutes a part of the specification validation: the functional specification respect.

However, it may be of use to examine the coherence of the specifications on a part independently of the mechanism in which the part is included. The proposed model permits to realize three kinds of specification analysis help.

5.1 Non consistence of a datum frame relative to a tolerance zone

Every specification for which the datum frame does not define the location of the tolerance zone has to be considered like incorrect. If an uncontrolled mobility leaves on the location of the tolerance zone, this one will be ambiguous.

In the proposed model, the undetermined components of the deviation and gap torsor which may stay in the deviation area show this lack of consistence. Indeed, the presence of one of this type of component shows that the deviation area may be deform without any control.

5.2 Extremum search of deviation component

The proposed model represents the geometric specifications by the way of a set of a non linear equations or of a set of linear constraint combinatory (in function of the contact configurations between the substitute element and the datum). Within this context and using a optimization algorithm, it is possible to search the extremum values of the select deviation components.

These values permit to check that all deviations are bounded (the set of geometric specification is suffisant), that the constraints relative to each specification are satisfied by at least one deviation value (the complete set of specification is necessary).

6. CONCLUSION

We have shown that the deviation torsor, the gap torsor and the part torsor associated to the undetermined components of the small displacements provide a coherent model of the geometric specification representation This systematic model gives the representation of the geometric specification as explicit mathematical.
This mathematical expression permits to tackle the questions relative to the specifications consistence with a point of view which ought to be developed. Indeed, the methods proposed to help the validation of the specifications on a part constitute some research axis which still need to be studied closely and completed.

7. REFERENCES

[Ballot 1995] Ballot E. "Lois de comportement géométrique des mécanismes pour le tolérancement" *Thèse de Doctorat de l'Ecole Normale Supérieure de Cachan*.

[Ballot et al., 1998] Ballot E., Bourdet P. "A computational method for the consequences of geometrical errors in mechanisms" *Geometric Design Tolerancing: theories, standards and application, edited by El Maraghy*. Chapman & Hall, ISBN0-412-83000-0.

[Bourdet et al. 1996] Bourdet P., Mathieu L., Lartigue C. and Ballu A. "The concept of small displacement torsor in metrology" *Advanced mathematical tool in metrology II in Series on advances in mathematics for applied sciences,* World Scientific, Vol. 40.

[Clément et al. 1991] Clement A., Desrochers A. and Riviere A. Theory and practice of 3-D tolerancing for assembly. *CIRP International Working Seminar on Tolerancing,* Penn State, pp. 25-55.

[Leveaux 1992] Leveaux F. "Contribution à la spécification géométrique des pièces mécaniques par simulation algorithmique du contrôle 3D" *Thèse de Doctorat de l'Ecole Normale Supérieure de Cachan*.

[Requicha 1983] Requicha A.A.G. Toward a Theory of Geometrical Tolerancing. *The International Journal of Robotics Research,* Vol. 2, No. 4, pp. 45-60.

[Turner 1990] Turner J. U. Relative positioning of parts in assemblies using mathematical programming. *Computer-aided design,* Vol. 22, No.7, pp. 394-400.

[Wolfram 1991] Wolfram S; "Mathematica : A system for doing mathematics by computer" *Addison-Wesley*.

Parametric tolerancing

J.M. Linares , S. Boukebbab , J.M. Sprauel
MECASURF, ENSAM / IUT, F13617 Aix en Provence
Tél: 33 442938153, Fax: 33 442938115
email: linares@iut.univ-aix.fr

Abstract: The objective of this work is to reduce scraps using a transcription optimization of the fitting functionality of mechanical parts. This paper brings to the fore the potentials of degrees of freedom present in mechanical assemblies.
The functional analysis lays down the functional conditions of global effect which are performed by the external tolerancing, the clearances interfaces and the internal tolerancing. The product analysis enable us to pull out the local functional conditions effect required by the internal tolerancing and the clearances interfaces. They represent stocks of liberties that aren't full used still nowadays.
The modeling of device interfaces allows us to have spaces of liberty. Those are generated by the assembly clearances and the variation of the internal tolerances. This notion of liberty space requires the settlement of a new description of the tolerance area which isn't currently define in the ISO standards.

Keywords: Tolerancing, Clearance, Degrees of Freedom, Tolerance zone, Design

1. INTRODUCTION

The world economy first translates into a decrease in profits and second into a need for competitiveness for all mechanical industries [Marty, 1995]. Within the past twenty years, efforts were focused on individualized processes. New words appeared, as well as new definitions for each craft (SPC, SMED...). This situation partly accounts for the failure of simultaneous engineering at its beginning. A systemic approach of industrialization problems brings answers at a wide scale for use. A new tool suggestion for design must absolutely account for the simplifications induced on the whole industrialization line. Any firm which is present in a sector open to competition must meet the customer's expectations concerning the technical characteristics and price of the product. Functional tolerancing emerges from this acknowledgement after a study phase. The tolerancing is described according to the tools suggested by the current standards. What concerns us is the realization of the global function of the product. The design engineer gives volume to the kinematics block diagram through a design study, then the tolerancing phase is activated. After this stage of the industrialization process, the biggest part of the product final cost is potentially defined. Mastering cost greatly depends on the foresight with which these implications will be understood and managed. With this research we suggest extending the use of tolerancing requirements (maximum material condition or MMC, least material condition, envelope requirement) by adaptive tolerancing in the industrialization process.

2. PROBLEM

Our tolerancing methodology is based on the notion of Functional Group [Linares, 1996]. A Functional group is defined as the set of surfaces which contribute together to a given elementary function. Such approach brings to the fore the duality which exists between the internal tolerancing and the clearances. The freedom spaces are composed of clearances and internal tolerancing.

The local functional flow is ensured by the tolerancing internal to the Functional Groups which are assembled and by the clearance deduced from the local functional conditions. The external tolerancing and the clearances characterize the external functional flow which must influence a global type functional condition. Optimizing the functional tolerancing comes to control the distribution of these spaces of freedom. This splits into local and global functional needs (Figure 1).

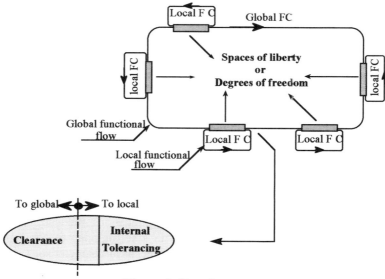

Figure 1; Freedom spaces

This notion of freedom space requires developing a new description of a tolerance zone [Boukebbab, 1998]. At present, with the tools supplied by ISO standards, only circular, cylindrical, plane, parallelogram-like zones or spherical zones can be described. The true spaces of freedom are of a complex form and can be described by implicit equations. In order to simplify and facilitate the use of design, we suggest to replace these implicit equations by vectorial functions (one sheet hyperboloid, ellipse, paraboloid ...).

Their coefficients will be given in a simplified form. To illustrate this operation, we shall deal with the example of two Functional Groups which consist in three holes or shafts and a plane.

3. PRESENTATION OF THE STUDIED EXAMPLE

To illustrate this freedom space concept, we shall deal with the example of two Functional Groups which consist in 3 holes (c1, c2, c3) or shafts (C1, C2, C3) and a plane (Figure 2). The clearance J1 between C1 and c1 is assumed to be lower than the clearance J2 between C2 and c2. The surfaces C3 and c3 complete the assembly. We will specially focus to the tolerance zone of C3 (or c3).

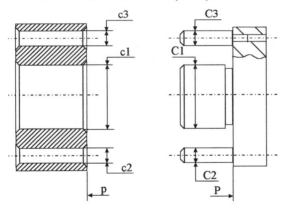

Figure 2; Example

This assembly problem can be solved by two different methods (figure 3). The first method is based on a graphic treatment by a Simplex. In this approach the 6 dimensions space of the small displacement vector coordinates is used [Giordano, 1992]. Two rigid parts assembled by theoretical perfect surfaces are considered. The real surfaces are assumed to be rigid and only the clearances between the two solids are taken into account (no form defect). The same kind of approach had already been proposed for tolerancing to describe feasibility spaces [Turner,1993]. However, no direct description of the space of freedom is obtained in the true physical 3 dimensions coordinates system.

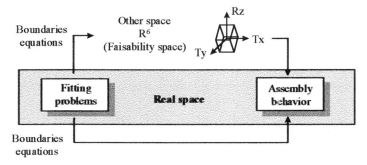

Figure 3; *different methods used to define the assembly behavior*

The second way which will be developed in our study, is based on the direct treatment of the boundaries equations in the 3 dimensions space. The assembly procedure imposes some hierarchy between the different surfaces which are connected. In our example it is identical for the two Functional Groups: the primary surface is the cylinder C1 (or c1) in MMC, the secondary surface is the cylinder C2 (or c2)) in MMC, tertiary surface of plane P or p. The assembly will behave in two different manners, depending on the size of the cylinders and on the clearances: the main orientation of the fitted part will be imposed either by the axis of C1 or by the normal to plane P.

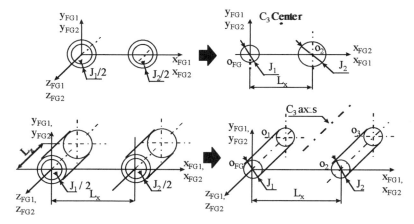

Figure 4; *Topology parameters*

The assembly is considered as perfectly rigid and without any form defect. Under these assumptions the boundaries conditions is greatly simplified. The conditions of contact are illustrated in figure 4. This leads to implicit equations which are difficult to solve analytically.

4. STATISTICAL APPROACH

The boundaries equations split into implicit equations for which two treatments may be used:
- An analytic treatment: The implicit equations are very complex and difficult to solve .
- An statistic treatment by a Monte Carlo method

The Monte Carlo simulation method has been published in many papers and books. It is well adapted to solve problems which depend on a great number of independent variables. In our single example, the number of internals and externals variables is greater than 10 (Figure 5).

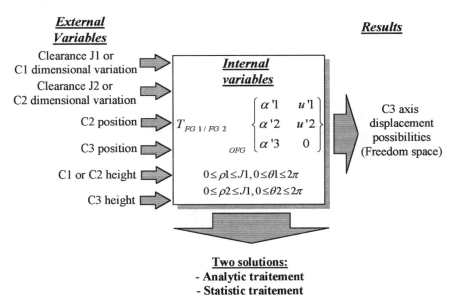

Figure 5; External and internal variables

Generally, all of the items of interest in these situations can be written in the form of equations, usually of considerable complexity and rarely amenable to explicit solution. The usual approach for handling these situations is the simulation.

The Monte Carlo method impose the knowledge of the distribution laws of each independent variable. It should be evident that although the mathematical definition of randomness can be precise, the practical definition (Rnd function) depends on the software and the application. A number of criteria have been proposed for judging the quality of these generators (serial correlation's of the pseudo random generator).

Such Rnd function should have has some qualities:
- **Stability.** The generator should pass all the statistical tests and have an extreme long period.
- **Efficiency.** Its Execution should be rapid and the storage requirement minimal
- **Repeatability.** A fixed starting condition (Randomize function) should generate the same sequence.
- **Simplicity.** The algorithm should be easy to implement and use.

5. SIMULATIONS RESULTS

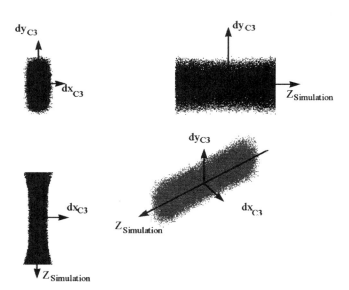

Figure 6, Freedom space in the reference frame used for the simulation.

A demonstration software has been developed for this example. At each iteration, a new position of the axis of cylinder C3 (or c3) is calculated. After a great number of simulations (more than 100 000 iterations), all the possible displacements of these axis C3 (or c3) can be plotted. The boundary of the domain which is thus obtained represents the freedom space of C3 (or c3).

The localization tolerance zone corresponds directly to this freedom space. However, the fitting of C1 into c1 requires that the parallelism defect between axis C3 (or c3) and C1(or c1) remains less than a given value which is derived from the clearance between C1 and c1 and the variation of the diameters of C1, c1, C3 and c3. This supplementary condition has to be added to the tolerancing.

6. LIBERTY SPACES

A great number of such simulations has been performed for different topologic conditions, the whole volume of the functional groups remaining less than 150x150x150 mm^3. Three different behaviors have thus been observed (Figure 7).

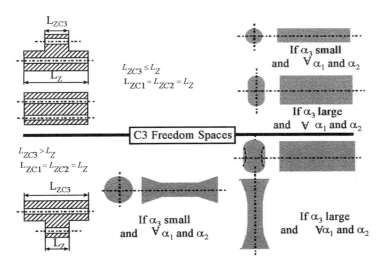

Figure 7; The three types of freedom space boundaries

These spaces of freedom are of complex form. To facilitate their description in order to use them in design, we suggest to replace the implicit equations by a vectorial function approximation: one sheet Hyperboloid when C1 and c1 are the primary surfaces of the functional groups or an Elliptical Cylinder when P1 and p1 are the top level surfaces.

The boundary of the freedom space depends on the height of cylinders C1 or C3 and on the angular rotation $\alpha'3$. Hover all the simulated cases can be described by our approximations. The parameters of the vectorial functions are adjusted in order to define a volume which bounds the freedom space.

7. COMPARISON WITH ISO TOLERANCE ZONE

The ISO standard 1101 defines different tolerance zones. In our example, the localization tolerance zones of C2, C3 (or c2, c3) are usually considered as cylindrical and are mainly derived from the clearance J1 .With such tolerancing mode the fitting condition is guaranteed in all cases. However, this increases the constraints for the manufacturing functions.

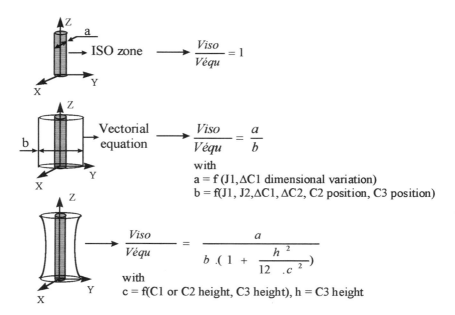

Figure 8; Comparison freedoms spaces boundaries with tolerance zone ISO

In order to reduce some scraps, the whole freedom space should taken into account. In fact, the fitting function can also be guaranteed by the simultaneous specification of location of C3 (zone defined by a vectorial function) and the parallelism between C3 et

C1 which enables the fitting of C1 to c1 (see end of §5). The difference between these two approaches is presented in figure 8. The benefit of the new method can be represented by the ratio between the volume of the ISO tolerance zone and the volume defined by the vectorial function.

8. CONCLUSION

In this paper it has been shown that vectorial functions could be used to improve the tolerancing of an assembly. This tolerancing mode allows to decrease some constrains for the manufacturing process. The designer should only account for the functional fitting needs of the assembly and should give to the manufacturer the greatest tolerance zones as possible. Simple vectorial functions which depend on the topology of the assembly have been proposed for that purpose to approximate the true spaces of freedom.

However, such approach requires to introduce mathematical definitions of dimensioning and tolerancing as it has already been proposed by the Y14.5.1 technical subcommittee [Requicha, 1983] [Srinivasan, 1993]. It brings to the fore the potentials of degrees of freedom present in mechanical assemblies.

REFERENCES

[Boukebbab, 1998] Boukebbab S., Contribution à l'étude des discontinuités dans les mécanismes en vue d'une ingénierie coopérative, Thèse de doctorat, ENSAM, 1998.

[Giordano, 1992] Giordano M., Duret D., Tichadou S., Arrieux R., Clearance Space in volumic dimensionning, Annals of the CIRP Vol. 41/1/1/1992.

[Linares., 1998] Linares JM., Boukebbab S., Sprauel JM., Co-operative engineering approach : Tolerancing, Control, Proceedings of the CIRP Seminar STC Design, New tools and workflow for product development, Berlin, May 14-15, 1998, pp 145-156.

[Linares, 1996] Linares, JM., Contribution à l'étude de la cotation fonctionnelle par une approche systémique, Thèse de doctorat, INSA de Lyon, 1996.

[Marty, 1995] Marty, C., Concurrent Engineering and economic effects of design decisions, Col. Int. INRIA, Grenoble, pp10-12, 1995.

[Requicha, 1983] Requicha A.G., Toward a theory of geometric tolerancing, international journal of robotic research, Vol. 25, n°4 p45-60.

[Srinivasan, 1993] Srinivasan V., Recent efforts in mathematization of ASME/ANSI Y14.5M Standard, proceedings of 3th CIRP Seminars on Computer Aided Tolerancing, Cachan, April 27-28 1993, pp 223-232.

[Turner, 1993] Turner J.U., A feasibility space approach for automated tolerancing, Journal of engineering for industry, Vol. 115, 1993, pp341-346.

Mathematical representation of Tolerance Zones

Max Giordano, Eric Pairel, Serge Samper
LMécA/CESALP (Laboratoire de Mécanique Appliquée)
Ecole Supérieure d'Ingénieurs d'Annecy, Université de Savoie,
BP 806, 74016 ANNECY Cedex, FRANCE
giordano@esia.univ-savoie.fr

Abstract : The notion of tolerance zone is fundamental for geometric tolerance analysis and synthesis. In this paper, a definition is proposed for the tolerance zones, slightly different from the ISO standard definition, but leading to a non ambiguous model. The tolerance zones are represented by inequations between the real variables that represent relative admissible displacements from the theoretical nominal positions. In the limit case of equality, the equations obtained are those of boundaries that limit a domain in the displacement space.

Some properties of these relations and of the corresponding volumes are investigated from their mathematical or geometrical representations, particularly the symmetry and transitivity.

The mathematical representation of tolerance zones allows to translate the standard concepts of projected tolerance zone and maximal material condition, and to analyse a given toleranced mechanism.

Keywords : tolerance modelling, geometrical tolerance, tolerance zone, mechanism theory.

1. INTRODUCTION

The notion of theoretical geometric elements associated to an actual feature is very much used in tolerancing modelling. J. Guilford and J. Turner use the term "virtual geometry" which can be derived from the actual feature [Guilford & Turner; 1993].

For F. Bennis, L. Pinot and C. Fortin, a tolerance zone corresponds to a set of transformations from a reference frame. The homogenous matrices are the mathematical representation of these transformations. The proposed algebraic method can help tolerance transfer, in particular [Bennis et al.; 1998].

Wirthz uses the Gauss criterion to associate a theoretical feature to an actual one. The position and the orientation of each feature are defined by means of vectors. Tolerances affected to the components of these vectors, define the position and orientation tolerances [Wirthz; 1993]. This model is simple but the main drawback, in our opinion, is that the shapes of tolerance zones are not built from functional requirements but depend on the vectors used to define the orientation and position : dimensioning and tolerancing are inter-dependant [Bialas et al.; 1997].

Clement and his team also use the term of vectorial tolerancing but the model is different. A theoretical surface is associated to actual one, but the position and orientation are toleranced by means of small displacement torsors (screw operators) [Gaunet 199-]. The concept of Technologically and Topologically Related Surfaces (TTRS) structures the tolerance definition and limits the number of combinations of relative tolerances of features. The tolerancing inequations are associated to tolerancing torsors [Temmerman 1998]

The aim of this paper is to analyse the properties of these tolerancing torsor associated to tolerance zones.

2.   A MODEL FOR TOLERANCE ZONES

2.1 Definition of a tolerance zone according to the model

An actual part is limited by a closed surface. The nominal geometry is defined by a set of theoretical features like plane, cylindrical, conic, spherical etc., called elementary features or surfaces. Supposing that the actual geometry is not too different from the nominal one, the type of theoretical feature corresponding to each actual point of the manufactured part is known.

Then another theoretical surface is associated to the actual one. It is the substitute element, not the real feature of the manufactured part. The substitute or associated surface element is identified by measuring many points of the actual surface. It is calculated by means of minimising a "distance" between these points and the associated surface. Different criteria may be used to define this "distance" such as the sum of the square euclidean distances (Gauss criterion) or the sum of distances, or the maximum distance, but with the theoretical surface on the free side of the material, etc. In the following, we call associated feature, this substitute feature whatever the criterion is.

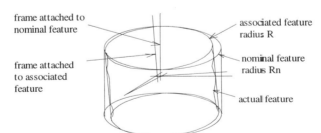

frame attached to
nominal feature

frame attached
to associated
feature

associated feature
radius R

nominal feature
radius Rn

actual feature

Figure 1; *Actual, associated and nominal features, and the frames attached to each of them.*

Then a frame is attached to each substitute element. For two elements A and B belonging to a part, the rigid displacement between the two frames is different from the displacement of the frame attached to the nominal part. If we note N_{AB} the homogenous

4x4 matrix relative to the nominal part and T_{AB} the matrix relative to the frames attached to associated elements, it is possible to define the deviation matrix by the relation : $T_{AB} = N_{AB} E_{AB}$ or $E_{AB} = N_{AB}^{-1} T_{AB}$

In the general cases, the maximum possible displacement deviation is small with respect to the nominal displacement so that we can define a deviation torsor (or screw) $\mathbf{E_{AB}}$.

This torsor is composed of two vectors, one defining the rotation and the other one defining the translation of the origin of the frame. For a given frame, this torsor is represented by 6 real numbers.

$$E = \begin{matrix} 1 & -\gamma & \beta & u \\ \gamma & 1 & -\alpha & v \\ -\beta & \alpha & 1 & w \\ 0 & 0 & 0 & 1 \end{matrix} \quad => \quad E = (\,\alpha\,,\beta\,,\gamma\,,u\,,v\,,w\,)$$

2.2 Analytic representation of a tolerance zone, deviation space

Assuming they are small, the deviations between the associated and nominal features, for two elements A and B are therefore characterised by the deviation torsor:

$$\mathbf{E_{AB}} = (\,\alpha\,,\beta\,,\gamma\,,u\,,v\,,w\,)$$

Choosing tolerances for the relative geometric locations between these two features consists in defining the set of possible limits of this torsor.

For example, for a parallelism and location tolerancing between a cylinder and a plane, parallel in nominal position, it is assumed that the B cylinder axis must be in the gap between two parallel planes at the distances (t) from each other. Thus the angle α and the small displacement v must be limited in accordance to the inequations (1) (Figure 2)

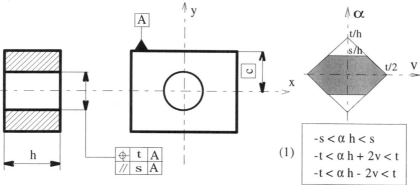

$$(1) \quad \begin{array}{l} -s < \alpha\,h < s \\ -t < \alpha\,h + 2v < t \\ -t < \alpha\,h - 2v < t \end{array}$$

***Figure 2**; Example of a deviation space for a cylinder and a plane.*

These inequalities may be represented by a region of the space torsor. This region is called "deviation space" [Giordano, Duret; 1993]. The torsors belong to a 6-dimension space thus it is only possible to represent a projection on two or three dimensions of this space. Some components of the deviation space are not limited, so the projections are done on the other components (figure 2). The number of limited variables is the dimension of the deviation space. It is equal to two in the previous example.

According to the model, a tolerance zone is a region of the 3D actual space where the points of the associated feature or the points of characteristic geometric elements of this feature such as an axis must be located.

So there are three kinds of representations for the geometric tolerances :
- by a tolerance zone,
- by a region of 6 dimension space,
- by inequalities.

It can be noticed that a set of compatible inequalities between small displacement components can always be represented by a region of the 6-D space. If the inequalities are not compatible with one another, this region is reduced to point O (no displacement). It is always possible to translate a tolerance zone into a set of inequalities or into a deviation space. On the other hand, it is not always possible to translate any deviation space into a tolerance zone. However, for the same feature, if different tolerance zones, each corresponding to a set of inequalities, are imposed, the complete inequality system finally obtained corresponds to a deviation space. And all the different combinations allow to build a great variety of possibilities for the example in figure 2. Nevertheless not all the possibilities are allowed, as will be seen the below.

2.3 Dimension of a deviation space

The dimension of this space has a maximum value of six and depends on the nature of toleranced features. A simple formula allows to determine this dimension. This formula is demonstrated in [Gaunet, 1994]. A part with (p) features to be toleranced with one another is considered. Each feature noted (i) is invariant for k_i independent motions. Each feature has k_i degrees of freedom. For a given frame, the number of small displacement components to be toleranced is : $\Sigma^p_{i=1}(6 - k_i)$. But, the part with its (p) features may have yet (m) degrees of freedom so that (6 - m) parameters can be arbitrary chosen to obtain the reference frame. For a part, the maximum number of variables to be toleranced is :

$$It = \Sigma^p_{i=1}(6 - k_i) - (6 - m)$$

For two features : (p=2) $It = 6 - k_1 - k_2 + m$
For a plane and a cylinder : $It = 6 - 3 - 2 + 1 = 2$

This number is a maximum because, from functional requirements, tolerances need not be attributed to all the components. For an orientation tolerance, only a maximum of three angles can be present in the inequations.

2.4 Examples of the three representations of geometric tolerances

Some simple examples are shown with their three possible representations : as zones in the real geometrical space, as volumes in the displacement space, and as inequalities associated to these volumes, figure 3 and 4.

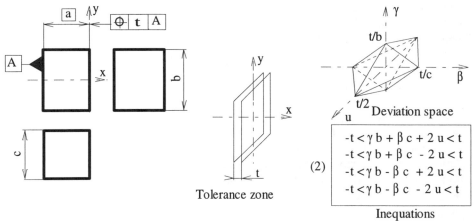

$$
\begin{aligned}
-t &< \gamma\,b + \beta\,c + 2\,u < t \\
-t &< \gamma\,b + \beta\,c - 2\,u < t \\
-t &< \gamma\,b - \beta\,c + 2\,u < t \\
-t &< \gamma\,b - \beta\,c - 2\,u < t
\end{aligned}
\tag{2}
$$

Figure 3; *Position tolerance of a plane, and the three representations of the tolerance zone.*

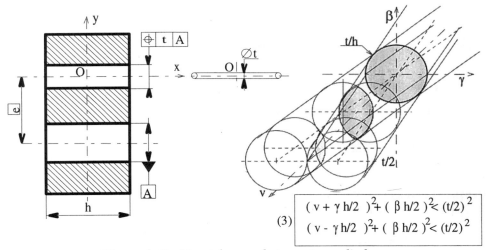

$$
\begin{aligned}
(v + \gamma\,h/2\ )^2 + (\ \beta\,h/2\ )^2 &< (t/2)^2 \\
(v - \gamma\,h/2\ )^2 + (\ \beta\,h/2\ )^2 &< (t/2)^2
\end{aligned}
\tag{3}
$$

Figure 4; *Position tolerance between two cylinders.*

The determination of variables that appear in inequalities can be obtained by a static analysis. We suppose that the part is in balance with given exterior forces. The unknowns are contact forces for each feature. The rank of the system is 6-m , where m is the degree of freedom of the part. The number of unknowns is $\Sigma^{p}_{i=1}(6 - k_{i})$, so the number of hyper-static unknowns is equal to the dimension of the deviation space (It). A small deviation component is then associated to each hyper-static unknown: angle for a torque and displacement for a force. For example, for two parallel cylinders, the static equation system has 8 unknowns and the rank is equal to 5. Frcm the static analysis, it is impossible to calculate the forces on the (y) axis and tcrques on (y) and (z) axes. Therefore, the tolerancing unknowns are v , β and γ (see figure 4).

3.   PROPERTIES OF TOLERANCE ZONES

3.1 Linear approximation, polytope associated to a deviation space

The assumption of small displacements leads to simple relations. For some tolerances, the relations are quadratic inequalities, but it is possible to replace them by some linear relations with an approximation. In particular, when the tolerance zone is a cylinder. For two parallel cylinders for example, the inequalities (3) are quadratic but inequalities (4) for different values of the parameter θ, are a linear approximation of (3). The deviation space, intersection of two oblique cylinders in figure 4, is translated into the polyhedron figure 5.

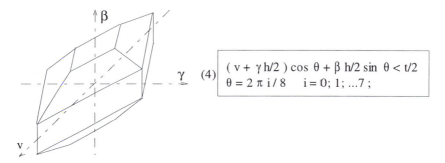

$$\text{(4)} \quad \begin{array}{l} (v + \gamma\, h/2\ )\cos\,\theta + \beta\ h/2\sin\ \theta < t/2 \\ \theta = 2\ \pi\ i\ /\ 8 \qquad i = 0;\ 1;\ ...7\ ; \end{array}$$

Figure 5; approximation of a space deviation.

Whatever the dimension of the deviation space is, a set of linear inequalities is translated into a convex polytope in this space : segment of a line for It =1, polygon for It=2 (figure 2), polyhedron for It=3 (fig. 3 and 5).

3.2 Symmetry

For a formal point of view, a geometric condition such as perpendicularity, parallelism or location is symmetrical. If the feature A must be perpendicular to the feature B, then, B must be perpendicular to A. But standard tolerancing is not symmetrical. However, for the considered model, different from the standard, and for a given geometric tolerance, it is possible to find a symmetric tolerance only in the particular case of a similar shape between the two features. For tb/h = ta/D the two deviation spaces are the same (figure 6).

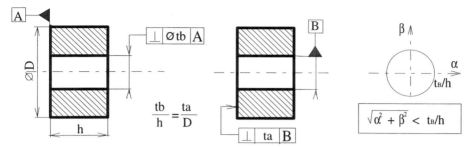

Figure 6; *Symmetry for two tolerance zones*

3.3 Transitivity

The transitivity leads to the notion of tolerance transfer generalised to the geometric tolerances. This problem is investigated by [Bennis et al.; 1998], without the use of linearity.

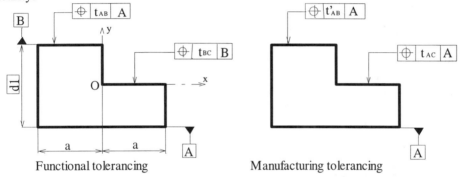

Functional tolerancing   Manufacturing tolerancing

figure 7; *Tolerance transfer for 3 parallel planes*

The problem can be formulated thus. Three features A, B and C belong to the same part. $\{E_{AB}\}$ is the set of deviation torsors of the toleranced feature B, in regard to the reference A. It includes all the possible values of torsor E_{AB}. It is assumed that $\{E_{AB}\}$ and $\{E_{BC}\}$ define the functional tolerances. The geometric tolerance transfer consists in substituting the tolerances $\{E'_{AB}\}$ and $\{E_{AC}\}$ for $\{E_{AB}\}$ and $\{E_{BC}\}$, (figure 7).

In practice the tolerance transfer is used for manufacturing. If E'_{AB} and E_{AC} deviations are consistent with the specifications, then the deviations E_{AB} and E_{BC} must be within the given set $\{E_{AB}\}$ and $\{E_{BC}\}$ respectively. The mathematical formulation is

$$\forall\ E_{AB} \in \{E'_{AB}\} \subset \{E_{AB}\} \text{ and } \forall\ E_{AC} \in \{E_{AC}\} \quad \text{then } E_{BC} \in \{E_{BC}\}$$

Now $E_{BC} = E_{AC} - E_{AB}$ therefore $\{E_{BC}\}' = \{E_{AC}\} + \{-E_{AB}\}'$ is the Minkowski sum of the two deviation spaces. Finally, $\{E_{BC}\}'$ must be included into $\{E_{BC}\}$.

184

For the example figure 8, the Minkowski sum has been built with t'$_{AB}$ = t$_{AC}$. In the assumption that t'$_{AB}$ < t$_{AB}$ the previous condition leads to t'$_{AB}$ = t$_{AC}$ = t$_{BC}$ /4

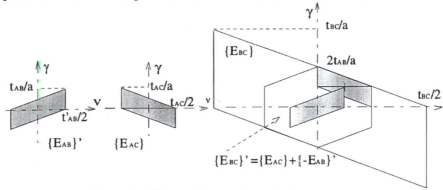

Figure 8; *Minkowski sum of two deviation spaces*

3.4 Projected tolerance zone

As shown in [Guiford and Turner 1993], the projected tolerance zone illustrates the concept of substitute features. A particular frame must be defined to express the components of the small displacement torsor. If the frame is at the centre of the tolerance zone, the deviation space is given by inequalities (a). But it is possible to obtain the inequalities (b) for the components of small displacements for the frame situated in the centre of the feature. (fig. 9)

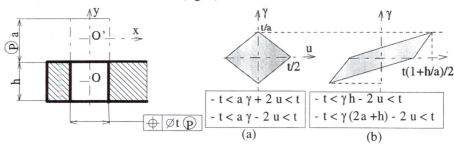

figure 9; *Projected tolerance zone*

3.5 Maximum material condition

According to the model, the maximal material condition is translated into a relation between geometrical and size tolerances : figure 10.

3.6 Application to closed loop mechanism analysis

The representation of tolerance zones by inequations on the components of small displacement torsors allows to determine the set of any relative positions of two features of the mechanism. However it is necessary to take into account the clearance between two features in each joint, and the associated clearance space [Giordano 98]. This method allows to analyse a given toleranced mechanism by combination of deviation and clearance spaces. It is necessary to compute the composition of these spaces by using their mathematical representation.

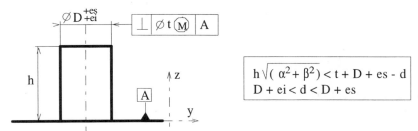

figure 10; *Maximum material condition*

CONCLUSIONS

The notion of theoretical substitute features associated to actual features is necessary to have a mathematical representation of tolerance zones. With the assumption of small deviations, the tolerance zones are translated into inequations on the components of the small displacement torsors. Non linear inequations will be translated into linear inequations for easier computation.

Then, the interpretation of some standard concepts can be analysed without any ambiguity such as projected tolerance zones, symmetry between two features, or tolerance transfer. The model allows to analyse any given complex toleranced mechanism. The number of inequalities increases quickly with the number of parts. Building an appropriate data structure and using mathematical representation of tolerance zones are necessary for a computer aided tolerancing software.

REFERENCES

[Bennis et al., 1998] F. Bennis, L. Pinot, C. Fortin «Geometric tolerance transfer for manufacturing by an algebric method»; In: *proceedings of International Conference on Integrated Design and Manufacturing in Mechanical Engineering, IDMME'98*; pp.713-720; Compiègne France, May 27-29; 1998.

[Bialas et al., 1997] S. Bialas, Z. Humienny, K Kiszka; "Relations Between ISO 1101 Geometrical Tolerances and Vectorial Tolerances - Conversion Problems"; In: *Proceedings of 5th CIRP, International Seminar on Computer-Aided-Tolerancing,* pp.37-48; Toronto; CANADA, April 27-29, 1997.

[Gaunet 1994] D. Gaunet, Modèle formel de tolérancement de position. Contributions à l'aide au tolérancement des mécanismes en CFAO. PHD thesis Ecole Normale Supérieure de Cachan fev. 94.

[Giordano, Duret, 1993] « Clearance Space and deviation Space » ; In : *Proceedings of 3ʳᵈ CIRP Seminars on Computer Aided Tolerancing* pp.179-196 April 27-28 1993 E.N.S. Cachan, Editor : EYROLLES Paris.

[Giordano 1998] M. Giordano «Modèle de détermination des tolérances géométriques»; In *Conception de produits mécaniques*, Chap. 13 Editor HERMES Paris 1998

[Guiford, Turner, 1993] J. Guiford and J. Turner ; « Representational primitives geometric tolerancing» ; In : *Computer-Aided Design*; pp. 577-586 ; Vol. 25 N° 9 Sept ; 1993.

[Temmerman 1998] M. Temmerman «Behavioral laws for tolerancing synthesis » ; In: *proceedings of International Conference on Integrated Design and Manufacturing in Mechanical Engineering, IDMME'98* ; pp.729-738; Compiègne France, May 27-29 ; 1998.

[Wirthz 1993] « Vectorial Tolerancing A Basic Element » ; In : *Proceedings of 3ʳᵈ CIRP Seminars on Computer Aided Tolerancing,* pp.115-128 ; April 27-28 1993 E.N.S. Cachan; Edt EYROLLES.

On the education of the real geometry of mechanical parts through the tolerancing and the tridimensional metrology

Claire LARTIGUE, Pierre BOURDET, Guy TIMON
Laboratoire Universitaire de Recherche en Production Automatisée
Ecole Normale Supérieure de Cachan
61 Av. du Président Wilson 94235 Cachan – France
lartigue@lurpa.ens-cachan.fr

Abstract: The aim of the paper is to present our experience of the education of tolerancing in University. The learning of tolerancing must give means to correctly read and understand standardised specifications, and must show how tolerancing allows to control the real geometry of the part surfaces in order to respect functional requirements. This approach is illustrated through a real case.
Keywords: education of tolerancing, geometrical specifications, functional requirements

1 INTRODUCTION

The education of tolerancing is more often limited to the reading of standardised specifications. Generally this learning relies on the numerous examples proposed by the ISO standard [ISO 5459, 1981]. Here, the standpoint is a designer standpoint, which leads to the design of an ideal nominal geometry.

On the other hand, the tridimensional metrology of parts using Coordinate Measuring Machines (CMM) allows to have a view of the real geometry. The real surfaces are known by point sets which are associated to geometrical models referred to *substitute surfaces*. The checking of the specifications is then based on calculations between substitute elements. As a result, such dimensions are not suitable with respect to the ISO standard.

Although the real part geometry is the common point between both types of education (tolerancing and tridimensional metrology), each education is taught independently and following its own point of view.

The paper presents our educational experience in University, which relies on the real geometry of mechanical parts through their standardised description and their tridimensional measurement. Taking the example of the tolerancing between two parallel planes, we show how tridimensional metrology, geometrical specifications and functional requirements are directly linked to the real geometry. The study emphasises the following phenomena:

- the scatter on the association of a geometrical model to a real geometry,
- the limits of tridimensional measurements *vis à vis* the geometrical specifications,
- the relevance of the geometrical specifications *vis à vis* the functional requirements.

Moreover, we propose a conceptual model of tolerancing, issued from ISO standard, which allows reading and interpreting simply all kinds of specifications

2   THE ISSUE

The objective of the education of tolerancing must be double. First, it must give means to correctly read and understand standardised specifications, but moreover, it must show how tolerancing allows to control the real geometry of the part surfaces in order to respect functional requirements.

To solve this issue, the teaching aid we use is the main part of a planing machine. Functional requirements are linked to the expected thickness of the shaving and to the quality of the planed board of wood.

So, the study concerns the determination of the geometrical specifications and the measurement of the studied surfaces in order to check specifications (figure 1). The study consists of two main steps:
- a critical analysis of a usual non-standardised specification,
- the expression of geometrical specifications from the functional requirements of the part.

The critical analysis relies on tests showing the influence of the choice of the measured points on the real geometry for the checking of geometrical specifications. In particular, we study the effect of various point numbers and different point distributions on the association of a geometrical model to a real surface.

In the second step, we propose to define geometrical specifications taking into account the functional requirements of the part and considering the real geometry. The description of the geometrical specifications is performed through a conceptual model of tolerancing.

3   FIRST EXPRESSION OF THE TOLERANCING

3.1   Introduction

The functions of the planing machine are to reduce with a given quantity the thickness of a wood board, and to ensure an acceptable flatness of the planed surface. With the hypothesis that during the rotation the cutting edge of the blade follows a cylinder which is tangent to the plane 2, the distance between the two planes corresponds to the thickness of the shaving. For instance, the thickness is chosen to 0.5mm, and as a result the designer imposes the planes to be parallel, distant of 0.5mm (in the exact position) and with an authorised distance variation of ± 0.1mm.

So the first expression of the functions as an ideal geometrical model corresponds to the specification proposed in figure 1.

This specification is obviously non-standardised but it is commonly used by most of the designers.

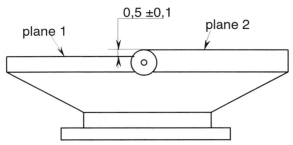

Figure 1 : *First expression of the tolerancing*

3.2 Checking of the specification

The first part of the study concerns the checking of the geometrical specification as indicated on the drawing, even though the specification is non-standardised.

Both specified surfaces of the planing machine are measured using a CMM, following a regular grid (x,y), covering in each case the whole surface. The representation of surface 1 is shown figure 2.

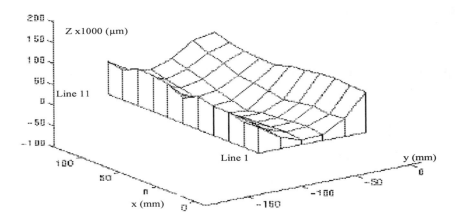

Figure 2 : *The distribution of the measured points*

A point set corresponding to a constant x defines a line (measured in 6 points) and the surface 1 is characterised by 11 point lines and the surface 2 by 12 point lines. Note that for both surfaces, the number of points is relatively high relative to the usual practice.

We can notice that the surface presents a hollow shape, which corresponds to the machining process. In practice, both surfaces were milled with a cylindrical tool for which the rotation axis was not perfectly in coincidence with the normal to the part movement.

As previously exposed, tridimensional metrology which allows the checking of geometrical specifications, relies on the association of geometrical models to point sets. Those models allow to define substitute elements and then, the checking is based on calculations between substitute elements. The association of geometrical models to point sets is performed using the concept of the Small Displacement Torsor [Bourdet et al., 1996]. In order to estimate the influence of the criterion, the association is realised using the Chebyshev criterion (specified in the standard) and using the least-square criterion (commonly used in metrology software).

So, tests will focus on factors influencing the geometrical model association: criterion, number and distribution of the measured points. A didactical software is developed under MATLAB and presents the following practicalities [Bourdet, 1993]:
    - visualisation of the real geometry through the measured points,
    - suppression of lines of points,
    - association of a geometrical model using the Chebyshev criterion,
    - association of a geometrical model using the least-square criterion,
    - calculation of the distance of all the points to a geometrical model,
    - calculation of the distance of the extremity points to a geometrical model.
Let us first consider the problem of the association of a geometrical model to a point set.

3.2.1   Modelling of the surfaces

Each surface is modelled by a plane. The ideal nominal element is a plane which normal is z, and which passes through the lowest point of all the measured points (figure 2). The components of the small displacement torsor correspond then to a small displacement along the z axis(w) and to small rotations around x (α) and around y (β). For each measured point the deviation, e_i, can be expressed by the equation:

$$e_i = \xi_i - (w + \alpha\, y_i - \beta\, x_i) \qquad (1)$$

with $\xi_i = z_i - z_0$

The association problem consists in:

    - minimising $\sum_1^n e_i^2$, using the least-square criterion,

    - minimising $[\max(e_i) - \min(e_i)]$, using the Chebyshev criterion.

The resulting form deviation is strongly linked to the used criterion. Moreover, the suppression of lines is also an influencing factor. Tests are performed with various numbers of points for the same distribution (corresponding to the whole coverage of the surface) and with various distributions for the same number of points (the choice of 18 points seems reasonable considering usual measurements of plane surfaces).

In order to emphasise the influencing factors, we consider the displacement of the points A,B,C,D, that materialise the extremities of the plane surface (figure 3). The moved points, during the association, characterise the *substitute plane*. Only the case of plane 1 is presented here, but the same approach can obviously be conducted for plane 2. Table I and table II summarise all the tests of the study. In our tests, the reference is the case including all the lines.

We notice that only the first three series of tests have sense for they correctly represent a large coverage of the plane surface. For the other cases, measured points are located on small portions of the real surface that obviously implies scatters on the form deviation, on the location and the orientation of the substitute element. This provides non-significant results.

Considering the three first tests and even though measured points are correctly distributed, we notice a variation of 30% on the value of the form deviation and a variation of 48% on the location of the substitute plane in function of the number of points, when the Chebyshev criterion is used.

In comparison, results obtained using the least-square criterion present a smaller variation interval for the substitute plane location, only 21%, which corresponds to a diminution of more than 50%.

	Line numbers	form deviation (mm)	upper point (z, in mm)	lower point (z, in mm)	location tolerance (mm)
Chebyshev	all, 1 to 11	0,097	0	0	0
	1-6-11	0,089	0,0011	-0,0192	0,020
	1-11	0,069	0,0144	-0,0322	0,0467
	9-10-11	0,064	0,0146	-0,0662	0,081
	1-2-3	0,073	0,00139	-0,0027	0,041
	1-3-6	0,081	0,0052	-0,0413	0,047
Least-square	all, 1 to 11	0,101	0	0	0
	1-6-11	0,098	0,0045	0,0116	-0,016
	1-11	0,069	0,0048	0,0156	0,02
	9-10-11	0,075	0,068	-0,0408	0,048
	1-2-3	0,074	0,0033	-0,0114	0,014
	1-3-6	0,089	0,0062	0,0318	0,038

Table I

Finally, the least-square criterion, non-standardised but commonly used in metrology software, gives the smallest scatters in the location of the substitute elements.

Results expressed in size of zone	3 first tests		all tests	
	Chebyshev	Least-square	Chebyshev	Least-square
maximum of the form deviation	0,0968	0,1005	0,0968	0,1006
variation on the form deviation	29%	31%	49%	45%
variation on the location of the upper point	0,032	0,016	0,066	0,041
variation on the plane location	0,047	0,020	0,0808	0,048
location error relative to form deviation	48%	21%	83%	49%

Table II

3.2.2   Interpretation of the specification 0,5 ± 0,1

As the associated planes to the surfaces are not parallel, the specification can be interpreted in various ways.

Generally, one of the two planes constitutes the datum. In our study, plane 1 is chosen as the datum and the checking of the specification is performed in two ways.

First, we consider the substitute element of plane 2, characterised by the moved points, A', B', C', D' (figure 3), and we express the distance as the distance of A', B', C', D' to the plane 1. The second way is to calculate the distance of all the measured points to plane 1. Results of the tests are presented in table III.

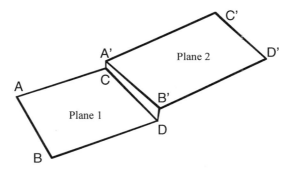

Figure 3 : Substitute planes

The various calculation methods of the distance involve results to be quite different, and the choice of the correct method turns out to be difficult.

calculation methods	all lines		lines 1,6,11		lines 1,11	
	dmax	dmin	dmax	dmin	dmax	dmin
distance of A', B', C', D' to plane 1	0,613	0,462				
distance of A', B' to plane 1 :	0,613	0,462				
distance of all the points to plane 1 :	0,666	0,425	0,653	0,425	0,616	0,400

Table III

To conclude this step, all the conducted tests show the obvious difficulty to correctly choose the number and the distribution of points for the checking of geometrical specifications. An a-priori knowledge of the manufacturing process of the part gives necessary information.

The last step of the work is the choice of the most suitable specifications to express the required functions of the part: thickness of the shaving and flatness of the planed surface.

4 EXPRESSION OF THE PART FUNCTIONS IN TERMS OF GEOMETRICAL SPECIFICATIONS

A geometrical tolerance allows to bound the variations of the real geometry into tolerance zones. Considering the ISO standards [ISO 5459, 1981] [ISO 1101, 1983], we have to define:

The real surfaces : corresponding to toleranced features and datum features

The ideal surfaces : corresponding to the datums (single or common), the datum-systems and the associated criteria

The tolerance zone

The tolerance zone is characterised by its shape. Its orientation and its location can be constrained by the datums or the datum systems.

For the example we have to consider the case of two nominally parallel planes. So, we identify :

- two real surfaces, referred as surface 1 and surface 2,
- a datum, for which the geometrical model is: *the outward tangent plane that minimise the form deviation*

Considering the direction of the wood board movement, the plane 1 is chosen as the datum. Surface 2 is thus the toleranced element.

As datum and toleranced element are clearly identified, the geometrical specifications proposed by the standard that can correspond to the functions are (figure 4):

- a specification of form deviation for the datum (flatness, for a plane surface)
- a specification of parallelism of the toleranced element relative to the datum,
- a specification of location of the toleranced element relative to the datum.

For all the geometrical specifications, the shape of the tolerance zones corresponds to two parallel planes.

194

For the specification of parallelism, the tolerance zone is constrained in orientation: the two parallel planes of the zone are parallel to the datum.

For the specification of location, the tolerance zone is constrained in location: the two parallel planes of the zone are parallel to the datum and located by means of exact dimensions.

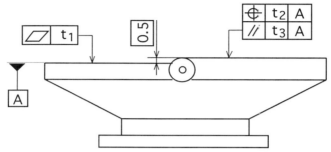

Figure 4 : *Geometrical specifications*

The values of t_1, t_2 and t_3 are defined considering the real geometry of the part and the functional requirements.

4.1   Function : thickness of the shaving, 0,5 ± 0,1

Both planes present a hollow shape, and we can make the hypothesis that the scattering on the association of the geometrical model is of the same order as the form deviation. Taking into account the machining process of the surfaces, the value of the form deviation is estimated to 0.1mm. This value gives a variation of 0.1mm on the thickness of the shaving. So, the dimension of the tolerance zone for the flatness is given by:
$t_1 = 0.1$mm.

Considering that the blade is tangent to the substitute plane cf surface 2, the location of the toleranced element is defined relative to the datum A (associated to the surface 1) with a value of t_2 such as $t_1 + t_2 = 0.2$mm, corresponding to the authorised variation of the shaving thickness. So, the dimension of the tolerance zone for the location is given by:
$t_2 = 0.1$mm.

If the machining of both surfaces is realised with the same quality, the form deviation is equal to the location deviation. In order to allow the adjustment of the machine tool, we have to reduce the value of the tolerance zone. This imposes the use of a machine tool and of a cylindrical tool of a better quality.

4.2   Function : flatness of the planed surface

The quality of the planed surface is defined through the specification of parallelism. The two surfaces are realised in one machining set up. So, the parallelism deviation is linked to the quality of the machine tool. Considering that the form deviation of surface 2 is of the same order as surface 1, the dimension of the tolerance zone for the location is given by: $t_1 < t_3 < t_2$

And taking into account the previous remarks : $t_1 = 0.05$mm, $t_2 = 0.15$mm, $t_3 = 0.08$mm, this leads to the geometrical specifications presented figure 5.

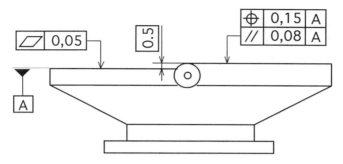

Figure 5: *Proposal of a tolerancing*

5   CONCLUSION

The teaching of tolerancing, which is generally limited to the learning of the reading of geometrical specifications, must show its link with the real geometry of surfaces and its possibilities to express the functional requirements of parts.

The experimentation we propose to the students consists in the checking of geometrical specifications that formalise a first expression of the functional requirements. Tests lead to bring out the influence of the real geometry of surfaces, in particular for the association of geometrical models to point sets (a plane, in the treated example). We show that, for the same surface, when number and distribution of points vary, the associated model may vary in a zone which size is of the same order as the form deviation of the surface. This effect is more VISIBLE when the association is conducted using the Chebyshev criterion. The use of the least-square criterion allows to bound the size of the incertitude zone, for the zone it is included in the incertitude zone resulting from the association with the Chebyshev criterion.

The numerous possibilities to express the distance between the two substitute planes show that the calculation of the distance must be directly linked to the functional requirements of the two plane surfaces. The use of geometrical specifications issued from the ISO standard show the possibilities to express functional requirements by limiting the variations of the real geometry using tolerance zones.

REFERENCES

[ISO 5459, 1981] ISO Standard – *Technical drawings, Geometrical Tolerancing, Datums and datum-systems for geometrical tolerances*, 1981

[Bourdet et al., 1996] Bourdet, P.; Mathieu, L; Lartigue C., Ballu, A. *"The concept of the small displacement torsor in metrology", Series on Advances in Mathematics for Applied Sciences*, vol 40, pp. 110-122, World Scientific, 1996

[Bourdet, 1993] Bourdet, P. *"Les outils d'aide à la spécification dimensionnelle et géométrique des pièces mécaniques", Polycopié de Cours de Licence de Technologie Mécanique* – Paris 6/ENS de Cachan, 1993

[ISO 1101, 1983] ISO Standard – *Technical drawings, Geometrical tolerancing, Tolerancing of form, orientation, location and run-out, Generalities, definitions, symbols, indications on drawings*, 1983

Choice of functional specifications using graphs within the framework of education

Alex BALLU
LMP - UPRES A 5469 CNRS - Université Bordeaux 1
351 cours de la Libération - 33405 Talence Cedex - France,
ballu@lmp.u-bordeaux.fr
Luc MATHIEU
LURPA - ENS de Cachan
61 Avenue du Président Wilson - 94235 Cachan Cedex - France,
mathieu@lurpa.ens-cachan.fr

Abstract : The teaching of three-dimensional tolerancing is relatively new and has to be improved. This paper develops the teaching of the choice of specifications. Two fundamental concepts are pointed out : the independence of the functional requirements and the influence of the parts, surfaces and deviations on a requirement

A manual method of tolerancing arises from these concepts. This method is based on representations in the form of graphs of the parts, surfaces, characteristics, functional requirements, specifications and on a set of rules for the choice of the specifications.

We have applied this method for a few years to our students and recently to the French teachers in design and manufacturing. This method is one of the bases of the education program in tolerancing developed at the national level.

Keywords : Education of tolerancing, Tolerance specification, Tolerance synthesis, 3D tolerancing

1. INTRODUCTION

For the last ten years, scientific knowledge on tolerancing strongly developed, but how is it taught ? What can be transmitted to the students and which pedagogy can be used ?

1.1. Education of Tolerancing

Having a double activity of research and teaching, we obviously put ourselves these questions. Public, in presence of which we practice, is under graduate student level, interested in the jobs of design and manufacturing (mechanical engineering and design department, manufacturing department). Up to now, the teaching usually dispensed to this public concerns :

- determination of tolerances by applying tolerance chains of only one dimension,
- elementary training in geometric specifications (by tolerance zones) with applications often limited to isolated parts.

197

The efforts which we carried out on teaching relate to :
- tolerancing standards with generic concepts [Ballu, Mathieu, 1993],
- choice of the specifications,
- tolerance analysis by space of small displacements [Giordano, Duret, 1993],
- rational use of 3D metrology,
- experimental activities on clearance [Ballu, Couetard, 1995],

The whole is presented with a three-dimensional view of the mechanisms.

1.2. Problem of the choice of the specifications

In this paper, a method to choose the geometric specifications is presented. Developed methods in research happen to be unsuitable to the students, because of, either the complexity of calculations [Ballot, Bourdet, 1997], or the great amount of specifications obtained [Clement et al., 1994]. Both methods are applied to all the mechanisms, but either the use of a calculator is compulsory or the specifications over constraint the parts.

The multiplicity of parameters and their interactions is difficult to apprehend and to process, even for simple cases. The use of data processing can be a solution, but that has the disadvantage of moving away the student from the physical phenomena concerned. Also, we choose a third way thanks to a simplified method, giving a solution of satisfactory quality but with the detriment of its universality. By this method, some mechanisms and some functional requirements cannot be treated.

1.3. Developed concepts

The main concepts pointed out are :
- **independence** of functional requirements,
- **influences** of parts, surfaces and geometrical deviations.

Each functional requirement of a mechanism must be studied independently from the others. The aim is not to specify a part independently of the mechanism, as it is unfortunately too often taught.

The second concept on which we lean on is the influence of the parts, surfaces and geometrical deviations on the requirement considered. Indeed, the geometrical specifications must relate to, and only to, these influential parameters.

The independence of the requirements and the influence of the parameters are really the two concepts pointed out by this teaching on the choice of the specifications. Our experience shows us their importance.

1.4. Example

The following of this paper presents the method used. This is illustrated by a simple mechanism which permits a linear movement of the carriage 6 (figure 1).

On the example, the base 1 is in permanent connection with the bearing 2 by a prismatic pair (surfaces a and b) and a tightening by the screw 3. Axes 4 and 5 are hooped on the carriage 6 so as to ensure two permanent connections. Axes 4 and 5 are both in cylindrical pair with the bearing 2. The centring device 7 is in revolute pair with the carriage 6 (surfaces g and h). The screw 8 ensures the tightening.

The parts 7 and 8 have no functions on the movement of 6, they are defined to illustrate the paragraph 3.1. E, F and K are particular points used at the paragraph 3.

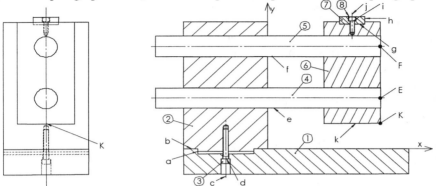

Figure 1 : *Studied mechanism.*

2. REPRESENTATION OF THE MECHANISM

The first stage of the method consists in understanding the structure of the mechanism to express the functional requirements. For that we lean on a graph.

2.1. Graph of the links

The structure of a mechanical product is usually modelized by a **graph of the links** (figure 2) in order to highlight the parts of the system and the links between the parts. For the needs of tolerancing, the links between two parts are decomposed in the elementary surfaces of type: plane, two parallel opposed planes, cylinder, cone, ... The corresponding graph is called **graph of the elementary links** (figure 2) and in this graph :

- each part is represented by a vertex,
- each surface is represented by a pole of the vertex of the corresponding part,
- each link is represented by an edge between two poles.

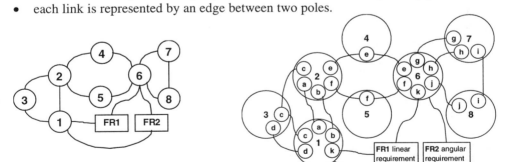

Figure 2 : *Graph of the links and graph of the elementary links*

2.2. Functional requirements

The graph of the elementary links being established, the study of the schedule of condition and of the process of assembly results in defining functional requirements. These requirements relate to one, two or several parts.

Within the framework of education we limit the study to the geometrical requirements on two parts. Those relate to :

- clearances or deviations,
- angular or linear.

A **deviation** requirement limits the variation of position with respect to a nominal position. A **clearance** requirement limits the extremum variation of position due to the clearance in the links of the mechanism.

On the graph of the links, a requirement is represented by a rectangular vertex connected by edges to the parts concerned (figure 2).

3. KEY PARTS, SURFACES AND DEVIATIONS

This section constitutes the second stage of the method. As we said in the introduction, we particularly endeavour to transmit the concept of influence of a part, a surface or a deviation on the functional requirement considered.

A key deviation means that the variation of position of a surface according to the direction of this variation involves a variation on the functional requirement. For example, the angular variation of the borings of the bearing 2 on z is a deviation which influences the requirement of linear deviation on y.

The research of the influences is not as obvious as that can appear. The large majority of the students has difficulties to identify these influences. These difficulties come especially from the number of parameters concerned. Also, our approach is to reduce the problem by the installation of several stages bringing gradually to the key deviations, by the resolution of elementary problems.

3.1. Key parts

At the level of the graph of the link, a requirement depends obviously on the parts connected to the vertex representing it, but it also depends on other parts of the mechanism. The parts on which the requirement depends, are called **key parts**.

The mathematical expression of displacements according to cycles of the graph makes it possible to show that the key parts have to be located on a cycle passing through the requirement. This type of cycle is called functional cycle for the requirement.

A **functional cycle** is a succession of vertices, poles and edges never passing twice by a same vertex.

But all the parts of all the functional cycles do not necessarily influence it.

For a given requirement, it is then possible to eliminate some parts. That is particularly interesting on a complex system. For the example, the functional cycles are : 1-3-2-4-6, 1-3-2-5-6, 1-2-4-6 and 1-2-5-6. The parts 7 and 8 do not belong to any functional cycle, so they are eliminated (figure 3).

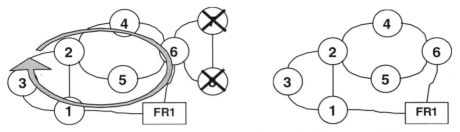

Figure 3 : Functional cycle 1-3-2-4-6 and graph of the potential key parts

3.2. Key surfaces

In the same way, at the level of the graph of the elementary links, surfaces on which the requirement depends, belong to the key parts and must be on a functional cycle. But all surfaces of a functional cycle do not influence necessarily the requirement. Surfaces on which the requirement depends, are called **key surfaces**.

For a given requirement, it is then possible to eliminate all surfaces in connection with the parts which were eliminated previously (figure 4).

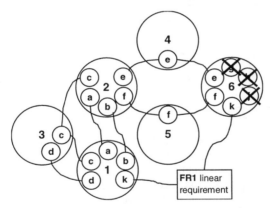

Figure 4 : Graph of the potential key surfaces

3.3. Reduction of the graph

The research of the key deviations for a functional requirement is difficult to realize manually when the number of potential key surfaces is important.

We suggest here a formal process of which only a principle is exposed and applied to the example.

The objective is to reduce the graph to simplify the analysis. The first task is to reduce the links placed in series and parallels. Figure 5 illustrates the types of reduction made. Figure 6 illustrates the reduction made for the example.

202

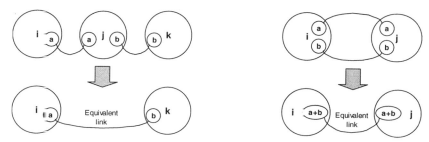

Figure 5 : *Reductions of the links in series and in parallels*

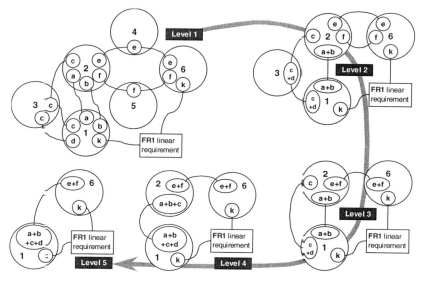

Figure 6 : *Reduction of the example*

For a great number of mechanisms, the process of reduction leads to a single link between the two parts to which the requirement relates. The process of reduction of the graph is specific to each couple of parts to which relates a functional requirement. The complete reduction is not always possible according to the mechanism and the couple of parts (figure 7).

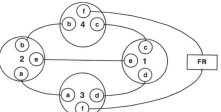

Figure 7 : *Irreducible graph*

Let us call graph of level 1 the most detailed graph, and graph of level k the graph reduced to its simpler expression.

3.4. Key deviations

Then, in order to determine the key deviations, it is necessary to go up from graph of level k to the graph of level 1 by analyzing the influence of the deviations progressively.

The first stage is to search for the deviations of the links of the graph of level k-1 which are of influence on the functional requirement. Then for any graph of level i, it is necessary to search for the deviations which are of influence on the deviations of the graph of level i+1.

The deviations are named α, β and γ for the angular deviations respectively on the axis x, y and z, and u_A, v_A and w_A for the linear deviations at the point A, respectively on the axis x, y and z.

It should be reminded that the linear deviation in a point can be influenced by an angular deviation by the effect of a " lever arm ". In the same way, an angular deviation can be influenced by several linear deviations involving the elimination of a degree of freedom in rotation.

Level 5	Level 4	Level 3	Level 2	Level 1	Level 0
6e+f /1a+b+c+d Prismatic of axis x V_K $\underline{\alpha}$	2a+b+c /1a+b+c+d Permanent v_K $\underline{\alpha}$	2a+b/1a+b : Prismatic of axis z v_K $\underline{\alpha}$		2a/1a : Plane normal to y $v_K, \underline{\alpha}$	2a : $v_K, \underline{\alpha}$
					1a : $v_K, \underline{\alpha}$
				2b/1b : Cylinder-and-plane of axis z, normal to x **No key deviations**	2b :
					1b :
		2c/1c+d : Ball and plane, normal to z **No key deviations**	2c/3c+d : Helical of axis y		2c :
					3c :
			3c+d/1c+d : Ball-and-cylinder of axis x	3c/1c : Ball-and-plane of axis z	3c :
					1c :
				3d/1d : Ball-and-plane of axis y	3d :
					1d :
	6e+f/2e+f : Prismatic of axisx v_K $\underline{\alpha}$		6e/2e : Cylindrical of axis x v_K $\underline{w}_F$	6e/4e : Permanent $v_K, \underline{w}_F$	6e : $v_K, \underline{w}_F$
					4e ; $v_K, \underline{w}_F$
				4e/2e : Cylindrical of axis x $v_K, \underline{w}_F$	4e : $v_K, \underline{w}_F$
					2e : $v_K, \underline{w}_F$
			6f/2f : Cylindrical of axis x v_K $\underline{w}_E$	6f/5f : Permanent $v_K, \underline{w}_E$	6f: $v_K, \underline{w}_E$
					5f: $v_K, \underline{w}_E$
				5f/2f : Cylindrical of axis x $v_K, \underline{w}_E$	5f: $v_K, \underline{w}_E$
					2f: $v_K, \underline{w}_E$

Figure 8 *: Table of the key deviations of FR1 and FR2*
(the key deviations on FR2 are underlined)

For instance, for the linear deviation requirement on the axis y between 1f and 6f :

- at the level 4, v_K of the link 2/1 and v_K of the link 6/2 are of influence on the requirement.
- at the level 3, v_K of the link 2a+b/1a+b is of influence on v_K of the link 2/1. The deviations of the link 2c/1c+d have no influence.
- ...

The summary table of figure 8 makes it possible to highlight the key deviations and the way in which they are propagated until the requirement. At the level 0, the key deviations of surfaces on the functional requirement are indicated.

Key surfaces are surfaces on which at least one deviation is a key deviation and the key parts are the parts on which at least one surface is a key surface. For the example key surfaces are those which are not in gray cells.

4. CHOICE OF THE SPECIFICATIONS

The choice of the specifications constitutes the last stage of the method. The geometrical specifications corresponding to a requirement are related to all the key surfaces and strictly to them and limit the key deviations and strictly them.

If a key deviation is not limited, then the definition of the part is under constrained with respect to the functional requirement from where functional problems appear. If a non-key deviation is limited, then the definition of the part is over constrained from where an overcost of manufacturing appears.

The specifications must be chosen requirement by requirement, and that in an independent way. Once each requirement has been considered, it is possible to eliminate the redundant specifications.

4.1. Characteristic specified

The key deviations being identified, the characteristic specified to be placed can be given. If the key deviations of a surface are angular, an orientation specification is adequate. If the key deviations of a surface are linear and/or angular, a location specification must limit them.

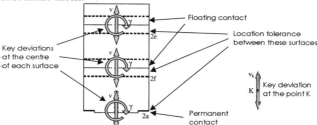

Figure 9 *: Key deviations and contacts of the bearing*

For the example and FR1, the key deviations are v_K at the point K or v and γ at the centre of each surface (due to the lever arm). Thus, the characteristic specified is a location (figure 9).

4.2. Datum

For the example, we have to choose the datum of the location specification : the plane or the cylinders (figure 9).

The choice of the datum must be realized according to the type of surface contact. Two principal types of contact are considered : the permanent contact (without clearance) and the floating contact (with clearance). The datum of a specification must first relate to surfaces of permanent contact. A datum or a datum system simulates the permanent positioning of a close part. For this reason, surfaces of a datum system must be, if possible, in contact with a same part. The hierarchy of the datum system must translate the setting into position of the close part, dominating bearing surface in primary datum, secondary bearing surface in secondary datum, ...

For the example, there are floating contacts on the borings, and a permanent contact on the plane (figure 9), so, the plane is chosen as the datum (figure 10).

If all the contacts are permanent, or if all the contacts are floating, the choice of the datum is based on the size of the surfaces.

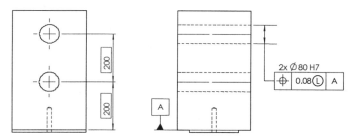

Figure 10 : Specifications of the bearing

4.3. Requirement

For the features of size (cylinder, 2 parallel planes, ...), requirements (envelope, maximum and least material) could be applied.

As for the datum, the type of contact of the surfaces is important for the requirement. A requirement is used on a surface if the contact is floating. Indeed, in this case, a requirement permits to limit the worst surface for a given functional requirement. According to the type of functional requirement, envelope, maximum material or least material must be used :

• maximum material for an assembly requirement,

• least material for a maximum clearance or a deviation requirement,

• maximum material for a minimal clearance requirement.

For the example, the contact on the plane is permanent, so there is no requirement to define on the datum. On the other hand, the contacts on the borings are floating, so there is a requirement to define on the toleranced surfaces. The worst case for the functional requirement is when the borings are at the state of least material (larger the clearance is, larger the deviation at the point K is). Thus, the least material requirement is used (figure 10).

5. NATIONAL PROGRAM OF EDUCATION

The tolerancing of a simple mechanism, as the example presented, appears simple, however, our six years experiment of teaching shows that tolerancing is far from being obvious for the students. This experiment gradually brought us towards the presented method. This one makes it possible to progress step by step. Moreover, this method highlights the physical phenomena intervening on a functional requirement.

We point out the fact that the concepts of independence of the functional requirements and of key parts/surfaces/deviations are fundamental.

After the validation of this type of teaching, a national action of education of the design and manufacturing teachers has been launched. There are 17000 French teachers to train to the tolerancing. Today, 40 voluntary professors coming from various regions have followed 2 sessions of 2 days. The introduced subjects were the standardized specifications and the choice of the specifications. Currently, these volunteers diffuse knowledge in their respective regions through many decentralized formations. A reference frame to be used by the teachers is under development.

REFERENCES

[**Ballu, Mathieu, 1993**] Ballu A. ; Mathieu L. ; "Analysis of dimensional and geometrical specifications: standards and models" ; In : *CIRP Computer Aided Tolerancing 3rd Seminar*, pp. 157-170 ; Cachan, France

[**Giordano, Duret, 1993**] Giordano M. ; Duret D. ; " Clearance space and deviation space : Application to the three-dimensional chains of dimensions and positions " ; In : *CIRP Computer Aided Tolerancing 3rd Seminar*, Cachan, France

[**Ballu, Couetard, 1995**] Ballu A. ; Couetard Y. ; "Dispositif de mesure spatiale des petits deplacements: application au comportement des liaisons avec jeux dans le cadre de travaux pratiques" ; In : *Seminaire Tolérancement et Chaînes de cotes*, ENS de Cachan, France

[**Ballot, Bourdet, 1997**] Ballot E. ; Bourdet P. ; " A computation method for the consequences of geometric errors in mechanisms " ; In : *CIRP Computer Aided Tolerancing 5 rd Seminar*, pp. 137-147 ; Toronto, Canada

[**Clement et al., 1994**] Clement A. ; Riviere A. ; Temmerman M. ; "Cotation tridimensionnelle des systèmes mécaniques: théorie et pratique", PYC Edition, France, ISBN 2-85330-132-X.

TASys: A Computer Aided Tolerance System based on a Variational Geometry formulation

A. Alvarez, J.I. Barbero, A.R. Carrillo, T. Gutiérrez and M. Aizpitarte
Cuesta de Olabeaga, 16 – 48013 Bilbao (SPAIN)
Email: angel@labein.es

Abstract: During recent years, the use of Variational Geometry has been widely adopted in Computer Aided Design systems. It is considered a key technology with application to a wide range of activities involved in design. An important application is the Tolerance Analysis.

This work presents *TAS*ys; a Tolerance Analysis System based on a non-manifold 3D variational solid modeller called *DAT*um, both developed by LABEIN. In this system, variations are controlled by dimensions. It is able to perform both worst-case and statistical tolerance analyses. As it is a Variational Geometry based Computer Aided Tolerancing System, the geometric tolerances have been translated into dimensions.

Besides, it has HPCN (High - Performance Computing and Networking) abilities, where the computation time is using several processors

Keywords: Computer Aided Tolerancing, Variational Geometry, Tolerance Analysis and High - Performance Computing and Networking

1. INTRODUCTION

During recent years, the use of Variational Geometry has been widely adopted in Computer Aided Design systems, and it is considered today a key technology with application to a wide range of activities involved in design. The central idea of the Variational Geometry is that dimensions, such as those which appear in a mechanical drawing, are the natural descriptors of geometry and provide the most appropriate means for altering a geometric model. So, it allows a simple representation to be used to describe the entire family of geometries that share a generic shape.

An important application of the Variational Geometry is Tolerance Analysis. It plays a very important role in reducing variations in manufacturing, thus improving the quality. The possibility of carrying out a tolerance analysis during the design process of a 3D mechanical assembly, allows these problems to be detected at an early stage, and thus it is possible to search for alternative design solutions or manufacturing methods.

This work presents *TAS*ys; a Tolerance Analysis System based on a non-manifold 3D variational solid modeller called *DAT*um, developed by LABEIN as well. In this system, dimensions control variations. It is able to perform both worst-case and statistical tolerance analyses, taking into account dimensional and geometric tolerances. Several post-processing tools were integrated in *TASys* for a better understanding of the tolerance behaviour and the influence of the different components.

Besides, *TASys* has HPCN (High - Performance Computing and Networking) abilities, where the computation time is reduced making the process more interactive by using several processors in the same computation. The possibility to use clusters of cheap Windows NT based PCs, allows to introduce this application in industries where big computing facilities are not available.

2. VARIATIONAL GEOMETRY FORMULATION

The basis of the present Variational Geometry formulation is related to the concept of "model variable", defined as a real-valued quantity that represents an elementary geometric property of a model. This definition is closely related to the definition given in **[Martino and Gabriele, 1989]**, where model variables represent variations from a nominal model, in the context of tolerance analysis.

A geometric entity (i.e., points, curves and surfaces) is defined in terms of a set of independent model variables that are identified with point coordinates, vector components and scalars representing angles and lengths. This definition can be either explicit or expressed through other geometric entities in terms of parametric equations

$$\mathbf{g} - \mathbf{g}(\mathbf{g}_1,...,\mathbf{g}_n,\mathbf{x}) = 0 \tag{1}$$

where $\mathbf{g}$, $\mathbf{g}_1...\mathbf{g}_n$ are geometric entities (points, curves, surfaces or length-angle scalars), and $\mathbf{x}$ is a set of model variables, that explicitly appear within the geometry equation.

Given **(1)** as the mathematical definition of a geometric entity, a constraint can be defined in terms of an equation that relates several geometric entities with a scalar value that can be interpreted as a dimension. It is expressed analytically by a non-linear equation involving dimensional and geometric entities in terms of

$$f_i \equiv f_i(\mathbf{g}_1,...,\mathbf{g}_{n(i)},L_i), i = 1,...,M \tag{2}$$

where $\mathbf{g}_1$, ..., $\mathbf{g}_{n(i)}$ are geometric entities, L_i is a value associated to the constraint equation and is called nominal value, $n(i)$ is the number of geometric entities involved in the constraint, and M is the total number of constraints attached to a given model. The value l_i for which the i-th constraint vanishes is called real value.

For given values of dimensions, the set of constraints **(2)** can be solved using a numerical method, evaluating corresponding values of geometric entities through solving constraints in terms of model variable values.

This VG formulation can be easily extended to handle assemblies. An assembly is defined as a set of geometric models or parts, each one with an associated transformation (rotation and translation), which is used to position the parts into assembly. The assembly relations are controlled by dimensions relating part geometric entities and the associated transformations.

The present formulation has been implemented in *DAT*um [**Gutiérrez et al., 1998**]. It is an object oriented variational non-manifold geometric modeller developed by LABEIN, with a STEP translator (compliant with ISO 10303-AP203) to exchange models with different CAD systems.

Its mathematical background [**Alvarez et al., 1998**] allows to support a wide range of dimensions in terms of general points, curves, surfaces and other dimensions: distances relating any combination of points, curves and surfaces, angles relating curves or surfaces and contacts. Besides, it deals with underconstrained dimensioning schemes; so, the designer can specify only the dimensions in which he/she is interested.

The classic assembly relations as mate, mate-offset, align, align-offset, insert and orient [**PTC, 1991**] can be easily derived from the implemented dimensions. Anyway, *DAT*um provides two assembly methods:

1)   "3-2-1" assembly method (**Figure 1**): Mate three surfaces with other three surfaces, so each surface of the first group touch one of the second group: coincident and facing each other.

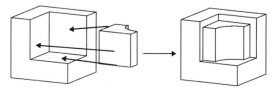

Figure 1: *"3-2-1" assembly method*

1)   "4-1-1" assembly method (**Figure 2**): Align a "male" revolved surface into a "female" revolved surface, mates two planar surfaces together so that they have an offset between the surfaces (although they can be coincident), and orients two planar surfaces to be parallel and facing in the same direction

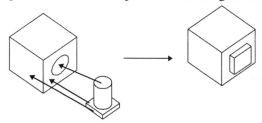

Figure 2: *"4-1-1" assembly method.*

3. TASYS: A V.G.-BASED TOLERANCE ANALYSYS SYSTEM

The Tolerance Design is defined as the process of selecting correct tolerances for a part's dimension that allow this part to be economically produced and functional and it has a great importance in mechanical design, because tolerances have a significant impact on product cost and quality.

The use of Variational Geometry allows the set of functional constraints $X_1,..., X_n$ to be combined with an objective or design function Y, to determine its variability, given the allowed tolerances for the dimensions involved in the constraints. A design function could be a dependent dimension or a cost function that depends on the model variables, so an explicit expression as functions of the constraints is usually unknown. The solution is found in the domain of the independent variables $X_1,..., X_n$.

A practical CAT system has to include both dimensional tolerances, which define the range of variation of the dimensions, and geometrical tolerances, which are defined as the zones within which the part features or their resolved geometries (centre-point, centreline and centre-plane) are constrained to lie.

In a VG-based Tolerance Analysis, dimensions control the variations. However, the "classical" dimensional tolerances are not enough to express all the possible variations in an assembly. So, the geometric tolerances defined in [**ASME, 1994**] and/or [**ISO, 1983**] have been translated into dimensions. As an example; the parallelism tolerance of a surface $\mathbf{s} \equiv \mathbf{s}(u,v)$, with reference to a datum plane with normal direction $\mathbf{n}$, (**Figure 3**) is expressed mathematically with the following equation:

$$f = f(\mathbf{s}(u,v),t) = t - \min(\mathbf{s}(u_1,v_1) - \mathbf{s}(u_2,v_2)^T \mathbf{n} \qquad (3)$$

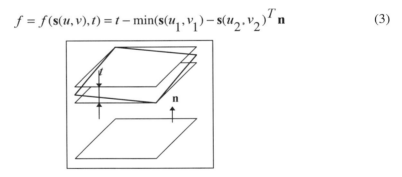

Figure 3: *Parallelism tolerance of a surface with reference to a datum plane*

There are two types of tolerancing analysis. A worst case tolerance analysis assumes that all fluctuations may occur in the worst possible fashion. This analysis finds the upper and lower limits of the response function $Y=F(X_1,...,X_n)$, giving the defined tolerances $z_i<X_i<Z_i$, where X_i is a set of given dimensions.

Mathematically, this problem is expressed as the minimisation of a non-linear function $G(Y)=Y-Z_Y$, where Z_Y means the upper limit of the design function Y, subject to the set of inequality linear constraints $z_i{<}X_i{<}Z_i$. It could be solved by means of linear programming methods [**Alvarez et al., 1998**]

The worst case analysis is usually excessively conservative, giving results that are very pessimistic. With a statistical analysis the low probability of a worst-case combination can be taken into account. Moreover, there is a growing trend to considerate the probabilistic behaviour of the response functions. The problem of statistical tolerance analysis is to estimate the moments of distribution (usually mean and standard deviation) of the response function giving the statistical distribution of the dimensions.

The statitiscal methods are very easy to implement in a VG-based CAT system, because the only evaluation of the response function at given values of the independent variables is needed.

Two methods are the most popular among the current CAT systems: the linear stack-up method, also known as root-sum-squares method [**Greenwood and Chase, 1990**] and the Monte Carlo simulation method [**DeDoncker and Spencer, 1987**]. The first one is a simple and inexpensive method, but it gives bad estimates if the response function is highly non-linear. Although the Monte Carlo is the simplest and the most used method, its main drawback is the high number of evaluations to achieve a good approximation of the moments of the response function.

Two statistical methods are becoming very promising: the Quadrature technique [**Evans, 1975**] and the modified Taguchi's method [**D'Errico and Zaino, 1988**]. Both methods are not available in commercial packages, and are easy to implement in a VG based Tolerance Analysis system.

Once an analysis has been performed, the user can obtain some information from this analysis. For example, after a worst-case tolerance analysis, the designer is able to see the worst-case tolerance condition. This feature has been extended to the statistical tolerance analysis, so, after a Monte Carlo simulation analysis, *TAS*ys is able to display the assembly at any value of the response function. Besides, *TAS*ys is able to estimate the effect of each input parameter or tolerance has on the assembly function.

4. EXAMPLE

The following example shows how *TAS*ys works. An air compressor for using in light industrial environments is shown as a case study (**Figure 4**). The process begins by creating and dimensioning the parts of the assembly (the base of the engine, the pillow block, the crankshaft, the connecting rod, the piston and the cylinder head). Once all the parts were defined, they are positioned in the assembly. The next step is to define all the tolerances along with the assembly response function. In this case, the position of the piston inside the cylinder head is selected to check interferences **Figure 5**)

Figure 4: Air compressor.   **Figure 5**: Tolerance problem.

Once, the problem is defined, the designer selects the desired type of tolerance analysis. For a worst-case tolerance analysis, it is possible to visualize the worst-case geometry of the assembly. As it can be seen from **Figure 6** there is interference. To evaluate the probabilistic behaviour of this interference, it is possible to perform a statistical tolerance analysis. The distribution function is shown in **Figure 7**, where the shadowed area corresponds to the probability of a defective assembly. This percentage is equal to 0.618534%.

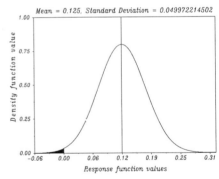

Figure 6: Worst case geometry. Interference problem.

Figure 7: Distribution function of the Response function

5. HPCN

High - Performance Computing and Networking (HPCN) is a powerful enabling technology that can significantly improve the competitiveness of businesses through its application to many diverse areas. Within the European ESPRIT programme LABEIN is participating as Technology Transfer Node (TTN) in the HPCN TTN Network as coordinator of the HIPERTTN project. The main objectives are:

- Raise awareness of HPCN use in industry
- Assess potential gains of HPCN use in SMEs
- Faster and accelerate take-up of HPCN
- Technology transfer by demonstration and best practice

In this sense, the Tolerance Analysis System *TAS*ys has been parallelized in a scalable cluster a Pentium Windows NT based PCs, using WPVM [**Alves, 1995**]. Currently, it is being applied to 3 Spanish SME industries and real cases with around 100 components are being tested. The efficiencies attained are between 0.8 and 1, and the speed-ups are close to the number of processors used.

6. CONCLUSIONS

Tolerance analysis plays a very important role in reducing variations in manufacturing, thus improving the quality. The possibility of carrying out a tolerance analysis during the design process of a 3D mechanical assembly allows these problems to be detected at an early stage, and thus it is possible to search for alternative design solutions or manufacturing methods.

HPCN technology reduces the computation time and makes the process more interactive. The use of clusters of Windows NT based PCs allows to introduce this application in industries where big computing facilities are not available.

A powerful CAT system called *TAS*ys has been presented in this work. It is based on a non-manifold 3D variational solid modeller called *DAT*um. It is able to perform both worst-case and statistical tolerance analyses, taking into account dimensional and geometric tolerances. Its implemented post-processing tools help the engineer to understand the tolerance behaviour and the influence of the different components.

7. ACKNOWLEDGEMENTS

This work has been partly supported by the following organisms and projects:
- UET/SPRI Basque Government. DAME project. Ref: CO97AE01.
- EC DGIII, ESPRIT HPCN programme. HIPERTTN, project number 24003.

214

REFERENCES

[**Alvarez et al., 1998**] Alvarez, A.; Longo, A.; Carrillo, A.R.; Gutiérrez, T.; Aizpitarte, M.; "Variational Geometry and its application to the Tolerance Analysis of 3D mechanical assemblies", In:*Proceedings of Mathematics & Design 98*, pp. 619-627.

[**Alves, 1995**] Alves, A.; Silva, L.M.; Carreira, L. and Gatriek, J.G.; "WPVM: Parallel Computing for the People", *In Proceedings of HPCN'95, High Performance Computing and Networking Europe*, Milano, Italy, Lecture Notes in Computer Science 918, pp. 582-587.

[**ASME, 1994**] ASME; "ASME Y14.5M-1994 Dimensioning and Tolerancing"

[**D'Errico and Zaino, 1988**] D'Errico, J.R. and Zaino, Jr N.; "Statitiscal Tolerancing using a modification of Taguchi's method", *Technometrics*, Vol. 30, No. 4, pp. 397-405

[**DeDoncker and Spencer, 1987**] DeDoncker, D. and Spencer, A.; "Assembly Tolerance Analysis with simulation and optimization techniques", *SAE Trans.*, Vol. 96, No. 1, pp. 1062-1067.

[**Evans, 1975**] Evans, D.H.; "Statitiscal Tolerancing: the state of the art. Part II: Methods for estimating moments",*J. Quality Tech.*, Vol. 7, No. 1, pp. 1-12.

[**Gutiérrez et al., 1998**] Gutiérrez, T.; Barbero, J.I.; Eguidazu, A.; Aizpitarte, M. and Carrillo, A.R.; "Virtual prototypes", In: *Proceedings of Mathematics & Design 98*, pp. 637-644.

[**Greenwood and Chase, 1990**] Greenwood, W.H. and Chase, K.W.; "Root sum squares Tolerance Analysis with non-linear problems", Trans. ASME J. Eng. Indust., Vol. 109, No. 2, pp. 382-384.

[**ISO, 1983**] ISO; "ISO 1101: Technical drawings – Geometrical tolerancing – Tolerancing of form, orientation, location and run-.out – Generalities, definitions, symbols, indications on drawings"

[**Martino and Gabriele, 1989**] Martino, P.M.; Gabriele, G.A.; "Estimating Jacobian and Constraint Matrices in Variational Geometry systems"; In: *ASME Design Automation Conference*, pp. 79-85.

[**PTC, 1991**] Parametric Technology Corporation; "Pro/Engineer ® User Guides Release 8.0".

Innovations in Integrating Tolerance Analysis Technologies for a Computer-Aided Tolerancing System

Charles G. Glancy and Timothy V. Bogard
Raytheon Systems Company
P.O. Box 660246 M/S 486
Dallas, Texas 75266
USA
c-glancy@ti.com

Abstract: CE/TOL 6σ is a commercial computer-aided tolerancing system for modeling, analyzing and optimizing tolerances in 3-D mechanical assemblies. CE/TOL 6σ is integrated with the Pro/ENGINEER CAD system and provides graphical tolerance model creation using the assembly geometry. The tolerance model is constructed using a versatile set of modeling elements. Graphical modeling eliminates the need to manually enter any equations. The tolerance analysis and optimization is performed using built-in worst case and statistical methods.

CE/TOL 6σ is a combination of many tolerance analysis technologies developed at Raytheon Systems Company and Brigham Young University. Because of the many different technologies that comprise CE/TOL 6σ, much attention was given to the system level design to ensure consistency and conceptual integrity. As a result of the attention to design, the CE/TOL 6σ tolerancing system demonstrates several innovations in the areas of tolerance model visualization, tolerance analysis capability, and automation.

Three innovations in the area of tolerance model visualization are presented: the interactive Assembly Network Diagram, the interactive Part Network Diagram, and the Tolerance Model Tree.

A new, second-order, general statistical analysis method is introduced. This new analysis method can perform RSS or DRSS analyses as well as consider non-Normal distributions.

The capability for automatic joint recognition is presented for automating tolerance modeling. A user interface element is described for displaying tolerance optimization objectives.

Keywords: computer-aided tolerancing system, tolerance analysis, tolerance modeling

1. INTRODUCTION

"Everything should be made as simple as possible, but not simpler."
— Albert Einstein

Fundamental business processes are driving the need for better computer automated engineering technologies. Rapid design capture, analysis and optimization processes are at a premium in most companies around the world. Minimizing non-value-added efforts to achieve an appropriate level of design robustness is a difficult re-engineering task. An inherent resistance to change within organizations severely limits rapid adoption of new processes, and has become a critical barrier to any new technology implementation. Flexibility and ease are two of the most important issues to be addressed. This paper is intended to discuss those technologies implemented in CE/TOL 6σ, a computer-aided tolerancing (CAT) system that allows for the rapid adoption of new tolerancing technologies necessary to achieve business objectives.

CAD companies and engineering management allow for a wide range of geometry creation processes. For the purposes of tolerance analysis, the appropriate CAD model should understand the true manufactured dimensions and assembly techniques. However, due to the complexity of modeling geometry, it is common practice to use dimensioning schemes that simplify creating geometry, not necessarily schemes that are needed for proper tolerance analysis. Confusion over industry standards further complicates the geometry capture process. As the design evolves, additional information that allows for a more precise tolerance analysis is desired, but typically not available. Tools that allow for rapid modeling, analysis, and optimization are very dependent then on user CAD modeling techniques, quality of manufacturing process data available, and methods for easily modifying a tolerance model to include other related variations. As a result, any tolerancing system must be flexible and easy to use.

2. CE/TOL 6σ TECHNOLOGIES

New technologies in tolerance modeling, analysis and optimization processes within CE/TOL 6σ have been implemented to directly address system flexibility. Taking tolerancing technology from research to final integration has required human factor innovations to achieve ease of use goals. Fundamentally, these innovations include intuitive graphical user interfaces, logical data representations, and simplified dialog boxes that are optimized for user interaction.

Simplicity must be a guiding principle of any good design. CE/TOL 6σ is a complex collection of diverse mathematical algorithms, statistical techniques and software technology. In CE/TOL 6σ the tolerance model is constructed using a versatile set of modeling elements. Graphical modeling eliminates the need to manually enter any equations. The tolerance analysis and optimization is performed using built-in worst case and statistical methods.

Three new technologies were introduced in CE/TOL 6σ for flexibility that required special attention to achieve ease of use for the system. Those technologies include:

1. surface based representation of tolerances directly associated to the CAD model,
2. assembly constraint tolerance model based on full 3D kinematic definitions, and
3. generalized statistical and worst case analysis engines.

Each of these technologies was developed to advance the capabilities of CE/TOL 6σ, but each had unique challenges with regard to user adoption.

For example, in conventional CAD systems using kinematic assembly constraints, the user must have significant knowledge and CAD skills to properly define, place, orient, and set the degrees of freedom of each joint each assembly contact location. This could take 40 to 50 unique steps in conventional kinematic modeling systems. The innovation in implementing this constraint system is an automated joint recognition and placement system requiring less than 4 unique steps. Suddenly, the burden on the user to get full advantage of the kinematic constraints has been radically reduced.

3. CE/TOL 6σ INNOVATIONS

The primary goal of CE/TOL 6σ is to make comprehensive tolerance modeling, analysis and optimization easier to use. This goal was achieved by simplifying the modeling, analysis and optimization processes. The design challenge was to modify each process enough to make it simple, while not compromising the flexibility of the system.

Six innovations in the areas of tolerance model visualization, tolerance modeling, analysis and optimization were required to properly integrate the three new tolerancing technologies in CE/TOL 6σ. The six innovations are:

1. Assembly Network Diagram
2. Part Network Diagram
3. Tolerance Model Tree
4. Automatic Joint Recognition
5. Simplified Tolerance Analysis Process
6. Simplified Tolerance Optimization Choices

3.1 Assembly Network Diagram

CE/TOL 6σ includes an interactive Assembly Network Diagram to aid in tolerance model visualization. The Assembly Network Diagram is a relationship graph of the tolerance model elements. The parts of an assembly are the nodes of the graph while the assembly constraints and assembly specifications are the arcs between the nodes. The Assembly Network Diagram is a valuable representation of both the state and extent of the tolerance model. The Assembly Network Diagram communicates what parts are included in the tolerance model, what assembly specifications are being analyzed and what constraints exist between the parts of the assembly.

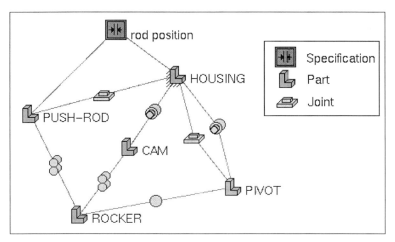

Figure 1: *Assembly Network Diagram*

Figure 1 shows the Assembly Network Diagram of a valve assembly. The specification symbol indicates a gap specification between the *PUSH-ROD* and *HOUSING* parts. The seven joint symbols illustrate the various constraint relationships between the five parts of the assembly. The difference joint symbols represent different constraint types. For example, there are two constraints connecting the *HOUSING* and *PIVOT* parts, one planar joint and one cylindrical joint.

The Assembly Network Diagram of CE/TOL 6σ is dynamic and interactive. For example, as the user creates an assembly specification between two parts, the symbols for the specification and the two parts are dynamically added to the Assembly Network Diagram. Once displayed, the symbol for the specification becomes interactive and can be selected in order to arrange the symbol in the diagram, to modify the specification, or to delete the specification entirely. The Assembly Network Diagram is an intuitive view for the CE/TOL 6σ user to create and manipulate the elements of the tolerance model.

3.2   Part Network Diagram

The Part Network Diagram of CE/TOL 6σ is a complementary view to the Assembly Network Diagram. The Part Network Diagram is a relationship graph of the part-level tolerance model elements. The nodes of this graph are datum features, datum reference frames, geometric tolerances and joints. The arcs between the nodes are dependencies. For example, a datum reference frame of the part may depend on three datum features. The Part Network Diagram communicates the dimension scheme of the part.

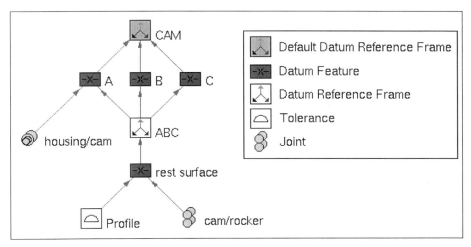

Figure 2: *Part Network Diagram*

Figure 2 is the Part Network Diagram for a cam part. The diagram illustrates the dimension scheme of the cam. For example, Figure 2 shows the profile tolerance of the *rest surface* is defined relative to datum reference frame *ABC*. The Part Network diagram is also useful in visualizing how variation propagates from one part joint to another. For example, the path from the *cam/rocker* joint to the *housing/cam* joint is easily identified as *rest surface* → *ABC* → *A*.

The Part Network Diagram, like the Assembly Network Diagram, dynamically updates as new elements of the tolerance model are created. Both the Part Network Diagram and the Assembly Network Diagram provide a new way for a user to visualize and interact with the tolerance model.

3.3 Tolerance Model Tree

The third tolerance model view of CE/TOL 6σ is the Tolerance Model Tree. The Tolerance Model Tree is a hierarchical view of every element of the tolerance model. The hierarchy of the tree contains the assembly, the parts of the assembly, the tolerances of the part, and the variables controlled by the tolerance. These four levels represent the most fundamental grouping of the tolerance model elements. This logical organization allows for any tolerance model element to be easily located and modified by the user. The four levels of the tree can be collapsed or expanded by the user, and allow a compact display of large numbers of model elements.

220

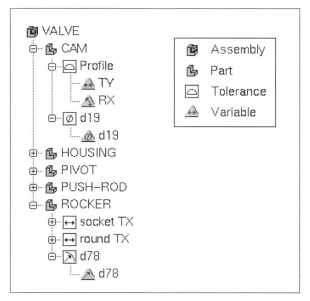

Figure 3: *Tolerance Model Tree*

Figure 3 is the Tolerance Model Tree for the valve assembly. In contrast to the Part Network Diagram, the Tolerance Model Tree includes every part tolerance that has been included in the tolerance model, not just the geometric tolerances. For example, the Tolerance Model Tree includes the diameter size tolerance *d19* for the *CAM* part in addition to the geometric tolerance named *Profile* that was visible in the Part Network Diagram (Figure 2).

The fourth level of the Tolerance Model Tree contains the variables. A CE/TOL 6σ tolerance model allows a tolerance to control up to three geometric variations where appropriate. The three possible geometric variations are size, location and orientation. A tolerance can have one variable for each of these three geometric variations. Figure 3 shows the *Profile* tolerance controlling two variables, a size variable, *TY*, and an orientation variable, *RX*.

3.4 Automatic Joint Recognition
Automatic joint recognition is a new technology in CE/TOL 6σ to simplify the tolerance modeling process. Kinematic joints are used in CE/TOL 6σ to define the constraint relationships between the component parts of an assembly. The joint recognition technology automates the identification of assembly constraint types during tolerance model creation. For example, to define the assembly constraint between two mating parts, the user simply selects the two contacting surfaces of the parts. The automatic joint recognition algorithms will then identify the mating surface characteristics and create the appropriate joint type with the correct position and orientation. This technology dramatically reduces the number of modeling steps.

3.5 Simplified Tolerance Analysis Process

CE/TOL 6σ now includes a second-order tolerance analysis method. This second-order method provides a more accurate approximation of a non-linear tolerance model than a first-order method. However, even with the addition of the new second-order method, CE/TOL 6σ was able to keep the analysis method choices very simple. The choices include two methods, worst case or statistical, and two accuracy settings, first-order or second-order. Both methods can be performed at either accuracy setting.

The statistical method of CE/TOL 6σ is a general method that allows the user to perform the popular analysis types of Root Sum Squares (RSS), Dynamic RSS (DRSS) and Static RSS (SRSS). Modifying specific settings of the input variables performs these distinct analysis types. For example, the distribution of an input variable defaults to a Normal distribution with a Standard Deviation equal to one third the symmetric tolerance. This initial variable state is the RSS state. The user can change this variable to a DRSS state by modifying the variable's Dynamic factor, D_k. Alternatively, the variable may be changed to a SRSS state by adding a static mean shift. Variables may also be modified to include non-Normal distributions. The generality of the CE/TOL 6σ statistical method keeps the analysis options simple and at the same time provides tremendous flexibility for the user.

3.6 Simplified Tolerance Optimization Choices

The CE/TOL 6σ tolerance optimization choices have also been simplified. CE/TOL 6σ includes sophisticated tolerance allocation and tolerance scaling algorithms to optimize tolerance values to a given assembly or part quality level. To help the users decide when to apply allocation or when to apply scaling, CE/TOL 6σ added the Optimize Dialog.

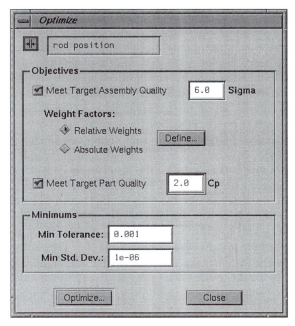

Figure 4: Optimize Dialog

The Optimize Dialog is shown in Figure 4. The dialog box contains check boxes and text fields for the user to define the specific objectives of the optimization. For example, the user can specify a target assembly quality of 6 sigma and target part C_p values of 2.0. Once the optimization objectives are defined, the user simply presses the *Optimize* button and CE/TOL 6σ then automatically determines which optimization algorithms to use. The Optimize Dialog greatly simplifies the user's need to understand the differences between tolerance allocation and tolerance scaling.

4. SUMMARY

Creating technological advances in core tolerance modeling, analysis, and optimization functions is required to advance tolerancing methods to meet continually increasing demands. Maintaining focus, from the user perspective, on how the new technologies can most easily be adopted is at the center of rapid technology deployment in the CAT market. Innovations in integrating the future technologies must keep pace with the increased desire for more intuitive and easy to use tools. CE/TOL 6σ has introduced innovations addressing these issues. Research continues in both technological advances for flexibility as well as innovative approaches to improve ease of use.

Representation of assembly and inspection processes for the tolerance analysis of automobile bodies

by Minho Chang, Jooho Kim, Hyung-Min Rho, and Sungdo Ha

CAD/CAM Research Center
Korea Institute of Science and Technology
Sungbuk, Hawolgok 39-1, 136-791
Seoul, Korea
minho@kist.re.kr

Abstract: An automobile body assembly is a complex system consisting of hundreds of compliant sheet metal parts. A number of locating schemes are used throughout the assembly and inspection processes. This paper presents a methodology to represent the assembly and inspection processes of an automobile body for tolerance analysis. The proposed representation methodology consistently describes dimensional variations with respect to various locating schemes that change through the assembly process.
Keywords: assembly, inspection, variation, tolerance, compliance

1. INTRODUCTION

Sheet metal parts have variations from their nominal shapes due to inherent variations in the stamping process. During the assembly process, compliant parts are further deformed by the relatively rigid assembly tooling. In recent years, various models are proposed to simulate the propagation of variations during the assembly of compliant, non-ideal parts [Chang, 1996] [Chang and Gossard, 1997] [Long and Hu, 1997] [Sellem and Riviere, 1998]. Part deformation during the assembly and inspection processes is calculated with the finite element method. The mechanical behavior of parts is modeled by the following constraints:

- Constitutive relations: Changes in the relative positions between the features of a part are constrained by the constitutive relation. If the changes are small and the material properties are linear, the constitutive relation is represented by a linear equation, $f=Ku$, where u is the position changes, and f and K are the corresponding forces and stiffness coefficients.

- Geometric compatibility: When features are in contact, the changes in their relative positions are constrained by the type of the contact. In a pin/hole contact, for example, only the translation along and the rotation around the axis of pin are permitted.

- Force equilibrium: When features are in contact, the summation of the force applied on each feature equals to zero.

2. ASSEMBLY AND INSPECTION PROCESSES

The assembly process is modeled as a process of placing, clamping, fastening, and releasing.

- Parts are placed in an assembly fixture by locator pins and blocks.
- Parts are clamped. Parts may deform in this process.
- Parts are welded by welding guns. Parts may further deform in this process.
- Finally, parts are released from the assembly fixture. In this process, parts spring back to minimize the strain energy stored in the assembly.

Similarly, the inspection process is modeled as a process of placing, clamping, measuring, and releasing.

- Parts are placed in an inspection fixture by locator pins and blocks.
- Parts are clamped.
- Parts are measured with rulers or measuring probes.
- Finally, parts are released from the inspection fixture.

The features that are used for positioning an assembly, such as locator holes and clamping surfaces, are called locators. Locators are the datum for an assembly, and variations of the same assembly can be described differently depending on the type and the number of locators. For the consistent interpretation of variations, it is preferred to use the same locator scheme during the assembly and the inspection processes. However, in practice, the locator scheme used for assembly often differs from that used for inspection. For example, Figure 1 (a) illustrates the locator scheme for a part during the inspection process, while Figure 1 (b) illustrates the locator scheme for the same part during the assembly process.

During the inspection process, parts are positioned with a minimum number of locators. This is because additional locators may cause part deformation. Usually two locator holes and three locator surfaces are used for positioning a sheet metal part[†], i.e., 3-2-1 locator scheme. Unlike the inspection process, various forces, such as welding gun forces, are applied during the assembly process. For secure positioning of parts subjected to forces, redundant locators are used during the assembly process.

[†] When a part is big, more than three locator surfaces are used in order to prevent the deformation caused by its own weight.

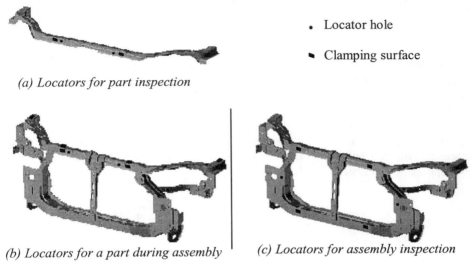

(a) Locators for part inspection

• Locator hole

↘ Clamping surface

(b) Locators for a part during assembly   (c) Locators for assembly inspection

Figure 1; Locators for the assembly and inspection processes

After parts are assembled together, parts cannot rigidly move relative to each other because of weld joints. Therefore, not all the locators used for positioning individual parts need to be used for positioning the assembly. An appropriate number of locators are selected and used during the subsequent assembly and inspection processes. For example, Figure 1 (c) illustrates the locator scheme for the inspection of the resulting assembly.

In addition to locators, welding surfaces play a significant role in the propagation of variations because variations in the welding surfaces may cause part deformation. When there are variations in a welding surface, the welding gun pushes the surface deforming parts. Then, the welding surfaces are welded together as they are deformed. Parts spring back when they are released from the fixture, but strain energy will remain in the assembly. Locators and welding surfaces are important for analyzing the propagation of variations, and they are collectively called assembly features in this paper.

3. REPRESENTATION OF ASSEMBLY AND INSPECTION PROCESSES

The assembly and inspection processes are modeled as a tree consisting of part, assembly, assembly station, and inspection fixture nodes as shown in Figure 2.

- A *part node* denotes a constituent part.
- An *assembly node* denotes an assembly consisting of a single or multiple

    part(s).
- An *assembly station node* denotes an assembly station where parts and assemblies are assembled.
- An *inspection fixture node* denotes an inspection fixture where parts and assemblies are measured.

An assembly station assembles multiple assemblies into a bigger assembly. This relation is represented in the process tree by an *assembly station node* having multiple *assembly nodes* corresponding to the incoming assemblies as its children and having the *assembly node* corresponding to the resulting assembly as its parent. An inspection fixture is the reference that is used for the description of variations of an assembly, and each *assembly node* has an *inspection fixture node* as its child. The assembly process proceeds from the bottom to the top of the tree, and the top-most node of the tree corresponds to the final assembly.

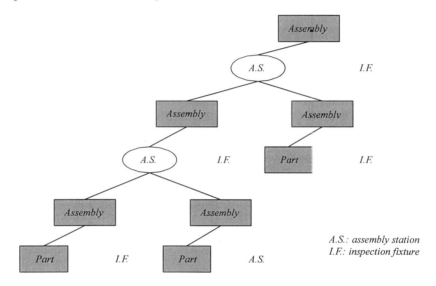

A.S.: assembly station
I.F.: inspection fixture

Figure 2; *Process tree*

    A node in the process tree is represented by an object that consists of variables and methods[†] (Figure 3).
- An *assembly station node* has the deviation of welding spots as member variables, and has a method to calculate the deviation.
- An *inspection fixture node* has a method to calculate the positions of the assembly features of the children assemblies when the deviations of welding

---

[†] The terms, member variables and methods, are typically used in the context of object oriented programming.

spots are given.
- An *assembly node* has, as member variables, the assembly features used in the parent assembly station and the assembly features used in the child inspection fixture. The *assembly node* has a method to calculate the positions of the assembly features of constituent parts.
- A *part node* has, as member variables, its geometry and all the assembly features on the part that are used at least once during the entire assembly and inspection processes. The *part node* has a method to calculate the positions of any point on the part.

The methods in the *part nodes* and *assembly nodes* can be obtained from the constitutive relations by using static condensation [Chang, 1996]. The methods in *assembly station nodes* and *inspection fixture nodes* can be obtained from the constitutive relations, geometric compatibility, and the force equilibrium conditions [Chang and Gossard, 1997]. Once the methods are obtained for each node in the assembly tree, the same methods can be used for the simulation of the assembly and inspection processes for various part and tooling variations.

Figure 3 illustrates the simulation procedure under the proposed representation methodology. First, the deviations of welding spots are calculated by using the method in the *assembly station node*. Then, the positions of the assembly features of the assemblies to be assembled are calculated by using the method in the *inspection fixture node*. Next, positions of the assembly features of the constituent parts are calculated by the method in the *assembly node*. Finally, variations of each part in the assembly are calculated by the method in the *part node*.

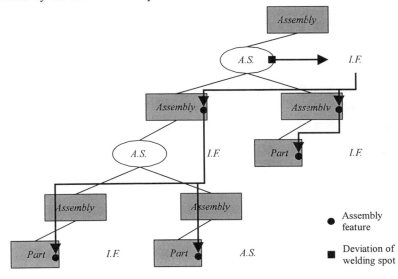

Figure 3; *Simulation of the assembly and inspection processes*

4. IMPLEMENTATION

A software program called *VariAns* is implemented based on the proposed methodology. The input of the program is the finite element mesh data of constituent parts, assembly sequence, locators, welding points, and the variations in incoming parts and assembly tooling. The program simulates the assembly and inspection processes to calculate the variations of intermediate subassemblies and the final assembly. *VariAns* consists of three modules: process editor, preprocessor, and the variation analysis module as shown in Figure 4. A user defines the assembly and inspection processes using the process editor. Then, the preprocessor automatically generates the methods of each node in the process tree. The users can observe the variations of intermediate subassemblies and the final assembly for various cases of part and tooling variations using the variation analysis module. The program is implemented in C++ with OpenGL graphics library on Windows NT. Figure 5 shows a screenshot of the program. The process tree is shown in the left-hand side. The variations of intermediate subassemblies are shown in the right-hand side. Currently, *VariAns* is being applied for validation at Korean automobile companies.

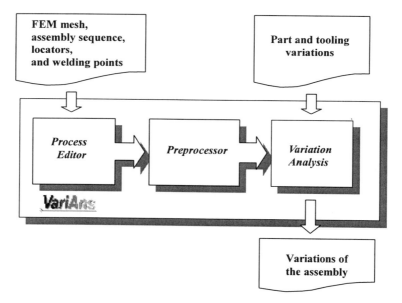

Figure 4; Overview of VariAns program

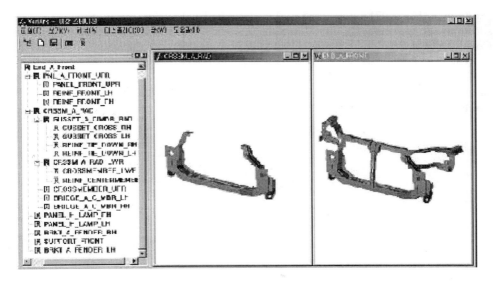

Figure 5; *Screenshots of VariAns program*

5. CONCLUSION

A new methodology to represent the assembly and inspection processes is proposed in this paper. The processes are modeled as a hierarchical tree consisting of part, assembly, assembly station, and inspection fixture nodes. Each node is modeled as an object consisting of member variables and methods. The proposed representation methodology facilitates the simulation of a large assembly and inspection process in several aspects. First, by encapsulating the part and tooling information for an assembly as a node in the process tree, modeling a large assembly process is conceptually straightforward. Second, this representation scheme facilitates the software implementation. Third, the tolerance analysis system implemented based upon this methodology is computationally efficient because once the methods in each node are determined, they can be reused for various cases of part and tooling variations.

REFERENCES

[Chang, 1996] Chang, M.; ¡Modeling the assembly of compliant, non-ideal parts,¡ ph.D. thesis dissertation, MIT, Cambridge, MA, U.S.A.

[Chang and Gossard, 1997] Chang, M.; Gossard, D.; ¡Modeling the assembly of compliant, non-ideal parts,¡ In: CAD vol. 29, no. 10, pp. 701 ? 708;

[Long and Hu, 1997] Long, Y.; Hu, J.; ¡A unified model for variation simulation of sheet metal assemblies,¡ In: Proceedings of 5th CIRP International Seminar on Computer-Aided Tolerancing, pp. 149 ? 160; Toronto, Ontario, Canada

[Sellem and Riviere, 1998] Sellem, E.; Riviere, A.; ¡Tolerance analysis of deformable assemblies,¡ In: Proceedings of DETC 98:1998 ASME Design Engineering Technical Conference; Atlanta, GA, U.S.A.

Two-Step Procedure for Robust Design Using CAT Technology

Rikard Söderberg, Ph. D. and Lars Lindkvist, Ph. D.
Chalmers University of Technology
Mechanical and Vehicular Engineering
S-41296 Göteborg, Sweden
rios@kanslim.chalmers.se
lali@mvd.chalmers.se

Abstract: This paper describes how computer aided tolerancing (CAT) techniques can be used in a two-step procedure to increase geometrical assembly robustness.

In the first step, in geometry *concept design*, the *general robustness* of a concept is increased by minimizing the number of controlling parameters for a critical characteristic and by maximizing the general ability to suppress variation. This step agrees with design philosophies put forward by Suh and Taguchi and uses CAT techniques to analyze and improve general assembly robustness with respect to locator positions and variation directions.

In the second step, in *detail design*, tolerances are allocated with respect to general robustness, manufacturing capability and manufacturing cost. At this stage, the final variation of the overall product characteristics may be simulated and part tolerances may be adjusted with respect to assembly sensitivity and cost. If loss functions are available, the total quality level of a concept may be analyzed.

Keywords: tolerancing, embodiment design, robust design, variation simulation, sensitivity analysis.

1. INTRODUCTION

Tolerance analysis has interested many authors over the years. In [Pheil, 1957] and [Mansoor, 1963], probability analysis was applied to component dimensions to analyze assembly dimensions. Cost minimization is treated in [Speckhart, 1972], [Spotts, 1973], [Parkingson, 1985], [Lee and Woo, 1989], [Cagan and Kurfess, 1992], [Kumar and Raman, 1992] and [Söderberg, 1993, 1994a, 1994b and 1995].

The research field of computer aided tolerancing (CAT) is highly focused on methods and tools for predicting variation in assemblies, statistically and with respect to cost and quality. Software tools, the representation of tolerances in CAD and different tolerance analysis methods have been discussed in [Bjørke, 1989], [Söderberg, 1994b], [Ianuzzi and Sandgren, 1994], [Chase and Parkingson, 1991], [Kumar and Raman,

1992] and [Nigam and Turner, 1995], [ElMaraghy et al., 1995], [Methieu et al., 1997] and [Clemént et al., 1997]. An evaluation of commerical systems for CAT is presented in [Salomonsen et al., 1997]. The incorporation of finite element analysis into variation analysis is discussed in [Long and Hu, 1997]. Assembly vector loops are treated in [Gao et al., 1996] and for geometric tolerances in [Chase et al., 1996]. In [Gao et al., 1995], the vector loop method (the direct linearization method) and the Monte Carlo simulation method are compared. Computer Aided Tolerance Management and the roles of the subcontractors are discussed in [Söderberg et al., 1998].

Trade-off analyses between cost and quality using loss functions, see [Taguchi et al., 1989], have been discussed by [Vasseur et al., 1992], [Krishnaswami and Mayne, 1994] and [Söderberg, 1993, 1994a, 1994b, 1995]. Spatial constraints, requirement decomposition and tolerance analysis in configuration design are described in [Söderberg and Johannesson, 1998].

2. COMPUTER AIDED TOLERANCING TECHNOLOGIES

A number of commercial computer aided tolerancing (CAT) tools are currently available on the market, see [Salomonsen et al., 1997], which provide a number of analysis types:

- **Sensitivity Analysis**: The influence of each geometrical feature is ranked for every specified critical assembly dimension. The analysis includes effects of feature *position* and variation *direction*.
- **Variation analysis**: Statistical data such as standard deviation, mean, tolerance range and acceptance rate for the total population are calculated for the specified critical assembly dimensions.
- **Contribution analysis**: The 3D influence of variation in each geometrical feature, according to specified tolerance and distribution, is ranked for specified critical dimension, including effects of feature *position*, *direction* and variation *magnitude*.

The analyses described in this paper are performed in the RD&T tool (Robust Design & Tolerancing) developed by the authors. Rigid bodies are then assumed.

3. EMBODIMENT DESIGN

In this paper, geometry *concept design* refers to early embodiment design where the overall geometry concept is evaluated with respect to geometrical robustness. Geometry *detail design* refers in this work to the late embodiment design phase, during which the final geometry is set and tolerances are selected with respect to product constraints, geometrical sensitivity and process variation.

The *robustness* and the *variation* represent two important characteristics of a design. Robustness is here defined as "the ability to suppress geometrical input variation". The variation is the actual variation in a particular surface or geometrical feature of the design. The relations and dependencies between variation, robustness, locators and tolerances can be formulated as:

$$\begin{bmatrix} robustness \\ variation \end{bmatrix} = \begin{bmatrix} x & 0 \\ x & x \end{bmatrix} \begin{bmatrix} locators \\ tolerances \end{bmatrix}$$

(1)

Equation 1 indicates a coupled behavior and that the robustness, controlled by the locators, should be treated first. On the basis of this and the information available in the different stages of the design process we will divide the *embodiment design* process into two steps:

1. During geometry *concept design* (early embodiment design), where functional requirements are fulfilled through the selection of geometry characteristics, very little manufacturing information is available and the focus should be on improving the robustness of the design.
2. During geometry *detail design* (late embodiment design), a design concept fulfilling the functional requirements of the product is to be adjusted to the actual manufacturing constraints. This means compensating the geometrical sensitivity of the concept by the selection of tolerances and minimizing the manufacturing cost of the product.

Chapters 4 and 5 will discuss concept and detail design in more depth.

4. CONCEPT DESIGN

During geometry *concept design*, the designer's task is to generate and evaluate a number of concept solutions. The situation is often characterized by the facts that:
- only rough sketches and/or CAD models exist
- several potential concept solutions are to be evaluated
- very little manufacture and assembly information is available

At this stage, the focus should be on optimizing the general robustness of the design, making it as insensitive to variation as possible.

4.1 Axiomatic Design

In the theory of *axiomatic design,* see [Suh, 1990], the *design equation* (equation 2) relates a group of parallel design parameters (DP:s) on the same hierarchical level to their corresponding functional requirements (FR:s). The matrix elements represent the partial derivatives at a specific design point.

$$\begin{bmatrix} FR_1 \\ FR_2 \\ FR_3 \end{bmatrix} = \begin{bmatrix} a_{11} & a_{12} & a_{13} \\ a_{21} & a_{22} & a_{23} \\ a_{31} & a_{32} & a_{33} \end{bmatrix} \begin{bmatrix} DP_1 \\ DP_2 \\ DP_3 \end{bmatrix}$$

(2)

An *uncoupled* design is characterized by the fact that each output is controlled by one input only. A design such as this, represented by a diagonal matrix with all non-diagonal elements equal to zero, has a very clear relation between input and output parameters and is very easy to tune. The situation can be compared to a set of tolerance chains, each with only one link.

We will in the following sections discuss how this philosophy may be applied to geometry problems in which the geometrical concept may be improved by decreasing the degree of geometrical coupling, i.e. the number of *inputs* controlling a certain *output*.

4.2 Robust Design

Generally, a robust design is a design that is insensitive to variation and disturbance. The important performance characteristics of the product are insensitive to manufacturing variation, temperature, wear etc, see [Taguchi, 1989]. Figures 1 and 2 show an example of a non linear relation between an input parameter, x, and an output characteristic, y. By shifting the nominal value, x_0, the sensitivity $\Delta y/\Delta x$ is decreased.

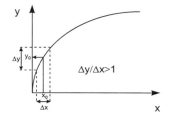

Figure 1; *Sensitive design*   **Figure 2**; *Robust design*

The main source of variation considered here is the manufacturing variation. By decreasing the sensitivity of the design, wider tolerances on input parameters may be used. For many cases, this results in reduced manufacturing costs.

4.3 CAT tools used in concept design

During geometry *concept design*, the focus is on optimizing the locating schemes for the assembly. Figures 3 and 4 present the evaluation of two different locating schemes. In figure 3, the locating scheme that has the primary locators (A1, A2, A3) on the vertical plane is analyzed, whereas figure 4 analyzes the locating scheme using the horizontal plane as primary locators. Discrete contact points are assumed. The output point to be analyzed is point MP.

Changing the locating scheme may in many cases be done without increasing the manufacturing cost, whereas tightening the tolerances often increases the manufacturing cost. Figures 5 and 6 show the sensitivity analysis results of the two cases:
- The stability matrix, showing the stability index, determined as the influence of the whole locating scheme (P-Frame) on the MP variation, in each direction as well as in total. For large assemblies, this matrix gives a good overview of the total

assembly sensitivity. Two special measures are then used to evaluate the total assembly robustness. This is described in [Söderberg and Lindkvist, 1998].

- The stability index list, showing the sensitivity coefficients for each locating point, determined as $\Delta output/\Delta input$ for the total magnitude of the output variation as well as the magnitude in each separate direction (X, Y and Z) in space.

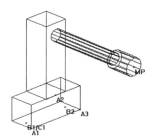

Figure 3; Vertical primary locators

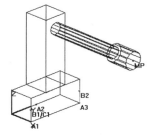

Figure 4; Horizontal primary locators

REF1.rdt ... — Unit Disturbance

Stability Index for One Entity

Measure Name: Z

P-Frame	Target Part	Point	Ref.Pt	Unit Dist.	Real Tol.
Part01	Ground	A2	A2	4.499982	0.899297
Part01	Ground	A3	A3	2.250002	0.450074
Part01	Ground	A1	A1	2.250002	0.450075
Part01	Ground	B2	E2	0.500004	0.100175
Part01	Ground	B1/C1	E1	0.500004	0.100175
Part01	Ground	B1/C1	C1	0.000000	0.000000

Measure	Part01
X	1.00
Y	1.25
Z	1.67
MAG	2.72

Figure 5; Vertical primary locators

CASE2.r... — Unit Disturbance

Stability Index for One Entity

Measure Name: Z

P-Frame	Target Part	Point	Ref.Pt	Unit Dist.	Real Tol.
Part01	Ground	A2	A2	3.999987	0.799498
Part01	Ground	A1	A1	2.500005	0.500205
Part01	Ground	A3	A3	2.500005	0.500204
Part01	Ground	B1/C1	C1	0.000000	0.000000
Part01	Ground	B1/C1	B1	0.000000	0.000000
Part01	Ground	B2	B2	0.000000	0.000000

Measure	Part01
X	1.17
Y	1.10
Z	1.50
MAG	2.61

Figure 6; Horizontal primary locators

It can be seen from the results that, in the Y and Z directions as well as in total (MAG), the right locating scheme, using the horizontal primary locators (figure 4), is the most robust scheme. In the Z direction, the contributors are only the A1, A2, A3 locating points, which result in a less coupled design solution that is easier to control than the left scheme where the B1 and B2 points also contribute to variation in MP. By changing the locating scheme from the one in figure 3 to the one in figure 4, two effects have been accomplished:

- The number of input parameters has been reduced, resulting in a design that is easier to adjust and control. This corresponds to the philosophy of *Axiomatic Design*.
- The sensitivity has been reduced, resulting in a design that better suppresses input variation. This corresponds to the philosophy of *Robust Design*.

An improvement in geometrical robustness as described here requires very little geometry information. For many types of early robustness evaluations, discrete points, representing locators, mating features and critical dimensions, along with individual feature variation direction, are sufficient to make robustness improvements.

5. DETAIL DESIGN

During geometry *detail design*, the design is adjusted to the manufacturing and assembly process. Geometrical concept sensitivity must be met by manufacturing and assembly precision and the final choice of distributors for the individual parts have to be made. The situation is often characterized by the facts that:

- geometry and CAD models are more highly developed
- product sensitivity is known (from concept design)
- manufacturing and assembly information is available

At this stage, tolerances may be allocated with respect to:

- product constraints
- assembly sensitivity
- manufacturing and assembly variation and cost

5.1 CAT tools in Detail Design

CAT technology provides two types of analyses that can be used during detail design to optimize the selection of tolerances, *variation* analysis and *contribution* analysis.

5.1.1 Variation Analysis

To determine the quality level and minimize the total loss for a product *before* production starts, variation in product key characteristics must be predicted. By using CAT tools for predicting variation, complex products may be analyzed and improved before the first physical prototype is built.

Figure 7 shows the RD&T Monte Carlo Distribution Analysis. In this case, the expected distribution for point MP is simulated for a set of tolerances and distributions on the two parts. On the basis of the number of Monte Carlo iterations, the simulation predicts the expected:

- mean value

- standard deviation, σ, and
- range

for the specified critical dimensions to be analyzed.

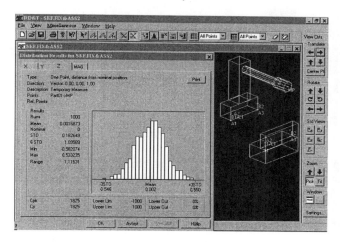

Figure 7; *Variation Analysis*

This information may then be used together with a loss function to calculate the expected loss for a particular set of tolerances on individual surfaces in the assembly. Different sets of input tolerances may then be evaluated against each other.

If information about manufacturing cost and a set of manufacturing alternatives are available, the total loss may be minimized by the optimal selection of tolerances, i.e. the optimal selection of manufacturing processes, see [Vasseur *et al.*, 1992], [Krishnaswami and Mayne, 1994] and [Söderberg, 1993, 1994a, 1994b, 1995].

5.1.2 Contribution Analysis

The RD&T Contribution Analysis (figure 8) presents the ranked contribution of each input feature variation with respect to position, variation direction and range (tolerance value). This information gives a hint of where the tolerances should be tightened in order to reduce output variation. It may also be compared with the sensitivity analysis which does not include the effect of tolerance value. Comparing the contribution analysis with the sensitivity analysis reflects how well the selected tolerances compensate for the sensitivity of the design.

Contribution analysis does not consider manufacturing cost. The economic effects of tightening or loosening tolerances must be considered by the user. The user must also decide what design philosophy to use. Basically, he can strive for:

- equal contribution, i.e. spreading the risk as much as possible, or
- a few major contributors, i.e. a clear and uncoupled design solution.

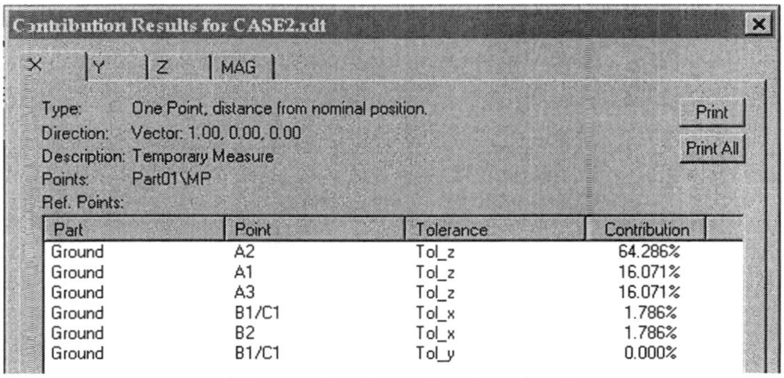

Figure 8; *Contribution Analysis*

The latter is preferred in the axiomatic design philosophy put forward by Suh since it makes the assembly design easier to tune, i.e. adjust to proper quality level.

6. SUMMARY

This paper describes how design robustness may be increased, and geometrical variation reduced, in a two-step procedure using CAT technology.

In geometry *concept design*, the *general robustness* of a concept is increased by minimizing the number of controlling parameters for a critical characteristic and by maximizing the general ability to suppress variation. This corresponds to philosophies put forward by Suh and Taguchi and uses CAT techniques to analyze and improve general assembly robustness with respect to locator positions and variation directions.

During *detail design*, tolerances are allocated with respect to general robustness, manufacturing capability and manufacturing cost. At this stage, the final variation of overall product characteristics may be simulated and part tolerances may be adjusted with respect to final sensitivity and cost. If loss functions are available, the total quality level of a concept may be analyzed.

For many types of early robustness evaluations, discrete points, representing locators, mating features and critical dimensions, along with individual feature variation direction, are sufficient to make robustness improvements.

REFERENCES

[Bjørke, 1989] Bjørke, O., *Computer Aided Tolerancing,* ASME PRESS New York, ISBN 0-7918-0010-5.

[Cagan and Kurfess, 1992] Cagan, J.; Kurfess, T. R.; "Optimal Tolerance Allocation over Multiple Manufacturing Alternatives", *Advances in Design Automation,* Vol. 2, ASME, DE-Vol. 44-2, pages 165-172, ISBN 0-85389-534-1.

[Chase and Parkinson, 1991] Chase, K. W.; Parkinson, A. R.; "A Survey of Research in the Application of Tolerance Analysis to the Design of Mechanical Assemblies", *Research in Engineering Design,* Vol. 3, pages 23-37.

[Clement *et al.*, 1997] Clement, A.; Rivière, A.; Serré, P.; Valade, C.; "The TTRS: 13 Constraints for Dimensioning and Tolerancing", 5th CIRP Conference on Computer Aided Tolerancing, Toronto, April 28-29 1997, p 73-83.

[ElMaraghy *et al.*, 1995] ElMaraghy, W. H.; Gadalla, M. A.; Valluri, S. R.; Skubnik, B. M.; "Relating ANSI GD&T Standards and Symbols to Variations on Primitives", Proceeding of the 1995 ASME International Mechanical Engineering Congress & Exposition, San Francisco, CA, USA Nov 12-17 1995.

[Gao *et al.*, 1995] Gao, J.; Chase, K.W.; Magleby, S.P.; "Comparision of Assembly Tolerance Analysis by the Direct Linearization and Modified Monte Carlo Simulation Methods", ASME Design Engineering Technical Conference, Boston, p 353-360.

[Gao *et al.*, 1996] Gao, J.; Chase, K.W.; Magleby, S.P.; "Generalized 3-D Tolerance Analysis of Mechanical Assemblies with Small Kinematic Adjustment", submitted to Journal of Design and Manufacturing.

[Hernla, 1993] Hernla, M., "Calculation of Measuring Uncertainty with CMMs under Industrial Conditions", Proceeding of the 3rd CIRP Conference on Computer Aided Tolerancing, Cachan, France, April 27-28 1993, p 171-178.

[Iannuzzi and Sandgren, 1994] Iannuzzi, M.; Sandgren, E.; "Optimal Tolerancing: The Link between Design and Manufacturing Productivity", ASME, DE-Vol. 68, Design Theory and Methodology - DTM 94, p 29-42.

[Krishnaswami and Mayne, 1994] Krishnaswami, M.; Mayne, R. W.; "Optimizing Tolerance Allocation Based on Manufacturing Cost and Quality Loss", *Advances in Design Automation,* Vol. 1, ASME, DE-Vol. 69-1, pages 211-217, ISBN 0-7918-1282-0.

[Kumar and Raman, 1992] Kumar, S.; Raman, S.; "Computer-Aided Tolerancing: The Past, the Present and the Future", *Journal of Design and Manufacturing,* Vol. 2, pages 29-41.

[Lee and Woo, 1989] Lee, W. J.; Woo, T. C; "Optimum Selection of Discrete Tolerances", *ASME Journal of Mechanisms, Transmissions, and Automation in Design, Vol.* I I 1, June, pages 243-25 1.

[Long and Hu, 1997] Long, Y.; Hu, J.; "A Unified Model for Variation Simulation of Sheet Metal Assemblies", 5th CIRP Conference on Computer Aided Tolerancing, Toronto, April 28-29 1997, p 149-160

[Mansoor, 1963] Mansoor, E. M.; "The Application of Probability to Tolerances Used in Engineering Designs", *Proceedings of the Institution of Mechanical Engineers,* Vol. 178, No. 1, pages 29-51.

[Methieu, 1997] Methieu, L.; Clement, A.; Bourdet, P.; "Modeling, Representation and Processing of Tolerances", Tolerance Inspection: a Survey of Current Hypothesis, 5th CIRP Conference on Computer Aided Tolerancing, Toronto, April 28-29 1997

[Nigam and Turner, 1995] Nigam, S. D.; Turner, J. U.; "Review of Statistical Approaches to Tolerance Analysis", *Computer-Aided Design*, Vol 27, p 6-15

[Parkinson, 1985] Parkinson, D. B.; "Assessment and Optimization of Dimensional Tolerances", *Computer Aided Design,* Vol. 17, No. 4, pages 191-199.

[Pheil, 1957] Pheil, G. D.; "Probability Applied to Assembly Fits", *Product Engineering,* Vol. 28, No. 21, page 88.

[Salomonsen *et al.*, 1997] Salomonsen, O., W.; van Houten, F.; Kals, H.; "Current Status of CAT Systems", 5th CIRP Conference on Computer Aided Tolerancing, Toronto, April 28-29 1997, p 345-359.

[Speckhart, 1972] Speckhart, F. H.; "Calculation of Tolerance Based on a Minimum Cost Approach", *ASME Journal of Engineering for Industry,* Vol. 94, No. 2, pages 447-453.

[Spotts, 1973] Spotts, M. F.; "Allocation of Tolerances to Minimize Cost of Assembly", *ASME Journal of Engineering for Industry,* Vol. 95, August, pages 762-764.

[Söderberg, 1993] Söderberg, R.; "Tolerance Allocation Considering Customer and Manufacturer Objectives", *Advances in Design Automation*, Vol. 2, ASME, DE-Vol. 65-2, pages 149-157, ISBN 0-7918-1181-6

[Söderberg, 1994a] Söderberg, R.; "Robust Design by Tolerance Allocation Considering Quality and Manufacturing Cost", *Advances in Design Automation*, Vol. 2, ASME, DE-Vol. 69-2, pages 219-226, ISBN 0-7918-1282-0.

[Söderberg, R., 1994b] Söderberg, R.; "Tolerance Allocation in a CAD Environment Considering Quality and Manufacturing Cost", *Lean Production: From Concept to Product*, Irish Manufacturing Committee, IMC-11, Belfast 31-2 September, pages 789-800, ISBN 0-85389-534-1.

[Söderberg, 1995] Söderberg, R.; "Optimal Tolerance Band and Manufacturing Target for Monotonic Loss Functions with Functional Limits", *Advances in Design Automation*, Vol. 1, ASME, DE-Vol. 82, pages 345-352, ISBN 0-7918-1716-4.

[Söderberg *et al.*, 1998] Söderberg, R.; Wandebäck, F.; Wahlborg, P-J,; "The Subcontractor's Role in Computer Aided Tolerance Management, ASME Design for Manufacturing Conference in Atlanta, September 13-16.

[Söderberg and Johannesson, 1998] Söderberg, R.; Johannesson, H. L.; "Spatial Incompatibility - Part Interaction and Tolerance Allocation in Configuration Design", ASME Design Theory and Methodology Conference in Atlanta, September 13-16,

Söderberg and Lindkvist, 1998] Söderberg, R.; Lindkvist L; "Computer Aided Assembly Robustness Evaluation and Geometrical Coupling Quantification", Submitted to the Journal of Engineering Design.

[Suh, 1990] Suh, N. P.; *The Principles of Design*, Oxford University Press, ISBN 0-19-504345-6.

[Taguchi et al., 1989] Taguchi, G.; Elsayed, E. A.; Hsiang, T. C.; *Quality Engineering in Production Systems,* , McGraw-Hill International Editions, ISBN 0-07-100358-4

[Vasseur *et al.*, 1992] Vasseur, H.; Kurfess, T. R.; Cagan, J.; "A Decision-Analytic Method for Competitive Design for Quality", *Advances* in Design Automation, ASME DE-Vol. 44-1, p 329-336

Detect the Unexpected –
How to Find and Avoid
Unexpected Tolerance Problems in Mechanisms

Ralf Schultheiss and Uwe Hinze
Ford Werke AG, D–50725 Cologne, Germany
rschulth@ford.com, uhinze@ford.com

Abstract: We discuss the current design process with tolerance analysis as a distinct process step and identify the deficiencies that have an adverse impact on design and product quality. Perhaps the greatest of them is the inability to detect unknown or unexpected tolerance problems.

As an alternative we present:

- A vision for a design process where tolerance analysis is no longer needed as a distinct activity.
- How the new process can be realized with a new computer tool.
- The enabling technologies named Variational Tolerance Envelopes and Kinematic Tolerance Space.

The Variational Tolerance Envelopes is a method to visualize worst case tolerances in mechanisms. This method is used to predict all interferences in mechanisms resulting from adverse part tolerance shapes. The Kinematic Tolerance Space method is used to detect the unknown kinematic problems in mechanisms. It offers a holistic view of the kinematics and complements the Variational Tolerance Envelope method. Thus we are able to realize an unmatched problem detection capability with respect to kinematic, geometric and tolerance problems in a design. Interference and collision problems can be detected as well as kinematic malfunctions. Most of them can be identified and highlighted automatically without human interaction.

We show how these methods are embedded in our ADAPT tolerance analysis system via Variational Geometry and Variational Assembly. We conclude with an outlook to the potential of these new technologies.

Keywords: tolerance problem detection, tolerance synthesis, tolerance envelopes, tolerance configuration space, kinematic tolerance analysis.

1. INTRODUCTION

Computer based tolerance analysis in Ford dates back to 1974 [1]. Since then we apply and develop a program named ADAPT. Its main application is the tolerance analysis of mechanisms.

Two of our major development objectives for this program package are:
- Detection of all possible tolerance problems in mechanisms.
- Development of an intuitive understanding of tolerance variation by design engineers.

The assignment of tolerances is mainly ruled by two, sometimes opposing factors which are:

Function: Functional Tolerances are the limits of geometric variation in a mechanism which still allow a correct function.

Production: Manufacturing Tolerances are the variations which result from the production process.

The scope of this paper is the analysis of Functional Tolerances.

2. THE TOLERANCE ANALYSIS PROCESS

Today the Auto industry faces fierce competition on world markets. Strong efforts are made to improve the design processes and tools. The objective is to improve the design quality and reduce the development time.

Tolerance analysis is still a relatively time consuming activity. To keep the analysis effort within affordable limits only areas of a design which are presumed to be critical will receive attention. Mostly these are safety items, important clearances or areas where interferences are expected. Perhaps the most demanding and time consuming tolerance analysis task is the analysis of kinematics. Often unproven assumptions and simplifications like linearization of nonlinear characteristics are applied to achieve a result within reasonable time.

This leads to the situation where, despite considerable analysis efforts, tolerances still are a factor of uncertainty and risk in a design. Things get even worse when many changes happen during the design process and when there is no time for a detailed analysis of their cross effects. The deficiencies of current tolerance analysis are most painful felt during the design of new automotive powertrain mechanisms. The orders to purchase the transfer lines, the machines and tools are placed at about the same time as the tolerance analysis starts. Should the tolerance analysis reveal a problem, modifications to the design may become very expensive. They usually require changes in the manufacturing process like modifications to a transfer line, machines and/or tools.

When this happens, the engineer often seeks help in statistics, hoping that the probability of the problem will remain within acceptable limits. If this fails too, he requests Manufacturing to produce with higher accuracy, what could result in higher costs per unit. In other cases the engineer takes the risk and accepts higher warranty costs and customer dissatisfaction with the argument: this is still cheaper than changing the transfer line.

After the production has started the downstream consequences of tolerance problems may cause all kinds of quality issues. In extreme cases a stop of production may be required until the problem is resolved.

Tolerance problems which occur during production somehow escaped previous tolerance analysis. Their cause often are areas or constellations in the design which were not expected or known to be critical. The reason might be that the engineer did not regard a certain feature as worth to be analyzed, or the complexity of the kinematic solution and the geometry in the mechanism, produced effects the engineer did not expect.

This tolerancing process is common practice today. It is fundamentally incapable to support the design of robust mechanisms. In a robust design concept the geometry and kinematics provide enough room for variation. Conceptual design problems cannot be resolved by tuning tolerance values or by the calculation of statistical probabilities.

Today the majority of tolerance analysis methods and tools are designed to support the tolerance analysis process as described above. They offer very limited support for the creation of robust designs with correct Functional Tolerances.

The capability of these tools to detect unexpected tolerance problems is very poor.

3. THE TOLERANCE SYNTHESIS PROCESS

In order to overcome this dilemma we developed a concept of a better process where an explicit tolerance analysis is no longer needed. It should be possible to determine the correct Functional Tolerances at any time during the design process. With a number of studies and pilots we tested the feasibility of this concept. We named this process the Tolerance Synthesis process.

At the time this concept was defined there were no tools on the market which came anywhere near the capabilities we required. Therefore, we decided to continue the development of our ADAPT tolerance analysis package to realize this new process. Today our research has reached a point where all the major theoretical problems are solved.

With this tool a design process will be possible where:
- Tolerance problems can be avoided, because the tolerance synthesis tool will be able to automatically detect interferences. The detection of other kinematic, geometric or tolerance problems is also supported through new visualization techniques.

- Functional Tolerances can be determined and optimized during the early design stages.
- Design concepts which are not robust can easily be identified. They can be corrected or alternative concepts can be pursued.
- Geometry and kinematic of a design can be optimized for best functional and manufacturing tolerances.

ADAPT has the following three key enablers for tolerance problem detection and the Tolerance Synthesis process:

- Variational Geometry
- Variational Tolerance Envelopes
- Tolerance Configuration Spaces enabling Variational Assembly

4. VARIATIONAL GEOMETRY

A variational geometry modeler allows to comfortably sketch and manipulate part models. The dimensions and tolerances are variables defining the geometry. By setting dimension values to their upper or lower limit, any worst case tolerance shape of a part can be generated.

The ADAPT variational geometry engine consists of a constraint analyzer, a geometry solver and a tolerance solver.

The constraint analyzer starts from an origin and recursively traverses through the constraint graph consisting of construction elements and constraints. It determines the definition state of all variational geometry elements and matches the definition patterns against its basic geometry rules.

The geometry solver executes the geometry rules previously determined by the constraint analyzer and propagates resulting geometry changes to other geometric elements [2, 3, 4].

5. TOLERANCE ENVELOPES

The constraint analyzer also assigns the rules for the calculation of Tolerance Envelopes. [5] It calculates a Tolerance Envelope around a part geometry. With Tolerance Envelopes we provide the engineer with a better more intuitive understanding of variation. They are the application of Euklidian Geometry on geometric elements with variation of size, location, and orientation.

With Variational Geometry it is only possible to calculate and display discrete worst case part shapes. Unlike this method Tolerance Envelopes visualize the full range of variation of a part geometry. Tolerance Envelopes are a holistic method to convey the complete potential of variation to the engineer. As Tolerance Envelopes are an integral part of the ADAPT variational geometry, the user can immediately see the effect of any modification resulting in a change of geometry and/or Tolerance Envelopes.

Figure 1 shows an ADAPT variational part model with all the constraints and

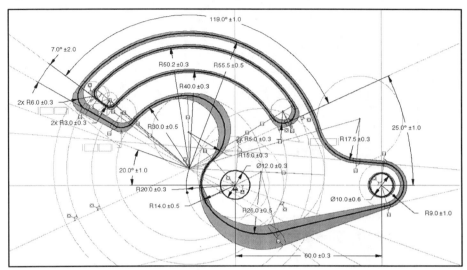

Figure 1

dimensions and the Tolerance Envelope as grey shade around its outline.

6. INTERFERENCE DETECTION WITH TOLERANCE ENVELOPES

How unexpected tolerance problems can be detected with the use of Tolerance Envelopes is shown in the following two figures.

Figure 2 is an overview of a mechanism with all its dimensions. An arrow points to a

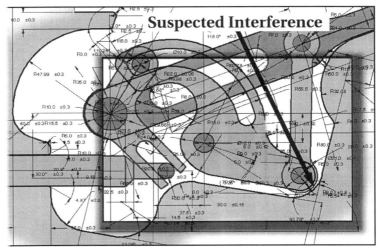

Figure 2

region where a human analyst might expect a tolerance problem.
Figure 3 is an enlarged view of the boxed detail shown in Figure 2. Here the Tolerance

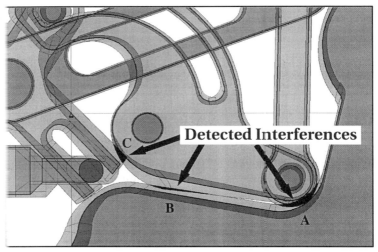

Figure 3

Envelopes of the mechanism can be seen as grey shades around the part outlines.
Arrows point to the areas marked A, B and C where tolerance envelopes interfere.

These are zones which may cause potential tolerance problems. The suspected interference shown in figure 2 is confirmed (zone A). But also two other problem zones (B and C) are shown which look completely unsuspicious in figure 2. Especially the large interference zone C of the two levers is as serious as the suspected zone A.

With the help of Tolerance Envelopes it is possible for the first time that the engineer can develop an intuitive understanding of variation. As variation becomes a visual design characteristic of the part geometries, engineers can create designs with high robustness and avoid or prevent tolerance problems.

7. KINEMATIC TOLERANCE SPACE

Variational Geometry can be used to model the kinematic relationships of the parts in a mechanism. With the variational geometry constraints, the various contact conditions of the parts can be described.

For the simulation of a mechanism this method has one big disadvantage. Changing contact conditions between the parts may require a frequent redefinition of the constraints. Thereby, a kinematic analysis of a mechanism can become a very tedious task [6].

The Kinematic Tolerance Space method developed by Leo Joskowicz and Elisha Sacks [7] offers a comprehensive solution to this problem.

It is a general worst case limit kinematic tolerance analysis method. It computes the range of variation in the kinematic function of a mechanism, based on its part tolerance specifications. It covers fixed and multiple contact mechanisms. A mechanism is modeled with pairwise part relations.

In contrast to our present approach of a kinematic analysis in discrete steps, it enables a holistic view of the kinematic and variational relation of parts. But it can also be used to define any discrete kinematic situation which is also needed for a complete analysis. It is the complement to our visualization of variation in discrete kinematic situations.

Figure 4 shows the principle of this configuration space method. The kinematic relation

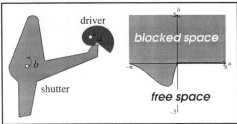

Figure 4

248

of two parts is captured in a diagram. The horizontal axis represents the rotation of part 'a' and the vertical axis that of part 'b'. The area shown in this diagram partitions into free space where the parts do not touch and into blocked space where they would interfere. The common boundary is called contact space. It contains all configurations where both part contours can touch each other.

This method allows the detection of unexpected kinematic, geometric and tolerance problems.

Figure 5 demonstrates the detection of kinematic problems caused by tolerances. Two

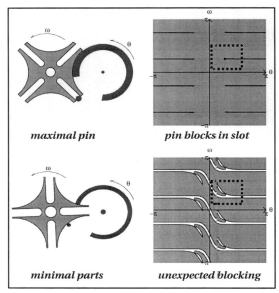

maximal pin   *pin blocks in slot*

minimal parts   *unexpected blocking*

Figure 5

extreme tolerance conditions of a geneva cross mechanism are shown there. In the upper example the pin is too big and cannot enter into the slot. The configuration space diagram shows for this situation an interrupted contact space. In the example with the minimal parts the diagram shows little pockets besides the intended path of the pin. This means that the pin cannot only move into the slot, but also slip along the side of the cross until it blocks the mechanism.

Tolerance Envelopes can also be calculated for the configuration space diagram. This allows to see in a single diagram, how variation affects the kinematics of a mechanism.

In figure 6 we show a detail of the Kinematic Tolerance Space for the geneva cross

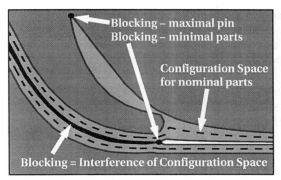

Blocking – maximal pin
Blocking – minimal parts

Configuration Space
for nominal parts

Blocking = Interference of Configuration Space

Figure 6

mechanism. It contains the two situations marked with the dotted boxes in the diagrams in figure 5. The dotted line in the diagram shows the configuration space for nominal parts.

The method allows a failure detection which is far more comprehensive than the pure simulation of a mechanism. Therefore we decided to combine our Tolerance Envelope method with the Kinematic Tolerance Space method. We will use the Kinematic Tolerance Space method for:

- The kinematic tolerance analysis of mechanisms via the configuration space diagrams.
- The calculation of part contact conditions in discrete kinematic situations.

The combination of these two methods allows us to realize a comprehensive detection capability for geometric and kinematic tolerance problems in mechanisms.

8. CONCLUSION AND OUTLOOK

Our research and development work focuses on the solution of the engineering problem. We therefore choose to reduce the complexity of the task by concentrating on a two dimensional solution first. During our long tolerance analysis practice it proved that about 90% of our automotive powertrain mechanisms can be analyzed in 2D planar models. However in many cases it may be necessary that several 2D projections from different viewing angles have to be combined.

To our knowledge ADAPT currently is the only production tolerance analysis tool which has some problem detection capabilities. It helped us to solve many design problems in the past. Our feasibility studies for the Tolerance Synthesis process were all carried out with ADAPT. Now, after the completion of our research, we are implementing these new features. Through that we will define a new state of the art for Dimensional Management tools and processes [8].

The need for the application of statistical methods in the design process will be reduced. This is possible through the visualization capabilities of Tolerance Envelopes and Tolerance Configuration Spaces. The engineer can provide enough space for variation by adjusting the part geometries accordingly. He can search for solutions where tolerances effectively can be compensated and instantaneously see the results of his design actions. Thus, he will create designs with optimum robustness. This will greatly improve product quality. At the same time it is also possible to optimize a mechanism for robustness and manufacturing costs.

REFERENCES

[1] **H. Petri, 1974**; "TTS – ein graphisches System zur Toleranzanalyse"; In: *Journal: Antriebstechnik,* Germany

[2] **M.Weiland, 1994**; "Parametric Design in ADAPT"; In: *Diplomacy thesis;* Fachbereich Informatik, University Dortmund, Germany

[3] **V.Trabandt, 1995**; "Konzepte und Implementierung eines Modellierers für Variational Geometry"; In: *Diplomacy thesis*; Fachbereich Informatik, University Dortmund, Germany

[4] **J.Weitkamp, 1995**; "Entwicklung eines Geometriemodellierers für Variational Geometry"; In:,*Diplomacy thesis*; Fachbereich Informatik, University Dortmund, Germany

[5] **Uwe C. Hinze, 1994**; "A Contribution to Optimal Tolerancing in 2–Dimensional Computer Aided Design"; In: *PhD thesis*; Department of Industrial Mathematics, University Linz, Austria

[6] **M.Aichinger, 1996**; "A Collision Problem for Worst Case Toleranced Objects"; In: *PhD thesis;* Department of Industrial Mathematics, University Linz, Austria

[7] **Leo Joskowicz**; **Elisha Sacks, 1997**; "Computer–Aided Mechanical Assembly Design Using Configuration Spaces"; In: *Proceedings of ICRA '98 Workshop*

[8] **Ralf Schultheiss**; Uwe Hinze, 1998; "From Tolerance Analysis to Tolerance Synthesis"; In: *Proceedings of ICRA '98 Workshop*

Application of TTRS Method in Industrial Practice Tolerance specification for Industrial Cooling Water Pumps

Salomons O.W., Begelinger R.E., Post E., Houten F.J.A.M. van

University of Twente, Mechanical Engineering Dept., Lab. of Production and Design Engineering, P.O. Box 217, 7500 AE Enschede, tel. X–31–53–4892532, e_mail: o.w.salomons@wb.utwente.nl

Abstract: This paper discusses the application of the TTRS method for tolerance specification in a company which designs and manufactures industrial cooling water pumps. The TTRS method has been applied by the engineering department of this company for cooling water pumps that are part of certain product families. No explicit use was made of commercial computer aided tolerancing tools since the company used Pro/Engineer as their CAD system for which the TTRS method is not yet commercially available. Only a spreadsheet program (for tolerance analysis) and a CAD system for generating drawings with tolerance specifications were used. Applying the TTRS approach gives a better insight in which faces are truly functional and which tolerances should (and should not) be specified. The trial showed that after a training in the basic concepts of the TTRS method, engineers were enthusiastic about the new method. In fact, the TTRS method has been adopted by the company as their internal standard tolerance specification methodology. Nevertheless, there are still some issues regarding the TTRS theory that came up during the industrial trial and that need to be solved. One of the most important problems is that tolerance value specification is not yet supported, especially in combination with complex shaped surfaces. The paper discusses these issues.

Keywords: tolerance specification, TTRS, tolerance management.

1. INTRODUCTION

Many companies do not have a clear methodology for tolerance specification. Therefore, most designers use their own experience, informal guidelines as well as their gut feeling in specifying tolerances. As a result, product designs may not contain all the required, functionally relevant, tolerances. In addition, tolerance values might be too tight, making it very costly to produce the product. Therefore, a clear and reliable method for tolerance specification is necessary as part of an overall tolerance management strategy. Tolerance management as a whole and it's necessity in industry has been treated extensively in [Söderberg 1998 a,b]. Tolerance management has been described by Söderberg as: "including all design, manufacture and inspection activities striving to minimize the effect of geometrical variation" [Söderberg 1998b]. In the authors perspective this definition should not always focus on minimization since in some cases increased geometric variation is desired e.g.

251

[Pegna et al. 1997]. Therefore, instead of minimization we would propose to focus on control and optimization of the effect of geometric variation (with respect to functioning, manufacturing, inspection etc.). Tolerance management can be supported partly by CAT tools. However, it is important to note that CAT tools in general focus only on a portion of the whole chain in which tolerance management should be applied. CAT tools can be used to verify the sensitivity of a design for geometric variations but they cannot generate a geometrically robust design in the first place. Söderberg proposes to use CAT tools in support of robust design in the following two steps [Söderberg 1998a]:

1. increase geometric robustness during concept design: CAT tools can be used to analyse the general stability and sensitivity to geometric variation.

2. assign tolerances in the detail design phase; CAT tools can be used here to verify the final sensitivity, manufacturing capability and cost.

The authors think these steps are very valuable since insight in these steps gives one a much better perspective on the items that play a role in tolerancing such as for instance the use of CAT tools. Thus, it can be ensured that the focus is not too much on the CAT tools but on the overall trajectory that is needed to achieve the goals in the field of tolerancing. The construction principles and the nominal geometry (geometric robustness) had already been verified for the presented industrial case. Hence the focus of this paper has been addressed towards the following step in tolerance management: tolerance assignment (specification) in detail design with a focus on the tolerance types.

With the advent of the TTRS method [Clément et al. 1991, 1994], a promising theoretical method for tolerance specification has become available. However, the TTRS method is still restricted to specifying tolerance types and datum systems. Tolerance values are not yet supported, although initial research has been performed in this direction. An example of such research is [Clément et al. 1994], describing an investigation into different functions (mainly applied to bearings) and their relation to tolerance values in several companies. Another example is [Salomons et al. 1998] in which it was proposed to derive tolerance values by means of several simulation runs of the functioning of the assembly (kinematics and dynamics). In the simulations, physical models are used which include tolerance information. The tolerance values should be varied and then the best optimum or trade off between functional behaviour and manufacturing should be proposed by the system. However, the problem is currently to obtain simulations which can include all the relevant aspects and provide reliable quantitative results [Salomons et al. 1998].

Although promising in theory, no extensive results from industrial practice on the usefulness of the TTRS method have been reported. This paper intends to set the first step to bridge this gap between theory and practice.

The paper discusses the application of the TTRS method for tolerance specification in a company which designs and manufactures industrial cooling water pumps. The work as described in this paper has been part of a larger project which involved the University of Twente and the company (Flowserve).

The outline of the remainder of the paper is as follows. An introduction will be given regarding the context of the project (section 2). In addition, tolerance specification for products with complex shapes in general is detailed by means of examples related to the cooling water pumps (section 3). The tolerance specification approach for the industrial cooling water pumps is addressed in section 4. Finally, some conclusions and recommendations are provided in section 5.

2.   CONTEXT

Flowserve, formerly known as Stork Pompen, is a world wide operating pump supplier, making high performance customer specific centrifugal pumps. Therefore, the centrifugal pumps by Flowserve are typical engineer to order products. Figure 1 provides an example

Pump shaft

Axial/Radial bearing
  construction

Bearing stool

Volute

Stuffing box cover

Radial bearing
  construction

Impeller

Suction bell

Intake

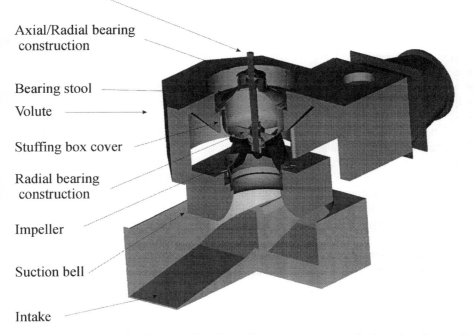

Figure 1; Example of a centrifugal cooling water pump as designed and manufactured by Flowserve.

of a centrifugal pump as produced by Flowserve. As in most market segments, but specifically in the market segment for engineer–to–order products, global competition imposes high requirements upon quality, delivery times and product price. This also holds for Flowserve's market segment in industrial cooling water pumps. To meet the demands mentioned, a complete re–engineering of the factory processes was seen as necessary. Therefore, the company started a project, with the Thermal Engineering and the Production and Design Engineering groups of the University of Twente. The project, which started in 1993, aimed at an integrated improvement of the design and manufacturing processes of Flowserve's centrifugal pumps. The project has recently been completed, and the achievements of the Production and Design Engineering group have been documented in [Begelinger 1998].

The initial situation in the company before the project began, resembled the traditional over–the–wall engineering approach. During the project, a concurrent engineering concept has been introduced. The project aimed at both the development of a computer support system that assists the designer in considering the various stages of the product life–cycle as well as the design of an organization that is able to support concurrent engineering, which is necessary to meet the high level of customization.

The centrifugal pumps of Flowserve can be regarded as belonging to product families. Most of the design work at Flowserve therefore can be regarded as re–design. This means that the basic layout of the different pump types is known in advance. During the project it was realized that in the old situation the amount of engineering hours per unit product was too high and that due to the nature of the Flowserve products, a certain level of design automation could be achieved. As a result of this automation, quotation generation can be performed much quicker and much more accurately and downstream processes can be integrated with more easily. The design tool that has been developed for Flowserve can be considered as an advanced product configurator for engineer to order products which sits on top of a parametric CAD system (in this case: Pro/Engineer). The system is called PROFIDT which is short for PROduct Family Integrated Design Tool [Begelinger 1998].

The most important module of the pump system as designed and manufactured by Flowserve consists of the hydraulic components because they take a big share in the quality, the cost and the delivery time. Therefore, the kernel of the PROFIDT system consists of a three dimensional parametric CAD–model for the basic hydraulic parts (impeller, diffusor and inlet configuration). The parametric CAD–models are driven by geometric parameters in order to make it possible to derive a customer specific hydraulic design quickly. The parametric models are built in such a way that certain hydraulic constraints are always maintained; the relationships determine the hydraulic performance. As an important part of the system, structural feedback loops will assure continuous improvement of the product realization process.

3.   TOLERANCING PRODUCTS WITH COMPLEX SURFACES

This section will elaborate on the tolerancing of products with complex surfaces. This is done by means of some examples related to the cooling water pumps dealt with during the project.

Inspection of products with functional complex surfaces will generally be essential since variations in geometry will affect functioning. One of the feedback loops as mentioned in the preceding section consists of a three dimensional photogrammetric measuring procedure for measuring the geometry of the impeller blades in combination with a best fit algorithm to monitor the differences between the as–built and the as designed geometry.

The feedback of hydraulic data in combination with accurate measurements of the geometry is very important. This feedback is necessary to get a better understanding of the relation between a specific geometric shape and the hydraulic performance. For products other than pumps a similar situation will be applicable. This way of working allows a more accurate prediction of the hydraulic performance, so the need for building prototypes will be reduced and the accuracy in all the downstream processes and the related manufacturing costs can be controlled better. For impeller pumps this work is currently still under way and is being performed by the University of Twente Thermal Engineering group. However, their models have not matured to the extent needed to incorporate them in PROFIDT or some future tolerancing module of PROFIDT.

In the current project, only the assignment of tolerances to surfaces which are part of the assembly has been performed: this is in conformance with the TTRS theory. The theory of TTRS assumes physical contact between the functional surfaces or almost physical contact (clearance). Thus, small clearances also involve functional surfaces. In the case of impeller pumps for example, there must be a small clearance between the impeller and the volute housing in which it resides. However, there are also surfaces which are not (almost) in contact with other surfaces of the assembly and that are still functional. An example are the impeller blade surfaces which are in contact with the water to be transported. This is beyond the scope of the TTRS theory and therefore, complex shapes such as the impeller blades surfaces have not been taken into account. In a way, this is not correct: functional surfaces need not always be in contact with other (functional) surfaces of the same assembly. Functional surfaces can for instance be in contact with a medium that has to be transported or guided via the assembly such as the water which is in contact with the impeller blades of the pump. Another example are the wings of an aeroplane. In order to be able to say something about the tolerances for such cases of complex surfaces, deep insight is necessary on the impact of geometric variation to the hydraulics or aerodynamics or other physical effects that influence the functioning of the assembly. This would then require even more extensive physical simulations than those proposed in [Salomons et al. 1998]. For the impeller pumps this work can only start when the hydraulic models are refined further.

In view of the above we can safely state that a pump's impeller blades are functional surfaces although they are not explicitly recognized as such in the functional surfaces determination method belonging to the TTRS theory. However since the tolerancing of these types of complex surfaces (which are functional but which are not in contact with other surfaces of the same assembly) is still a research issue that is not yet fully resolved, it has not been taken into account in the current study. Functioning, manufacturing as well as inspection aspects need to be taken into account together with the tolerances on complex functional surfaces whenever tolerance management is the final goal.

4. TOLERANCE SPECIFICATION OF IMPELLER PUMPS

From the previous section it will have become clear that tolerancing plays an important role for Flowserve's centrifugal pumps. This holds for the whole trajectory from tolerance specification all the way to inspection. In this paper however, the focus will be on tolerance specification with a focus on tolerance type specification. The first sub–section will address the situation of tolerance specification at Flowserve before the start of the project (section 4.1). Section 4.2 will address some details on the considerations regarding the possible use of CAT tools within Flowserve's tolerance management strategy. In section 4.3 the TTRS tolerance specification methodology that has been used is summarized. In section 4.4 the situation after the completion of the project is described.

4.1 Situation before the start of the project

Prior to the start of the project, no particular tolerancing methodology was used at Flowserve. Tolerances were specified as much as possible in conformance with ISO/ANSI standards and as annotations to the 2D CAD drawings. However, there was a large dependence on the individual designer's attitude towards tolerances. Therefore, inconsistent and incomplete tolerancing schemes have been encountered in the drawings. This sometimes lead to problems in downstream processes.

4.2 Use of CAT tools within Flowserve's tolerance management strategy

With reference to the tolerance management and robust design approach as proposed by [Söderberg 1998 a,b,], it can be said that the robustness of the nominal design of the cooling water pumps could hardly be improved. The next step would then be to improve tolerance specification. It was apparent that it was desirable to improve the tolerance specification approach at Flowserve; there was no methodology applied and errors occurred too often. The introduction of a CAT (Computer Aided Tolerancing) system has been considered as well. However, most of these systems do not yet offer a tolerance specification methodology which supports the user in tolerance specification [Salomons 1997]. They only offer an 'a posteriori' tolerance check. Using a CAT system for a (traditional) tolerance analysis was also not considered to be a very cost–effective solution. Due to the simple geometry of the parts (rotation symmetric), a worst case tolerance analysis could already be performed by the use of a simple spreadsheet program such as Excel. Except for the impeller blades, most parts are rotation symmetric, and hence most tolerance analysis tasks are simply 2D or 1D. The spreadsheet could be linked with the nominal dimensions generated by the PROFIDT system, allowing semi–automatic tolerance analysis for each new variant generated. Therefore, it was decided that the introduction of a commercial CAT system had little to offer at that stage.

Hence it was proposed to introduce a generic tolerance specification methodology to be carried out "manually" by the engineers. With this approach, already a lot of improvement could be made. This can be regarded as the starting point in a tolerance management process as was mentioned in section 1.

4.3 The TTRS based tolerance specification methodology

On the basis of the research at the laboratory of Production & Design Engineering in the field of tolerancing e.g. [Salomons 1995], the TTRS approach as proposed by Clément et al. [Clément et al. 1991, 1994] was selected for the project.

TTRS stands for Technologically and Topologically Related Surfaces. The TTRS approach by Clément et al. basically comes down to extracting functional (hence the technological aspect) surfaces from a nominal design model and tolerancing those functional surfaces which lie on a single part (the topological aspect). Functional surfaces are surfaces that have assembly relations with other surfaces. As we have seen previously, impeller blades, aircraft wings etc. are exceptions to this rule: they are in contact with some medium such as water or air and not with other surfaces of the assembly. In a CAD system the assembly relations are often explicitly available such as relations like mate, align etc.. In an automated tolerance specification system the order of tolerancing faces depends on the loops found in a graph representing all components with their faces involved in assembly relations. The loops represent overconstraints on the kinematic degrees of freedom of the nominal geometry and thus tolerances are regarded as a way to relax these overconstraints [Dufossé 1993]. Clement et al. [Clément et al. 1991, 1994] have provided some rules which determine the sequence in which loops should be processed. The basic rule is to process short loops of which the tolerances work in one direction first [Clément et al. 1991]. Some additional rules have been proposed later in [Salomons 1995]. In summary, the method has the following steps:
1. Determine the parts
2. Determine functional surfaces on the basis of assembly relations
3. Derive assembly graph (parts connected by assembly relations on functional surfaces)
4. Detect kinematic loops in assembly graph
5. Determine TTRS per part based on rules on which loop to process first
6. Infer tolerance types from TTRS

In the case of Flowserve the assembly relations are explicitly available in PROFIDT/ Pro/Engineer. In the case of Flowserve, since no CAT tool was used, the composition of a graph based on the assembly relationships was done manually.

Since Flowserve has a limited number of product families with relatively little variation in functional surfaces, it is worthwhile to (manually) create the assembly graph, detect the loops and infer the required tolerance types. Figure 2 provides an example of a graph of a BSV/BCV product family of an impeller pump. Figure 2 also provides the result after 8 rounds of loop detection. The tolerance values have to be added manually but can be verified by using the spreadsheet based tolerance analysis approach. Also, the tolerance types that belong to certain product families and that resulted from applying the TTRS approach have been programmed to automatically appear in drawings that are generated from the 3D CAD models. This is shown in figure 3. More details can be found in [Post 1997].

4.4 Situation after completing the project

The situation after completing the project was that the involved engineers had had a training on how to specify tolerances using the TTRS approach and now use this approach in prac-

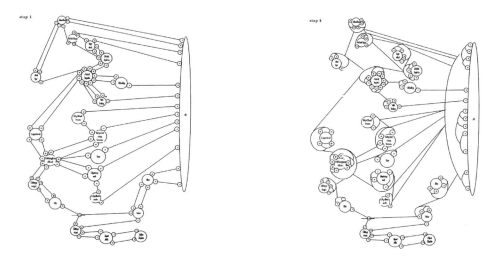

Figure 2; *Assembly graph of the pump before (left) and after (right) TTRS determination using 8 loop detection steps [Post 1997]*

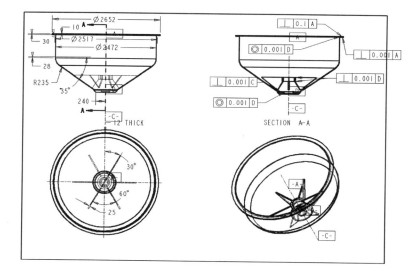

Figure 3; *Drawing of the stuffing box cover with tolerances generated by Pro/Engineer.*

tice. The engineers have embraced the TTRS approach enthusiastically; it provides them some clear guidelines by which to conduct their tolerance specification tasks which they did not have before. No quantitative results are available on how much improvement tolerancing schemes have undergone since the introduction of the new approach and how much this has improved down stream processes. However there are strong indications that a clear improvement has taken place. The approach has been documented with reference to specific Flowserve product examples [Post 1997]. Thus, new engineers will easily be made familiar with the tolerance specification approach.

5. CONCLUSIONS & RECOMMENDATIONS

Introducing the TTRS approach as an aid in the tolerance specification process for the parts of impeller pumps has been completed successfully in industrial practice at Flowserve. This can be seen as a first step towards tolerance management, or, a first step to achieve global consistency of tolerances in an industrial environment. For product family based products such as Flowserve's impeller pumps, it seems worthwhile to introduce the TTRS based tolerance specification approach even without any computer support. However, possible future computer support may smoothen and accelerate the tolerance specification process further. Especially this holds when seen from the broader context of tolerance management.

An issue that remains is the specification of tolerances on the impeller blades. These blades are functional surfaces since they are in contact with the water and since their shape determines the proper functioning of the pump to a great extent. However, in the current TTRS approach only functional surfaces are regarded that are related to surfaces within the product itself. In addition, since the impeller blades represent complex shaped surfaces and since the shape and tolerances affect the actual functioning of the pump, tolerancing the blades is not a trivial matter. Therefore, analyzing the effect of tolerances on the efficiency of the pump would be an interesting exercise. In such an analysis the issue of specifying the tolerance values also will have to be addressed, probably by performing many simulation runs for different tolerance values. For Flowserve the next step in tolerance management is recommended to be taken: looking at the tolerance values both from the viewpoint of functioning but also including manufacturing capability and cost in the tolerance specification task.

Summarizing, and irrespective of a particular product or product family, we propose to research tolerance specification of (functional) complex shaped surfaces. This needs to be done in relation to their functioning, manufacturing and inspection. Especially, the specification of tolerance values needs to be addressed more than in the few references available today. Finally, more feedback is needed from companies who try to implement some kind of tolerance management approach so that tolerancing research can be directed to the real needs in industry. In the end this will help achieve global consistency of tolerances from both an industrial as well as from an academic perspective.

260

ACKNOWLEDGEMENTS

Herbert Bult from Flowserve is acknowledged for his support during the project.

REFERENCES

[**Begelinger 1998**] Begelinger R., Computer support in the design of product families, PhD thesis University of Twente, 1998.

[**Clément et al. 1991**] Clément , A., Desrochers, A., Rivière A., Theory and Practice of 3–D Tolerancing for Assembly, 2nd CIRP International Work. Seminar on CAT, Penn State University, May 1991, USA.

[**Clément et al. 1994**] Clément , A., Rivière, A., Temmerman, M., Cotation Tridimensionelle des Systèmes Mécaniques, Théorie & Pratique, PYC Edition, Yvry–Sur–Seine Cedex, ISBN 2–85330–132–X), in French, 1994

[**Dufossé 1993**] Dufossé P., Automatic dimensioning and tolerancing, 3rd. CIRP seminar on Computer Aided Tolerancing, Cachan (F), April 1993, 1–10.

[**Pegna et al. 1997**] Pegna J., Fortin C., J. Mayer, 1997, Teaching tolerances, Proc. CIRP Int. Seminar on computer Aided tolerancing, Toronto.

[**Post 1997**] Post E., Automatische tolerantie generatie/Implementatie BSV/BCV pomp, internal document, in dutch, Flowserve/University of Twente, 1997.

[**Salomons 1995**] Salomons, O.W., "Computer support in the design of mechanical products, constraint specification and satisfaction in feature based design for manufacturing", Ph.D. Thesis, University of Twente, Enschede (NL)

[**Salomons et al. 1997**] Salomons O.W., Houten F.J.A.M. van, Kals H.J.J., Current status of CAT tools, proceedings int. CIRP Seminar on Computer Aided Tolerancing, Toronto, April, 1997.

[**Salomons et al. 1998**] Salomons O.W. , Zijlstra J., Zwaag J. van der, Houten F.J.A.M. van, Towards dynamic tolerance analysis using bondgraphs, proceedings ASME Design Engineering Technical Conferences DETC '98, Design Automation Conference, Atlanta, September 1998.

[**Söderberg 1998a**] Söderberg, R., Robust Design of CAT Tools, Proceedings of DETC98: 1998 ASME Design Engineering Technical Conference, 13–16 September 1998b, Atlanta, USA.

[**Söderberg 1998b**] Söderberg, R., Wandebäck, F., Wahlborg, P., The Subcontractors role in Computer Aided Tolerance Management , Proceedings of DETC98: 1998 ASME Design Engineering Technical Conference, 13–16 September 1998b, Atlanta, USA.

The configuration space method for kinematic tolerance analysis

Leo Joskowicz[1] and Elisha Sacks[2]

[1] *Institute of Computer Science*
The Hebrew University, Jerusalem 91904, Israel
[2] *Computer Science Department*
Purdue University, West Lafayette, IN 47907, USA
E-mail: josko@cs.huji.ac.il, eps@cs.purdue.edu

Abstract: We present an overview of our kinematic tolerance analysis algorithm for mechanical systems with parametric part tolerances. The algorithm constructs a variation model for the system, derives worst-case bounds on the variation, and helps designers find unexpected failure modes, such as jamming and blocking. The variation model is a generalization of the configuration space representation of the part contacts in the nominal system. The algorithm handles general planar systems of curved parts with contact changes, including open and closed kinematic chains. It constructs a variation model for each interacting pair of parts then derives the overall system variation at a given configuration by composing the pairwise variation models via sensitivity analysis and linear programming. Our implementation analyzes systems with up to 100 parameters in under a minute on a workstation.
Keywords: tolerance analysis, parametric tolerancing, kinematics, planar mechanical systems, configuration space.

1. INTRODUCTION

This paper describes our research in kinematic tolerance analysis of mechanical systems. Tolerance analysis is the task of estimating the worst-case or average error in critical system parameters due to manufacturing variation. The major steps in tolerance analysis are tolerance specification, variation modeling, and sensitivity analysis. Tolerance specification defines the allowable variation in the shapes and configurations of the parts of a system. The most common are parametric and geometric tolerance specifications [Voelcker, 93], [Requicha, 93]. Variation modeling produces mathematical models that map tolerance specifications to system variations. Sensitivity analysis estimates the worst-case and statistical variations of critical properties in the model for given part variations. Designers iterate through these steps to synthesize systems that work reliably and that optimize other design criteria, such as cost.

The critical system parameters effect the system function and capacity to be assembled. In assembly tolerancing, very general part variations must be modeled, so geometric tolerance specifications are the norm. Statistical sensitivity analysis is appropriate because guaranteed assembly is more expensive than discarding a few defective products. Most algorithms perform tolerance analysis on the final assembled configuration [Chase and Parkinson, 91] although recent research explores toleranced assembly sequencing [Latombe and Wilson, 95]. In functional tolerancing, the relevant part variations occur in functional features whose descriptions are parametric. Parametric tolerances, which are simpler than geometric tolerances, are best suited to capture these variations. Worst-case analysis is most appropriate because functional failures that occur after product delivery can be unacceptable.

Our research addresses functional kinematic tolerance analysis of mechanical systems. Kinematic tolerancing is the most important form of functional tolerancing because kinematic function, which is described by motion constraints due to part contacts, largely determines mechanical system function. The task is to compute the variation in the part motions due to variations in the tolerance parameters. Variation modeling derives the functional relationship between the tolerance parameters and the system kinematic function. Sensitivity analysis determines the variation of this function over the allowable parameter values.

We illustrate kinematic tolerance analysis on an intermittent gear mechanism (Figure 1). The mechanism consists of a constant-breath cam, a follower with two pawls, and a gear with inner teeth. The cam and the gear are mounted on a fixed frame and rotate around their center; the follower is free. Rotating the cam causes the follower to rotate in step and reciprocate along its length (the horizontal axis in the figure). The follower engages a gear tooth with one pawl (snapshot a), rotates the gear 57 degrees, disengages, rotates independently for 5 degrees while the gear dwells (snapshot b), then engages the gear with its opposite pawl and repeats the cycle (snapshot c). The mechanical function is conversion of rotary motion into alternate rotation and dwell. The dwell time is determined by the gear tooth spacing. Quantitative tolerance analysis bounds the variation in the gear rotation, which is the critical parameter in precision indexing. Qualitative analysis detects failure modes, such as jamming, when one pawl cannot disengage because the other prematurely touches the gear.

Creating a variation model is the limiting factor in kinematic tolerance analysis. In most cases, the analyst has to formulate and solve systems of algebraic equations to obtain the relationship between the tolerance parameters and the kinematic function. The analysis grows much harder when the system topology changes, that is when different parts interact at various stages of the work cycle, such as in the example above. Contact changes represent qualitative changes in the system function. They occur in the nominal function of higher pairs, such as gears, cams, clutches, and ratchets. Part variation produces unintended contact changes in systems whose nominal designs prescribe permanent contacts, such as joint play in linkages. The analysis has to determine which contacts occur at each stage of the work cycle, to derive the resulting kinematic functions, and to identify potential failure modes

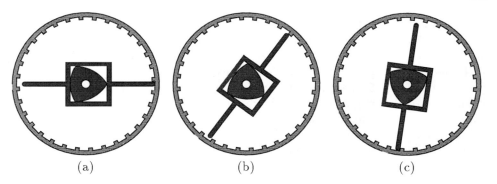

Figure 1: *Intermittent gear mechanism: (a) upper follower pawl engaged, (b) follower disengaged, (c) lower pawl engaged.*

due to unintended contact changes, such as play, under-cutting, interference, and jamming. Once the variation model is obtained, sensitivity analysis can be performed by linearization, statistical analysis, or Monte Carlo simulation to quantify the variation in each mode [Chase and Parkinson, 91].

We have developed a general kinematic tolerance analysis algorithm that addresses these issues. The algorithm constructs a variation model for the system, derives worst-case bounds on the variation, and helps designers find failure modes. The variation model is a generalization of our configuration space representation of the part contacts in the nominal system. The algorithm handles general planar systems of curved parts with contact changes, including open and closed kinematic chains. It analyzes systems with 50 to 100 parameters in under a minute, which permits interactive tolerancing of detailed functional models.

In this paper, we summarize our research on kinematic tolerance analysis [Joskowicz and Sacks, 96], [Sacks and Joskowicz, 98 and 98b]. We describe the configuration space representation, explain its role in tolerance analysis, and outline the analysis algorithm for general planar systems. We use the intermittent gear mechanism as a case study to illustrate the ideas. We conclude with our recent experience on industrial examples. In doing so, we present for the first time the big picture behind our methodology and identify future research issues.

2. PREVIOUS WORK

Previous work on kinematic tolerance analysis of mechanical systems falls into three increasingly general categories: static (small displacement) analysis, kinematic (large displacement) analysis of fixed contact systems, and kinematic analysis of systems with contact changes. Static analysis of fixed contacts, also referred to as tolerance chain or stack-up analysis, is the most common. It consists of identifying a critical dimensional parameter (a gap, clearance, or play), building a tolerance chain based on part configurations and contacts, and determining the parameter variability range using vectors, torsors, or matrix transforms [Clément et al., 97], [Whitney et al., 94]. Recent research explores static analysis with contact changes [Inui and Miura, 95], [Ballot and Bourdet, 97]. Because they require the user to identify systems configurations, configurations where unexpected failures occur can easily be missed. Kinematic analysis of fixed contact mechanical systems, such as linkages, has been thoroughly studied in mechanical engineering [Erdman, 93]. It consists of defining kinematic relations between parts and studying their kinematic variation [Chase et al., 97]. Most commercial CAT systems include this capability for planar and spatial mechanism [Solomons et al., 97]. These methods are impractical for systems with many contact changes, such as a chain drive, and can miss failure modes due to unforeseen contact changes. Our method overcomes these limitations by automating variational contact model derivation and analysis for general planar systems.

3. CONFIGURATION SPACE

We model nominal kinematic function within the configuration space representation of rigid body interaction [Joskowicz and Sacks, 91], [Sacks and Joskowicz, 95]. Configuration space is a general representation for systems of rigid parts that is widely used in robot motion planning [Latombe, 91]. We model the interactions of pairs of planar parts with three-dimensional spaces whose points specify the relative configuration (position and orientation) of one part with respect to the other. We perform contact analysis by computing a configuration space for each pair of parts.

Configuration space partitions into three disjoint sets that characterize part interaction: blocked space where the parts overlap, free space where they do not touch, and contact space where they touch without overlap. Blocked space represents unrealizable configurations, free space represents independent part motions, and contact space represents motion constraints due to part contacts. The spaces have useful topological properties. Free and blocked space are open sets whose common boundary is contact space. Contact space is a closed set comprised of algebraic patches that represent contacts between pairs of part features. Patch boundary curves represent simultaneous contacts between two pairs of part features.

We illustrate these concepts on the gear/follower pair of the intermittent gear mechanism (Figure 2). In the first configuration space, the gear translates at the fixed orientation $\psi = 0$ degrees. The uv coordinates specify the position of the gear center relative to the follower center (also the cam center). The gear orientation and the cam configuration are those shown in Figure 1a. Free space is the three white regions, blocked space is the grey region, and contact space is the black curves that separate them. If the pair starts in one region, it stays there forever. The

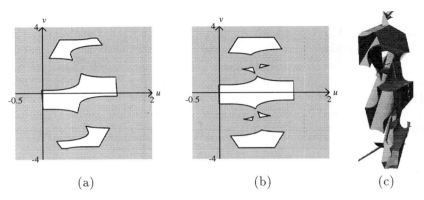

(a)   (b)   (c)

Figure 2: *Gear/follower configuration space: (a) slice at $\psi = 0$ degrees, (b) slice at $\psi = 1$ degrees, (c) full space.*

complex shape of the contact space encodes the way that the gear can slide along the follower pawls. In the second configuration space, the gear translates at the fixed orientation $\psi = 1$ degree. The free space consists of seven regions with different shapes than before, since additional gear teeth can touch the follower pawls at this orientation. The third configuration space represents translation and rotation of the gear relative to the follower. It is three dimensional with coordinates (u, v, ψ). This is the configuration space that we use for tolerance analysis. The first two, which are planar slices, help us understand the full space.

The configuration space of a pair is a complete representation of the part contacts. Contacts between pairs of features correspond to contact patches (curve segments in two dimensions and surface patches in three). The patch geometry encodes the motion constraint and the patch boundary encodes the contact change conditions. Part motions correspond to paths in configuration space. A path is legal if it lies in free and contact space, but illegal if it intersects blocked space. Contacts occur at configurations where the path crosses from free to contact space, break where it crosses from contact to free space, and change where it crosses between neighboring contact patches.

The configuration space representation generalizes from pairs of parts to systems with more than two parts. A system of n planar parts has a $3n$-dimensional configuration space whose points specify the n part configurations. A system configuration is free when no parts touch, is blocked when two parts overlap, and is in contact when two parts touch and no parts overlap.

We have developed a configuration space computation program for planar pairs whose part boundaries consist of line segments and circular arcs [Sacks and Joskowicz, 95], [Sacks, 98]. These features suffice for most engineering applications with the exception of involute gears and precision cams, which are best handled by specialized methods [Gonzales-Palacios and Angeles, 93], [Litvin, 94]. The program computes an exact representation of contact space: a graph whose nodes represent contact patches and whose arcs represent patch adjacencies. Each node contains a contact function that evaluates to zero on the patch, is positive in nearby free configurations, and is negative in nearby blocked configurations. Each graph arc contains a parametric representation of the boundary curve between its incident patches. After constructing configuration spaces for the pairs in a mechanical system, we analyze the system mechanical function in the system configuration space. We construct the relevant portion of the system configuration space from the pairwise spaces.

4. KINEMATIC VARIATION

We model kinematic variation by generalizing the configuration space representation to toleranced parts. The contact patches of a pair are parameterized by the touching features, which depend on the tolerance parameters. As the parameters vary around their nominal values, the contact patches vary in a band around the nominal contact space, which we call the contact zone. The contact zone defines the kinematic variation in each contact configuration: every pair that satisfies the part tolerances generates a contact space that lies in the contact zone. Kinematic variations do not occur in free configurations because the parts do not interact.

We illustrate contact zones on the gear/follower pair. The gear is parameterized by the inner radius $r_i = 15.7mm$, the outer radius $r_o = 16.4mm$, and the ratio $\mu = 0.3$ between the angular width of a tooth and the angular spacing between teeth. The follower is parameterized by the cam radius $r_c = 5mm$ (which determines the size of the inner square), the pawl thickness $w = 0.5mm$, and the length $l = 9.8mm$ of the arms. The worst-case parameter variations are $\pm 0.01mm$. The full configuration space has a three-dimensional contact zone that is hard to visualize. Instead, we examine configuration space slices, which have planar contact zones (Figure 3). The contact zone is bounded by the curves that surround the nominal contact curves. Its width varies with the sensitivity of the nominal contact configuration to the tolerance parameters.

Each contact patch generates a region in the contact zone that represents the kinematic variation in the corresponding feature contact. The region boundaries encode the worst-case kinematic variation over the allowable parameter variations. They are smooth functions of the tolerance parameters and of the part configurations in each region. They are typically discontinuous at patch boundaries because the adjacent patches depend on different parameters, as can be seen in Figure 3 where the boundary curves of adjacent zones cross. The variation at boundary configurations is the maximum over the neighboring patch variations. The contact zone regions represent the quantitative kinematic variation, while the relations among regions represent qualitative variations such as jamming, under-cutting, and interference.

The contact zone is obtained from the parametric part models and the nominal

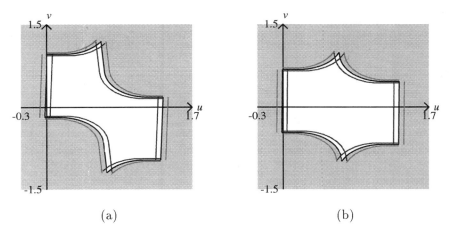

Figure 3: *Contact zones for gear/follower configuration space slices (a)* $\psi = 0$ *degrees and (b)* $\psi = 1$ *degrees.*

contact patches. Each patch satisfies a contact equation $g(u, v, \psi) = 0$, which we rewrite as $g(u, v, \psi, \mathbf{p}) = 0$ to make explicit the dependence on the vector $\mathbf{p}$ of tolerance parameters. A parameter perturbation of $\delta\mathbf{p}$ leads to a perturbed patch that satisfies

$$g(u + \delta u, v + \delta v, \psi + \delta\psi, \mathbf{p} + \delta\mathbf{p}) = 0. \tag{1}$$

Following the standard tolerancing approximation which considers only the first-order effects on kinematic variation, we obtain the linear expression

$$\frac{\partial g}{\partial u}\delta u + \frac{\partial g}{\partial v}\delta v + \frac{\partial g}{\partial \psi}\delta\psi = -\sum_i \frac{\partial g}{\partial p_i}\delta p_i \tag{2}$$

where the partial derivatives of g are evaluated at (u, v, ψ) and δp_i is the ith element of $\delta\mathbf{p}$. This equation approximates the portion of the perturbed patch near the configuration (u, v, ψ) with a plane.

The left side of Equation 2 specifies the normal direction of the perturbed contact patch, which is independent of the parameter variations. The right side specifies the distance between the perturbed and the nominal patch, which is the kinematic variation, for any allowable parameter variation $l_i \leq \delta p_i \leq u_i$ with $l_i \leq 0$ and $u_i \geq 0$. The worst-case kinematic variation (largest distance) occurs when the right side is maximal or minimal. It is maximal when every term is maximal, i.e., when $\delta p_i = l_i$ when $g_{pi} > 0$ and $\delta p_i = u_i$ otherwise. Switching u_i and l_i yields the minimal value.

The derivatives $\partial g / \partial p_i$ measure the sensitivity of the contact configuration to the tolerance parameters. Designers can make small changes in the kinematic function by changing the parameters in accordance with the sensitivities. The following table shows the sensitivities of the gear/follower pair in the $\psi = 0$ slice:

parameter	maximum	minimum	average
r_c	0.87	0.00	0.58
w	3.00	1.00	2.33
l	1.00	0.00	0.67
r_i	1.00	0.00	0.39
r_o	1.00	0.00	0.31
μ	1.54	0.05	0.68

The values are from 84 sample contact configurations. Linear interpolation between these configurations is accurate to two significant digits. The average pawl thickness sensitivity is four times that of the other parameters, which makes it a good candidate for small design changes.

The contact zone model generalizes from pairs to systems. The contact space is a semi-algebraic set in configuration space: a collection of points, curves, surfaces, and higher dimensional components. As the tolerance parameters vary around their nominal values, the components vary in a band around the nominal contact space, which is a higher-dimensional analog of the three-dimensional contact zone of a pair. We avoid general algebraic methods, performing kinematic tolerance analysis on individual operating modes.

System operating modes are defined by driving forces and initial conditions. We can perform the analysis for any number of modes, but cannot analyze the sensitivity to the continuously infinite space of all possible modes. Given the forces and initial conditions, the laws of physics determine the time evolution of the system state (part configurations and velocities). We can compute a nominal sequence of states by simulation [Sacks and Joskowicz, 93], [Sacks and Joskowicz, 98] or by physical measurement. This yields a nominal path in the system configuration space. We perform kinematic tolerance analysis by computing the kinematic variation at sampled configurations along the nominal path. For example, we derive the combined effects of cam, follower, and gear tolerances on the indexing accuracy of the mechanism and test for jamming due to simultaneous variations in all three parts.

We compute the system variation at each configuration along the nominal path by determining which pairs of parts are in contact, obtaining the corresponding parameterized contact equations from the pairwise configuration spaces, and solving a linear optimization problem. The variables are the part coordinate variations $(\delta x_i, \delta y_i, \delta \theta_i)$ and the tolerance parameters p_i. The constraints come from the tolerances and from the contact patches. The tolerances provide two constraints per parameter $l_i \leq p_i \leq u_i$. We collect the contact equations into a vector equation

$$\mathbf{g}(\mathbf{x}, \mathbf{p}) = 0 \tag{3}$$

with $\mathbf{x}$ part coordinates and $\mathbf{p}$ tolerance parameters. We linearize the contact equations around the current configuration and the nominal parameter values to get

$$D_{\mathbf{x}}\mathbf{g}\delta\mathbf{x} + D_{\mathbf{p}}\mathbf{g}\delta\mathbf{p} = 0 \tag{4}$$

with $D_{\mathbf{x}}\mathbf{g}$ the Jacobian matrix with respect to $\mathbf{x}$ and $D_{\mathbf{p}}\mathbf{f}$ the Jacobian matrix with respect to $\mathbf{p}$. This equation is the system analog of Equation 2. It approximates the portion of the perturbed configuration space near $\mathbf{x}$ with a hyper-plane. The objective functions are the maxima and minima of the coordinate variations. We solve one linear program for each function to obtain the system variation.

5. CONCLUSION

We have implemented our kinematic tolerance analysis algorithm and applied it to detailed parametric models of a variety mechanisms including a geneva cam pair, a 35mm camera shutter, a movie camera film advance, and a micro-mechanical gear discriminator. In all cases, the program produced interesting qualitative and quantitative results beyond the scope of current systems and not derivable by hand. We are currently analyzing more industrial examples, such as a transmission gear shift mechanism, a double chain transmission, and other micro-mechanical designs. We plan to extend the scope of our algorithm to spatial tolerance analysis of planar systems and of fixed-axes spatial systems. We also plan to study tolerance synthesis within the configuration space representation.

REFERENCES

[**Ballot and Bourdet, 97**] Ballot, E. and Bourdet, P. A computation method for the consequences of geometric errors in mechanisms. in: *Proc. 5th CIRP Int. Seminar on Computer-Aided Tolerancing*, Toronto, 1997.

[**Chase et al., 97**] Chase, K., Magleby, S., and Glancy, C. A comprehensive system for computer-aided tolerance analysis of 2D and 3D mechanical assemblies. in: *Proc. 5th CIRP Int. Seminar on Computer-Aided Tolerancing*, 1997.

[**Chase and Parkinson, 91**] Chase, K. W. and Parkinson, A. R. A survey of research in the application of tolerance analysis to the design of mechanical assemblies. *Research in Engineering Design* **3**, 1991.

[**Clemént et al., 97**] Clemént, A., Rivière, A., Serré, P., et al. The ttrs: 13 constraints for dimensioning and tolerancing. in: *Proc. 5th CIRP Int. Seminar on Computer-Aided Tolerancing*, Toronto, 1997.

[**Erdman, 93**] Erdman, Arthur, G. *Modern Kinematics: developments in the last forty years.* (John Wiley and Sons, 1993).

[**Gonzales-Palacios and Angeles, 93**] Gonzales-Palacios, M. and Angeles, J. *Cam Synthesis.* Kluwer Academic Publishers, 1993.

[**Inui and Miura, 95**] Inui, M. and Miura, M. Configuration space based analysis of position uncertainties of parts in an assembly. in: *Proc. of the 4th CIRP*

Int. Seminar on Computer-Aided Tolerancing, 1995.

[**Joskowicz and Sacks, 91**] Joskowicz, L. and Sacks, E. Computational kinematics. *Artificial Intelligence* **51**, 1991.

[**Joskowicz and Sacks, 96**] Joskowicz, L., Sacks, E., and Srinivasan, V. Kinematic tolerance analysis. *Computer-Aided Design* **29**, 1997.

[**Latombe, 91**] Latombe, J.-C. *Robot Motion Planning*. Kluwer Academic Publishers, 1991.

[**Latombe and Wilson, 95**] Latombe, J.-C. and Wilson, R. Assembly sequencing with toleranced parts. in: *Third ACM Symposium on Solid Modeling and Applications*, 1995.

[**Litvin, 94**] Litvin, F. L. *Gear Geometry and Applied Theory*. Prentice Hall, 1994.

[**Requicha, 93**] Requicha, A. A. G. Mathematical definition of tolerance specifications. *Manufacturing Review* **6**, 1993.

[**Sacks, 98**] Sacks, E. Practical sliced configuration spaces for curved planar pairs. *International Journal of Robotics Research* **17**, 1998.

[**Sacks and Joskowicz, 93**] Sacks, E. and Joskowicz, L. Automated modeling and kinematic simulation of mechanisms. *Computer-Aided Design* **25**, 1993.

[**Sacks and Joskowicz, 95**] Sacks, E. and Joskowicz, L. Computational kinematic analysis of higher pairs with multiple contacts. *Journal of Mechanical Design* **117**, 1995.

[**Sacks and Joskowicz, 97**] Sacks, E. and Joskowicz, L. Parametric kinematic tolerance analysis of planar mechanisms. *Computer-Aided Design* **29**. 1997.

[**Sacks and Joskowicz, 98**] Sacks, E. and Joskowicz, L. Dynamical simulation of planar systems with changing contacts using configuration spaces. *Journal of Mechanical Design* **120**, 1998.

[**Sacks and Joskowicz, 98b**] Sacks, E. and Joskowicz, L. Parametric kinematic tolerance analysis of general planar systems. *Computer-Aided Design* **30**, 1998.

[**Solomons et al., 97**] Solomons, O., van Houten, F., and Kals, H. Current status of cat systems. in: *Proc. of the 5th CIRP Int. Seminar on Computer-Aided Tolerancing*, Toronto, 1997.

[**Voelcker, 93**] Voelcker, H. A current perspective on tolerancing and metrology. *Manufacturing Review* **6**, 1993.

[**Whitney et al., 94**] Whitney, D., Gilbert, O., and Jastrzebski, M. Representation of geometric variations using matrix transforms for statistical tolerance analysis. *Research in Engineering Design* **6**, 1994.

Dynamic Tolerance Analysis, part I:
A theoretical framework using bondgraphs

Salomons O.W., Zwaag J.A. van der, Zijlstra J., Houten F.J.A.M. van,
University of Twente, Department of Mechanical Engineering
Laboratory of Production and Design Engineering
Enschede, 7500 AE, The Netherlands, tel. X–31–53–4892532/2551
fax: X–31–53–4893631, e_mail: o.w.salomons@wb.utwente.nl

Abstract: A theoretical framework is proposed by which the effect of tolerances in combination with physical effects such as wear can be analysed on the dynamic behavior of mechanisms. The framework uses bondgraphs in order to simulate the dynamic behavior under the influence of tolerances and other physical effects. The paper describes how a geometric model, including it's tolerances, can be transformed into a corresponding bondgraph model that on it's part consists of submodels. Based on this bondgraph model, simulations could be performed which could provide insight in the dynamic behavior of the mechanism. The paper details on how the different geometric tolerances such as form, orientation, position as well as size and clearances can be accounted for in a bondgraph model.
Keywords: dynamic tolerance analysis, bondgraphs, tolerance value specification.

1. INTRODUCTION

When striving for tolerance management [Söderberg 1998a,b], care should be taken that the nominal design is robust to geometric variations. One can achieve this by keeping the kinematic degrees of freedom in view and by the application of good construction principles. Once a robust nominal geometric assembly has been generated, tolerance specification can start. Tolerance type specification could be supported by means of the TTRS approach [Clément et al. 1994]. This method ensures that functional surfaces in the assembly will be toleranced properly. The authors have indeed applied the TTRS approach to tolerance type specification in their FROOM prototype system e.g. [Salomons 1995], [Salomons et al. 1996]. Based on this research the next step is computer support in tolerance value specification. Tolerance value specification in practice is all too often based on experience or a guess. What can be done to this situation is that the designer provides upper and lower limits for the tolerance values or just a guess. Based on this information several types of tolerance analyses can be started. The results of the tolerance analysis can be used to improve tolerance value specification (see figure 1). Future CAT tools might automate large portions of the iterative tolerance value specification approach of figure 1.

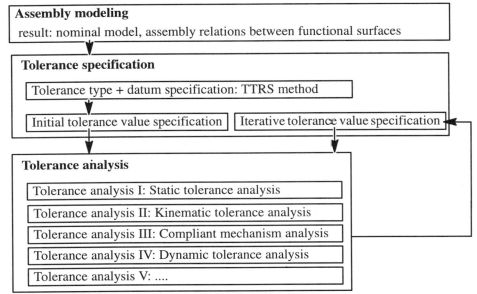

Figure 1; *Scenario of iterative tolerance value specification supported by various types of tolerance analysis: a possible automatic mode for future CAT tools?*

This paper will mainly focus on some tolerance analysis functionality that still is miss-ing in most CAT systems today and which can be used to support tolerance value specifica-tion. Therefore, a brief survey of tolerance analysis approaches is provided.

Tolerance analysis is generally considered as a means to verify the proper functioning of a mechanism after tolerances have been specified. In practice, whether using commer-cially available Computer Aided Tolerancing (CAT) systems or not, tolerance analysis most often comes down to checking two aspects:

- checking the feasibility of assembly
- checking the quality of assembly

These checks are not at all concerned with the actual functioning of the mechanism. They specify necessary but not sufficient conditions for the proper functioning of the as-sembly. Therefore, since designers have no sufficient insight into the functioning conse-quences of the tolerance values they specify, often too tight tolerances are the result. If the effect of tolerances on the functioning of an assembly is known in advance then this could be a great help to tolerance specification. The functioning of assemblies may depend on a wide range of physical effects such as kinematics, hydraulics, dynamics, aerodynamics etc.. Analysing the effect of the tolerances with the applicable physical effects could then help in determining the tolerance values (see figure 1). Assuming that kinematics and dy-namics are most important for the mechanical domain, this paper focuses on these effects. We will refer to this as dynamic tolerance analysis where the final goal is to assist the designer in his functional tolerance (value) specification by using the results from a dynamic simulation of the assembly under influence of its tolerances and other physical aspects.

This series of two papers is a follow–up on [Salomons et al. 1998]. Part I mainly summarizes the main items from the first paper, complemented with more information on the place of dynamic tolerance analysis in it's broader context, and some of the particular submodels used. Part II mainly discusses new simulation results as well as model refinements based on these results.

The organization of the remainder of this paper is as follows. Section 2 briefly summarizes related literature. In section 3 a motivation is given for using bondgraphs in dynamic tolerance analysis. Section 4 elaborates on the bondgraph submodels which include tolerances and which should be used in the actual simulations. The actual simulations, the problems encountered and further model refinements are discussed in part II of this paper. In section 5 some preliminary conclusions and recommendations are presented regarding the theoretical framework that is provided in this paper.

2. RELATED LITERATURE

A distinction is made between static tolerance analysis, kinematic tolerance analysis and dynamic tolerance analysis.

The static tolerance analysis approach is the oldest approach to tolerance analysis, probably first proposed by Bjørke [Bjørke 1978]. However, this approach not only assumed a static assembly but was limited to 2D and did not support for geometric tolerances. Only recently these disadvantages have been overcome. Representative examples are the papers by Chase et al. [Chase et al. 1996] and Rivière et al. [Rivière et al. 1994]. Similar to these references, most commercially available tolerance analysis systems assume a static assembly when performing their tolerance analysis task [Salomons et al. 1997]. Static tolerance analysis assumes rigid bodies which recently has been overcome in compliant mechanism tolerance analysis, e.g. [Sellem 1998], [Liu & Hu 1995].

Tolerance analysis which not only includes the geometric variations but also the kinematics in combination with tolerances is the next step in tolerance analysis. Examples are [Funk and Stolzenberg 1995] and [Joskowicz et al. 1997].

Some of the static tolerance approaches, e.g. [Chase et al. 1996], did take into account variations due to kinematics together with dimensional and geometric variations. However, they were not concerned with checking whether the global kinematic function of the mechanism is still acceptable under increased part variation. Moreover, they were focused at calculating clearances between parts (quality of assembly) which may give some indication of overall kinematic function but should not be taken as a reliable measure.

In [Funk and Stolzenberg 1995] and [Joskowicz et al. 1997] the overall kinematic function under increased part variation is indeed considered. Funk and Stolzenberg only focused on planar mechanisms and size tolerances and therefore did not take into account clearances and geometric tolerances. Thus, 3D effects and 3D mechanisms were not taken into account. The new aspect in the work by Funk and Stolzenberg is the use of FEM techniques as part of the kinematic tolerance analysis. The work by Joskowicz et al. can be seen as an extension to this. Many different types of mechanisms can be covered including mechanisms with higher pairs, varying topology and changing contacts. However, in both the work by Joskowicz et al. and that by Funk and Stolzenberg, the effect of tolerances on the dynamic behavior has not been taken into account.

Dynamic tolerance analysis approaches have been reported either for a number of specific mechanisms (hence not generic) or as a more generic approach but with certain limitations. We will first treat the more specific approaches and finally the more generic ones.

In [Wisniewski & Gomer 1997], the commercial VSA–3D software has been reported to be used for the dynamic tolerance analysis of an engine balance problem. The method is based on combining a conventional statistical tolerance analysis (static) with optimizing explicit equations for dynamic force balance. Therefore, we do not consider this as a generic approach. Some other work that has been reported so far is also not generic and often omits certain tolerance types; either clearances or geometric tolerances, e.g. [Townsend and Mansour 1975], [Miedema and Mansour 1976], [Funabashi et al. 1978, 1980], [Sakamoto et al. 1990], [Cleghorn et al. 1993].

In his thesis, Srinivasan [Srinivasan 1994], states that especially form tolerances will have an effect on the behavior of mechanisms. It is stated that zone based methods do not describe the frequency of surface profiles whereas these do have a significant effect on functioning. Fractal based tolerancing is proposed in order to describe the form tolerances of manufactured profiles. The scope is however limited to 1D form tolerances.

Suzuki et al. propose the notion of physically based modelling in order to get an idea of the relation between behavior of an assembly (function) and variations in part's shapes [Suzuki et al. 1996]. The method employs simulations, in particular those for calculating contact states of an assembly of simple parts in motion. The simulations adopt simplified physical models since accurate, quantitative models were seen to be infeasible.

3. BONDGRAPHS USED IN DYNAMIC TOLERANCE ANALYSIS

This section deals on how dynamic tolerance analysis could be realized. Section 3.1 details on transforming the geometric model into a generic dynamic analysis model. In section 3.2 bondgraphs and finite element formulations are discussed as most interesting alternatives for realizing dynamic tolerance analysis. It is argued that for the time being bondgraphs seem to have most appeal. In section 3.3 the transfer of geometric model data to bondgraph model data is discussed.

3.1 From geometric product model to dynamic analysis model

There are roughly two ways in which to transfer geometric information (including tolerances) to a dynamic simulation environment:
– inferring the (dynamics) equations directly from the geometric model
– inferring the (dynamics) equations indirectly from the geometric model

The first option is hard to implement since the method has to be generic. We would like to transfer any geometric product model into a dynamic model. Using just the B–rep, CSG or any other plain geometry representation format won't give much help in making a generic, automated information transfer. In fact, in the case of some of the reported previous work in dynamic tolerance analysis a direct transfer of specific geometry to specific dynamics equations was made often requiring a human to make the transfer work. Therefore, the second option seemed more attractive. Here both the geometric and dynamic models have higher levels of semantics than just a face in a B–rep or a compliance in a bondgraph. The information transfer is then made to work at this higher aggregation level.

3.2. Bondgraphs vs. a finite element formulation

In the information transfer on a higher semantic level as described above, bondgraphs seemed particularly attractive, since they provide a means to model physical systems and offer means to be simulated easily in a dynamic simulation environment. Moreover, bondgraphs are domain independent since they are based on the generic principle of conservation of energy [Karnopp et al. 1990]. Related to this, it may be interesting to note that Bjørke, one of the founders of the research in tolerance analysis, recently also has applied a bondgraph like approach to his manufacturing systems theory [Bjørke 1995].

A possible alternative to bondgraphs, when the focus is restricted to the mechanical domain, is a finite element formulation. Finite element formulations have now been applied in tolerance analysis of compliant mechanisms [Sellem & Rivière 1998], [Liu & Hu 1995]. Also in the domain of mechanism dynamics, on which this series of two papers is focusing, a finite element formulation could be applied. However, it would be considerably different from the compliant mechanism application. Because of the multi–domain nature of a bondgraph formulation, initial preference was given to a bondgraph formulation. In part II we will get back to the bondgraph formulation versus the finite element formulation.

3.3 Information transfer from geometric model to bondgraph model

The two main types of information that need to be transferred are: (1) model topology information, so that bondgraph model topology can be inferred from the geometric model and (2) relevant parameters for the bondgraph elements, so that dynamic simulations can be performed. We will first try to resolve the model topology issue followed by the parameters issue.

In a geometric model of an assembly, the topology of the geometric model is determined by its parts (solids) and the assembly constraints defined between the parts. In a physical bondgraph modeling environment however, the topology of the corresponding bondgraph model may be built according to two alternatives. The first alternative is to build the bondgraph topology directly from the constituting bondgraph elements such as compliances, which are connected by means of bonds and 0– or 1–junctions. The second alter-

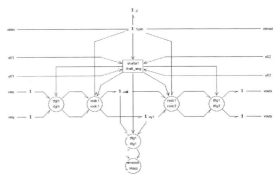

Figure 2; *Iconic diagram (IPM) of a single* bar

Figure 3; *Bondgraph sub–model belonging t the bar IPM of figure 2*

native is to build the bondgraph topology indirectly by using sub–models, possibly in combination with iconic diagrams. Before further resolving the issue of how to infer bondgraph topology from a geometric model, we will first detail some more on bondgraph sub–models and iconic diagrams since these play an important role in solving the topology issue.

Iconic diagrams depict Idealized Physical Models (IPMs) to which bondgraph models can be attached [Vries 1994]. Iconic diagrams are connected by ports. Ports are used to transfer the energy related 'through' and 'cross' variables. Within an iconic diagram three types of connections may occur: power ports, signal ports and pose ports. The pose ports have been introduced as a new element in order to account for positions and orientations [Zijlstra 1997]. In the mechanical domain the pose ports are especially important. An example of an iconic diagram of a single bar is provided in figure 2. The bar consists of two springs connected by means of a mass. The black dots on the left and right of the two springs indicate the ports of the bar: in this case there are three incoming and three outgoing power ports (velocities: vinx, voutx, viny, vouty, omin, omout) and two incoming and outgoing pose ports (positions: x01, x02, y01, y02). The corresponding bondgraph sub–model is shown in figure 3. Note that the ports can be identified from this figure as well. Note that the pose ports are in fact implemented as signal ports.

Knowing that bondgraphs can be built using either iconic diagrams (e.g. figure 2) with bondgraph sub–models underneath (e.g. figure 3) or using bondgraph elements directly, we have roughly two alternatives for inferring the bondgraph model topology from a geometric model. In the first alternative, the bondgraph model is built indirectly from the geometric model, using bondgraph sub–models. In the second alternative, the bondgraph model is created directly from the geometric model, without the use of sub–models. A mapping from geometric elements directly to bondgraph elements is not trivial and would have to be implemented separately for each case. A mapping from geometric models to (connected) sub–models (represented by iconic diagrams) is considered much easier, and offers the potential of genericity since sub–models can be re–used in other cases. Therefore, a mapping from a geometric model to connected sub–models has been preferred over a direct mapping of geometric model to bondgraph model.

In the domain of mechanisms, two main types of sub–models (iconic diagrams) will occur. The first sub–model type is directly related to a single discrete part in the geometric assembly model. For example, an iconic diagram of a bar in a physical model of a four bar mechanism is directly related to the geometric model of a bar part in a geometric assembly model of this mechanism. All physical parameters of the physical bar sub–model, such as compliance, moments of inertia, can be inferred directly from the geometric model. The second type of sub–model has no direct geometric counterpart in the geometric model and serves to connect sub–models of the first type. For example, an iconic diagram for a kinematic joint of a four bar mechanism will connect two different bars and cannot directly be inferred from the geometry of one single part in the geometric model. The kinematic joint type and its parameters can only be inferred from using geometric information from the two parts it connects, including the assembly constraints between them.

Thus, the geometric model has to be transferred to a model consisting of corresponding iconic diagrams with power, signal and pose ports. This can be achieved by a 1:1 conversion of each geometric component model to its corresponding iconic model and by pruning the assembly constraints in the geometric assembly to infer the proper type of kinematic joint and transfer these into the corresponding iconic diagrams that will connect the other iconic diagrams.

The relevant parameters for the bondgraph elements as part of the sub–models corresponding to the iconic diagrams can be computed from the geometry and material properties and include: mass, center of mass, compliance, moments of inertia, orientation/position, functional dimensions and tolerances.

All the above part (link) related parameters, except for the tolerances, can be assigned to the corresponding bondgraph submodels once the mapping between geometric model and iconic diagram has been established [Zijlstra 1997]. In the following section we will detail the mapping of the tolerances to the bondgraph sub–models that correspond with iconic diagrams. As we will see, often tolerances map from geometric objects (links) to bondgraph submodels of joints.

4. MODELING TOLERANCES WITH BONDGRAPHS

This section details on the generic bondgraph sub–models of the iconic diagrams that need to be present in the library of a physical modeling system when the influence of tolerances is to be simulated.

In a geometric model of individual parts (links) we have to deal with size tolerances and geometric tolerances (position, orientation and form tolerances). In an assembly these individual part tolerances give rise to clearances. Clearances involve surfaces of two different geometric objects. Clearances are therefore related to joints. The same holds for form tolerances, when they should have an effect on the dynamic behavior. For dynamic tolerance analysis clearances are of special interest. Clearances not only provide for micro–degrees of freedom in addition to the macro–degrees of freedom resulting from the kinematics, but also may give rise to collisions with an impact on the overall dynamic behavior.

4.1 Joints with Form Tolerances

Figure 4 shows a conceptual model of a surface with a form tolerance and roughness. The form tolerance specifies the amplitude while the roughness specifies the frequency of a sig-

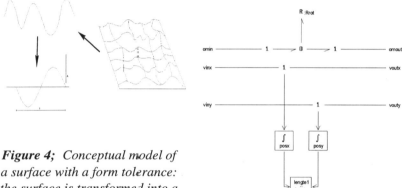

*Figure 4; Conceptual model of
a surface with a form tolerance:
the surface is transformed into a
1D signal*

*Figure 5; Bondgraph sub–model of an almost
ideal 2D revolute joint (only friction)*

nal that corresponds with the surface. This 2D signal can be translated into a 1D signal. When two surfaces interact when they form a kinematic joint, the interaction can be modeled by mixing the 1D signals from corresponding surfaces which participate in the kinematic joint. This is similar to [Srinivasan 1994]. Taking a simple 2D revolute joint as an example, we can model the effect of non–ideal, interacting functional surfaces with only form deviations as a disturbing noise on the resulting speeds in the x– and y–directions of outgoing pose ports of the joint.

An ideal bondgraph sub–model of a 2D revolute joint without form tolerances has been depicted in figure 5. This 2D revolute joint model contains a resistance (R) element to take friction into account. However, this does not yet represent in–depth knowledge of tribological processes that take place. The model may be extended in the future to include such refinements. The integration symbols are used to infer possible variations in bar length and were used in order to verify the accuracy of the model. This issue will be addressed in part II. The 2D revolute joint has been extended to a 2D revolute joint with form tolerance

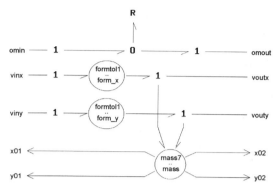

Figure 6; *Bondgraph sub–model of a 2D revolute joint with form tolerance, friction as well as a (small) mass*

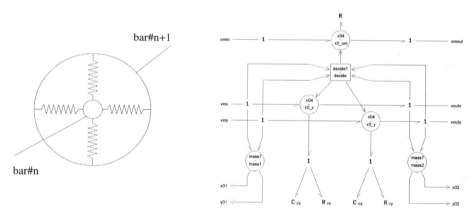

Figure 7; *Conceptual model of a 2D revolute joint with clearance modeled with springs*

Figure 8; *Bondgraph sub–model of a joint with clearance modeled by means of springs and dampers which are controlled by switching 0–junctions*

in figure 6. Note that in contrast to figure 5, this 2D revolute joint contains a (small) mass. Although conceptually a kinematic joint should not have a mass, it has been part of some sub–models in order to overcome numeric problems that occurred in the case of non–rigid bar models that were used in conjunction with the specific 2D revolute joint model with form tolerances. In part II the issue of parasitic elements is revisited.

4.2 Joints with clearances
For joints with clearances there are two ways on how to incorporate these in a bondgraph model of a joint. Let us again take a 2D revolute joint as an example.

The first possibility is to model a clearance conceptually using springs. This conceptual idea has been depicted in figure 7. In the actual bondgraph sub–model, that is based on such a conceptual model, apart from springs, there will also be dampers in order to overcome numeric problems. The corresponding bondgraph model has been depicted in figure 8. In this model, two switching 0–junctions can be observed which switch the spring compliance to low when the pin contacts the hole wall and to high otherwise. Note the presence of dampers parallel to the springs. The dampers are switched similar to the springs. The disadvantage of this model is that it does not reflect reality very well.

The second possibility is to approach reality closer by modeling a clearance using a discrete state model. The discrete states that one can observe in a joint with clearance are [Miedema and Mansour 1976]: impact mode, free–flight mode and following mode. In part II, the discrete state models will be addressed in more detail.

4.3 Links with size and position tolerances
Size and position tolerances will have an effect on the lengths of the links and thereby have a translatory effect on the positions of the joints related to the particular link. This means that size and position tolerances can be taken into account relatively easily by adding them to the model at one place: by changing a link's submodel parameters (like length) or by changing the initial conditions for a joint.

4.4 Links with orientation tolerances
Orientation tolerances will affect the orientation of the joints. If in dealing with 2D planar mechanisms, as we have done so far, the orientation of the joint changes within the plane, then the effect is similar to a size or position tolerance and can be taken into account similarly as described in the previous section. However, if the orientation tolerance works perpendicular to the plane of the mechanism, then the effect of an orientation tolerance in a nominal 2D mechanism is 3D and will propagate in 3D. In that case 2D submodels will not suffice anymore. However once 3D models are used, the inclusion of the effect of orientation tolerances can relatively easily be included in the submodels similar to the modifications as suggested in the previous paragraph.

5.   CONCLUSIONS AND RECOMMENDATIONS

Dynamic tolerance analysis seems a very useful extension to the traditional static tolerance analysis approaches in that it may truly help analyse the influence of tolerances on functional behavior. Thus, it may enhance tolerance value specification. A bondgraph formulation has been selected as a means to simulate a mechanism's dynamic behavior. Bondgraphs were favored over a finite element formulation due to the fact that bondgraphs have more extensive multi–domain properties. A generic method has been proposed in order to automatically transfer geometric product model data into submodel based bondgraph models.

Generic bondgraph based submodels have been constructed for the several cases of tolerances: (revolute) joints with form tolerances and clearances and links with position, size and orientation tolerances. Simulation results are necessary to verify the usefulness of the bondgraph based approach.

ACKNOWLEDGEMENTS

dr. ir. Theo de Vries and dr ir. Jan Broenink are gratefully acknowledged for their advice and support in building the bondgraph models.

REFERENCES

[Bjørke 1978] Bjørke Ø., 1978, *Computer Aided Tolerancing*, Tapir Publishers.

[Bjørke 1995] Bjørke Ø., 1995, *Manufacturing Systems Theory*, Tapir Publishers.

[Chase et al. 1996] Chase K.W., Gao J., Magleby S., Sorenson C.D., 1996, Including geometric feature variations in tolerance analysis of mechanical assemblies, accepted for publication in *IIE transactions*.

[Cleghorn et al. 1993] Cleghorn W.L., Fenton R.G., Wu J–F., 1993, Optimum tolerancing of planar mechanisms based on an error sensitivity analysis, *J. of Mechanical Design*, Vol.115, 307–313.

[Clément et al. 1994] Clément A., Rivière A., Temmerman M., 1994, *Cotation tridimensionelle des systèmes mécaniques, théorie & pratique*, PYC Edition, Yvry–Sur–Seine Cedex (ISBN 2–85330–132–X), in French (English version: *Three dimensional tolerancing of mechanical systems*, Addison Wesley Editions).

[Funabashi et al. 1978] Funabashi H., Ogawa K., Horie M., 1978, A Dynamic analysis of mechanisms with clearances, *Bulletin of the JSME*, Vol. 21, no 161, 1652–1659.

[Funabashi et al. 1980] Funabashi H., Ogawa K., Horie M., Iida H., 1980, A Dynamic analysis of the plane crank– and rocker mechanisms with clearances, *Bulletin of the JSME*, Vol. 23, no 177, 446–452.

[Funk & Stolzenberg 1995] Funk W., Stolzenberg J., 1995, Rechnerunterstützte Toleranzanalyse ungleichmäßig übersetzender Getriebe, *Konstruktion*, Vol. 47, 363–369

[Joskowicz et al. 1997] Joskowicz L., Sacks E., Srinivasan V., 1997, Kinematic tolerance analysis, *CAD*, Vol. 29, No. 2, 147–157

[Karnopp et al. 1990] Karnopp D.C., Margolis D.C., Rosenberg R.C., 1990, *System dynamics: a unified approach*, 2nd edition, John Wiley & Sons, New York.

[Liu & Hu 1995] Liu S.C., Hu S.J., On offset finite element model and its applications in predicting sheet metal assembly variation, Int. J. Mach. tools Manufact., Vol. 35, No. 11, 1995, 1545–1557.

[Miedema & Mansour 1976] Miedema B., Mansour W.M., 1976, Mechanical joints with clearance: a three mode model, *ASME Journal of engineering for Industry*, Vol. 98, 1319–1323

[Rivière et al 1994] Rivière A., Gaunet D., Dubé I., Desrochers A., 1994, "Une approche matricielle pour la représentation des zones de tolérance et des jeux", *Proceedings FORUM 1994 de la SCGM* (Société Canadienne de Génie Mécanique), McGill University, 27–29 June.

[Sakamoto et al 1990] Sakamoto Y., Funabashi H., Horie M., Ogawa K., 1990, A synthesis of planar mechanisms with pairs of optimum tolerances, *JSME Int. Journal*, Series III, Vol.33. No.2, 139–144.

[Salomons 1995] Salomons, O.W., 1995, *"Computer support in the design of mechanical products constraint specification and satisfaction in feature based design for manufacturing"*, Ph.D. Thesis, University of Twente, Enschede (NL).

[Salomons et al. 1996] Salomons, O.W., Jonge Poerink, H.J ,Haalboom, F.J, Slooten, F. van, Houten, F.J.A.M. van, Kals H.J.J. 1996a, A Computer Aided Tolerancing Tool I: Tolerance Specification, *Computers in Industry*, Vol. 31, No.2, 1996, 161–174.

[Salomons et al. 1997] Salomons O.W., Houten F.J.A.M. var, Kals H.J.J., 1997, Current Status of CAT systems, *proc. Int. CIRP Seminar on Computer Aided Tolerancing*, Toronto, (to be published by Chapman & Hall).

[Salomons et al. 1998] Salomons O.W., Zijlstra J., Zwaag J.A. van der, Houten F.J.A.M. van, Towards dynamic tolerance analysis using bondgraphs, proc. of DETC'98, Design Automation Conference, no. DETC98/DAC–5631, September, Atlanta, 1998.

[Sellem & Rivière 1998] Sellem E., Rivière A., Tolerance analysis of deformable assemblies, Proceedings of DETC98: ASME Design engineering Technical Conferences, September 13–16, 1998, Atlanta, GA.

[Söderberg 1998a] Söderberg, R., Robust Design of CAT Tools, Proc. of DETC98: 1998 ASME Design Engineering Technical Conference, September 1998, Atlanta, USA.

[Söderberg 1998b] Söderberg, R., Wandebäck, F., Wahlborg, P., The Subcontractors role in Computer Aided Tolerance Management , Proceedings of DETC98: 1998 ASME Design Engineering Technical Conference, September 1998, Atlanta, USA.

[Srinivasan 1994] Srinivasan R.S., 1994, *A theoretical framework for functional form tolerances in design form manufacturing*, PhD thesis, University of Texas at Austin, available via WWW: http://shimano.me.utexas.edu.srini.html

[Suzuki et al. 1996] Suzuki H., Kase K., Kato K., Kimura F., 1996, Physically based modelling for evaluating shape variations, in: *Computer–Aided Tolerancing*, ed. Kimura F., Chapman & Hall.

[Townsend & Mansour 1975] Townsend M.A., Mansour W.M., 1975, A pendulating model for mechanisms with clearances in the revolutes, *ASME J. for Engineering for Industry*, Vol. 97, 354–358.

[Vries 1994] Vries T.J.A. de, 1994, *Conceptual design of controlled electro–mechanical systems, a modeling perspective*, PhD thesis, University of Twente, Enschede, NL.

[Wisniewski & Gomer 1997] Wisniewski D.M., Gomer P., 1997, Tolerance analysis using VSA–3D for engine applications, *proc. 5th. Int. CIRP Seminar on Computer Aided Tolerancing*, Toronto, April, 361–372

[Zijlstra 1997] Zijlstra J., 1997, Dynamic tolerance analysis using FROOM & 20–sim, MSc thesis, PT–616, University of Twente, Enschede

Dynamic Tolerance Analysis, part II:
Issues to be resolved when applying bondgraphs

Salomons O.W., Zwaag J. van der, Zijlstra J., Houten F.J.A.M. van,
University of Twente, Department of Mechanical Engineering
Laboratory of Production and Design Engineering
Enschede, 7500 AE, The Netherlands, tel. X–31–53–4892532/2551,
fax: X–31–53–4893631, e_mail: o.w.salomons@wb.utwente.nl

Abstract Based on the theoretical framework that has been discussed in part I, the implementation of part of the theoretical framework for dynamic tolerance analysis is discussed. The implementation has been based on two tools: an academic prototype of a geometric modeling system (FROOM) and a modeling and simulation system for physical systems based on bondgraphs (20–Sim). Based on the implementation, some issues have emerged which are discussed as well. Some of these issues are related to the particular implementation environment that was used and are therefore conceptually less demanding. However, there were also some issues which are not related to the implementation environment and that are less easy to solve. The main non–implementation related problems which obstruct the proper application of dynamic tolerance analysis using bondgraphs are the following: submodel library incompleteness, incompatibility of submodels and inadequate accuracy. The paper proposes some ways to resolve these issues. It also describes initial research that is currently being performed in addressing the issues mentioned.
Keywords: dynamic tolerance analysis, bondgraphs, tolerance value specification

1. INTRODUCTION

In order to verify the theoretical framework that has been provided in part I of this series of two papers, this paper discusses the partial implementation of the theoretical framework, initial simulations as well as model refinements based on the simulation results. The tests have been performed on a four bar mechanism of which the parameters have been chosen to be the same as much as possible with those used in the papers [Miedema & Mansour 1976] and [Townsend & Mansour 1975]. Hence these papers were used as a reference. More details of the work described in this paper are to be found in [Zwaag 1998 a,b].

The outline of the paper is as follows: section 2 discusses the partial implementation that has been realized so far focusing on planar mechanisms in general and the four bar mechanism in particular. Section 3 discusses initial simulation results. In section 4 some model refinements are proposed such as a joint clearance model with discrete states, and links with distributed masses. Section 5 provides a discussion. In section 6 conclusions and recommendations are provided.

283

2. PARTIAL IMPLEMENTATION OF THE THEORETICAL FRAMEWORK

In the current, partial implementation, two systems are being used: FROOM [Salomons 1995], [Salomons et al. 1995,1996a,b] and 20–Sim [Broenink 1997], [Controlllab 1998]. FROOM focuses on geometric modeling while 20–Sim focuses on dynamic modeling, simulation and analysis.

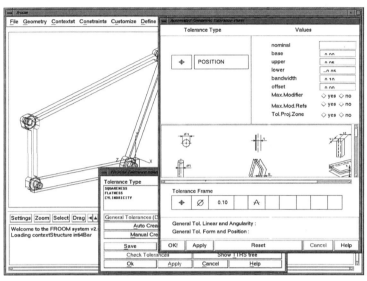

Figure 1; *Automatic tolerance type specification in the FROOM tolerancing module.*

In the FROOM system the nominal geometric model of the assembly is built first. Then the user can enter the FROOM tolerancing module. The user can specify tolerances both manually and semi–automatically. In the latter case, tolerance types are proposed by the system automatically while the user still has to provide the tolerance values. This is in accordance with the scenario as depicted in figure 1 of part I (Initial tolerance value specification). Tolerance types can be proposed automatically based on the kinematic loops that can be found in the conceptual graph, which is used to represent the assembly internally. This is also illustrated in figure 1 of part I. The FROOM system captures the tolerances in a TTRS–tree, which is a data structure according to the TTRS model, proposed in for instance [Clément & Rivière 1993] and [Clément et al. 1994].

Once the tolerances have been specified, the user may perform a static tolerance analysis in order to verify the feasibility of assembly and the quality of assembly. As a next step he may choose to export all the relevant data to 20–Sim. This option is illustrated in figure 2. The transfer of this data is based on a file transfer format based on an extension of the VHDL–A language (VHDL–A = VHSIC Hardware Description LAnguage – Analog extension). The actual implementation is described in more detail in [Zijlstra 1997].

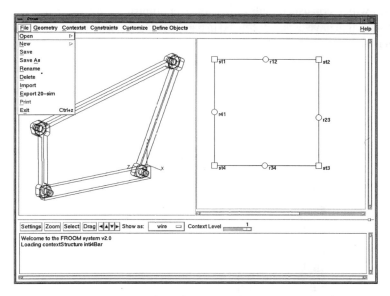

Figure 2; *Exporting the information from FROOM to 20–Sim*

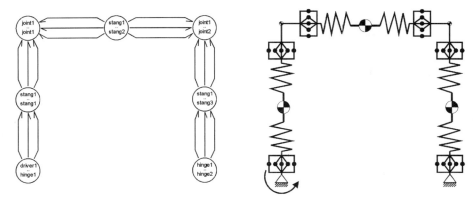

Figure 3; *Top–level "word"*
bondgraph of four bar mechanism

Figure 4; *IPM of the four bar mecha-*
nism using models for bar, hinge and joint

In 20–Sim the VHDL–A input from FROOM can be read in and be transformed into an Ideal Physical Model of iconic diagrams. However, since the 20–Sim 2.1 version, with which major parts of the research have been carried out, does not support iconic diagrams yet, this was not fully implemented and use had to be made of word–bondgraphs, sub–models and their corresponding bondgraphs. Figure 3 shows the top level word bondgraph belonging to the FROOM four bar mechanism of the previous figures. In the future apart from the word–bondgraph, an IPM will be present. Figure 4 shows the IPM for the four bar

286

mechanism which consists of sub–IPMs for bars, joints and hinges. When combining all the submodels that have been introduced in part I, we can get a model of the four bar mechanism on a much lower abstraction level. An example is provided in figure 5. When combin-

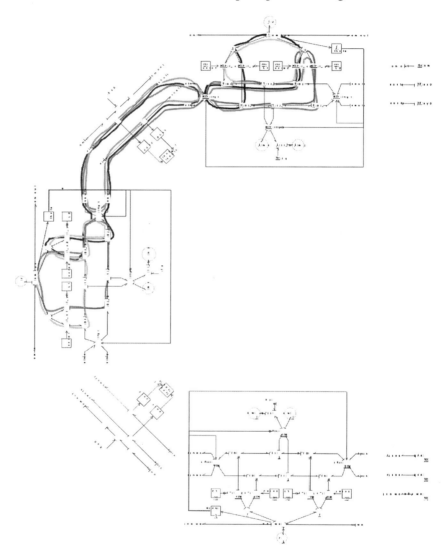

Figure 5; *Total bondgraph model consisting of submodels for frictionless joints and ideal hinge points. Storage ports with derivative causality are encircled and causal cycles are drawn.*

ing submodels, causality assignment problems may arise. If causality assignment is problematic we will refer to the submodels to be combined as being incompatible. Of course, when striving for a generic submodel library, one would like to see all useful model combinations to be compatible. Also note the 9 causal cycles in this figure and the 9 derivative storage ports, indicating possible numerical problems during simulation. This may mean that we will not be able to get simulation results at all or that one is forced to use integration methods with variable time steps (Zwaag 1998a). Some of the problems may be solved by enhancing the model, using parasitic elements, for example. We have already seen an example of this in part I, figure 5 where a small mass was added to the model of a revolute joint. However, in some cases it seemed that this may harm the requirement of genericity and thereby reusability of the sub–models. In the case of the four bar mechanism for example, in order to reduce the causality problems, it sometimes resulted in different bar sub–models to be used in the overall model of a four bar.

3. INITIAL SIMULATION RESULTS

Tests were mainly applied to the mechanism shown in figure 6. The main parameters are:

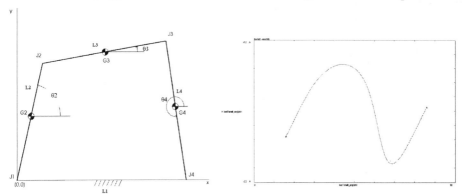

Figure 6; *Definition of the four bar mechanism under study.*

Figure 7; *Input–output relation of angle of driver vs. rocker bar of the nominal mechanism (Θ_2 vs Θ_4 of nominal mechanism).*

L1 = 180 mm, L2 = 100 mm, L3 = 197 mm, L4 = 180 mm.

An example of initial simulation results focusing on the nominal mechanism only is indicated in figure 7. In other plots drift was observed in the lengths of the bars using the integrators shown in figure 5 of part I. We will revisit this issue later. The next step would be to simulate a mechanism with tolerances, e.g. with one clearance joint using the bond-graph submodel that was introduced in part I (figure 8). However, due to the causal cycles that came into existence, at first hand this was not possible. Therefore and because the spring based clearance model of part I is conceptually not correct, a more refined joint clearance model was developed. This model has discrete states, something which is hard to accomplish in a continuous flow based formulation as bondgraphs are. In addition, particular technical problems in 20–Sim made it hard to overcome. In the following paragraph we will discuss this finite state model for clearance joints.

4. MODEL REFINEMENTS

4.1 Joint clearance models with discrete states

By using the law of momentum conservation the discrete state model can be implemented using bondgraphs, although this is much more difficult than in the spring based model of clearances that was presented in part I. This is due to the problems that discrete events may cause in a bondgraph based simulation environment [Breedveld 1996]. The 20–Sim system did not provide a standard facility for using discrete state events. Therefore, use has been made of equation models in combination with a reset event [Zwaag 1998b]. These have been inferred manually. Bondgraph models normally also result in equation models, but these can be inferred automatically from the bondgraph model. Hence the genericity of the bondgraph model is lost in favor of a specific equation model including discrete states. The model described in [Townsend & Mansour 1975] therefore has been transferred into an equation model. Because of the need to correctly cope with the reset function in 20–Sim, only the Euler integration method could be used. Simulation results regarding the input–output angles of nominal model versus clearance model are shown in figure 8. In the lower

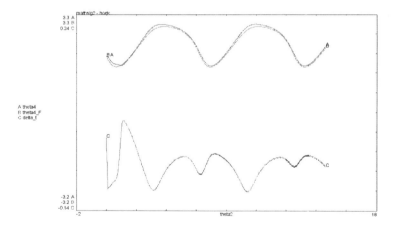

Figure 8; Input–output relation of driver and rocker bar: nominal vs. clearance mo-del (clearance = 0.005m)

plot, showing the difference between input/output angle ($\Delta\Theta$) of the nominal vs. the toleranced mechanism, already some noise can be discerned. It can be attributed to the small clearance and the Euler integration method (step size 1.10^{-5}). If the step size is chosen smaller, then the noise will be less. The first few peaks the lower part of figure 8 can be attributed to transients due to initial values of simulation parameters. Also results have been generated for the position of the pin with respect to the hole in the clearance joint (J3 in figure 6). Simulation results for various rpm rates are shown in figure 9. The results seem

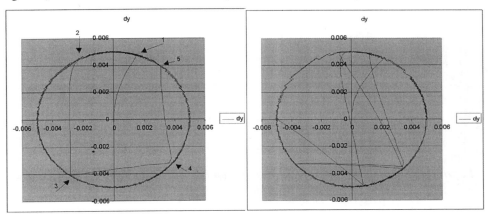

Figure 9; Crossovers at 50 rpm (left) and 40 rpm (right).

to be in conformance with [Funabashi 1978, 1980].

4.2 Links with distributed mass
The link submodel as was introduced in part I has one point mass in it's center (part I, figure 2). In order to counter the problem of the increasing bar length it was verified whether a model with distributed masses would have a positive influence on this undesired effect. Figure 10 shows one of the distributed mass models that has been tested: 3 point masses

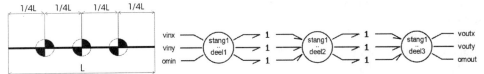

Figure 10; Three distributed masses and corresponding word bondgraph.

equally distributed over the link's length. The figure also shows the corresponding bond-graph model. In figure 11 the simulation results are shown. With respect to the link model with one point mass, the drift error seemed to be rather stable (lower plot). However the amplitude still is rather large (0.2 mm).

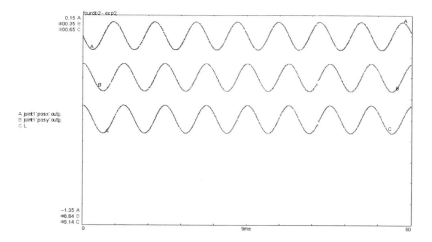

Figure 11; *Simulation result of four bar mechanism with distributed mass links: A = joint 1 position x, B= joint 1 position y, C = link length*

5. DISCUSSION

Since only a small set of 2D submodels was needed in order to get simulation results for the four bar mechanism, no complete 2D sub model library is available yet for use in 20–Sim. However, a partial 2D sub–model library has become available. Because of this, the compatibility of all possible submodels cannot yet be guaranteed. Also, 20–Sim offered some technical problems regarding the implementation of the discrete state models for joint clearances. For instance using equation models instead of bondgraph based submodels in the last simulation rounds was a result of this. Some of these problems still need to be re-solved. Therefore, this functionality has not yet been completed fully. Currently work is being performed to solve these and other problems such as the discrete state clearance mod-el in combination with bondgraph submodels instead of equation models.

The accuracy of the simulation results will strongly influence the usability, especially in our case of dynamic tolerance analysis where both small and large numbers are impor-tant. One of the most important problems that were encountered during simulation was the fact of drift in the length of the bars of the mechanism. The drift was noted by using the two integrators in the sub–models of kinematic joints as shown in figure 5 of part I. The drift already occurs for the nominal models. This is a serious problem because the inaccura-cies as a result of drift in the bar lengths blurs the perception on the effects of variations as a result of the tolerances. The drift problem has already been observed previously for mechanical systems that are modeled by means of bondgraphs [Dijk 1994]. The drift problem can be attributed to the fact that two open integrator loops cannot be expected to provide zero length difference between the actual bar length and the calculated bar length. The calculated bar length is inferred using the open integrator loops of figure 5 (part I) and Pythagoras' formula. The drift is due to numeric inaccuracies that are blown up as a result of the integration.

Because of the above mentioned problems, the issue whether bondgraphs or a finite element formulation is better, reoccurs (see also part I). The authors still think bondgraphs have a big appeal especially in conceptual design where a lot of geometric details are not present but where some physical parameters are, or can be estimated. Moreover, in conceptual design, geometric parameters don't matter much but the ability to switch between and combine domains (e.g. mechanics, hydraulics, electronics etc.) is an important advantage provided by bondgraphs. However, in detail design within the mechanism domain, where geometry does matter, a finite element formulation might be better. Therefore, with regard to dynamic tolerance analysis initial research has started to investigate the possible use of a finite element formulation, possibly in combination with a bondgraph based formulation, when integration of conceptual and detailed design tasks has to be achieved.

6. CONCLUSIONS AND RECOMMENDATIONS

In order to be able to validate the genericity of the proposed approach, the set of generic sub–models should be extended from being limited to a few planar mechanisms – as is currently the case – to the full range of both 2D and 3D mechanisms. In addition to extending the number of sub–models, the sub–models themselves need to be verified thoroughly in order to allow them to be compatible and therefore reusable in various types of different mechanisms. Especially causal problems need to be avoided. Also the accuracy problems need to be eliminated. In view of the problems encountered using bondgraphs for dynamic tolerance analysis, it may be worthwhile to investigate other formulations such as a finite elements formulation. Initial research in this direction is currently under way.

ACKNOWLEDGEMENTS

dr. ir. Theo de Vries and dr ir. Jan Broenink are gratefully acknowledged for their advice and support in building the bondgraph models.

REFERENCES

[Breedveld 1996] Breedveld P.C., 1996, The context–dependent trade–off between conceptual and computational complexity illustrated by the modelling and simulation of colliding objects, *proceedings IEEE–SMC CESA'96 conference*, Lille, 48–54
[Broenink 1997] Broenink J.F., 1997, Modelling, simulation and analysis with 20–Sim, *The Journal A*, Vol. 38, no.3, 22–25.

[Clément & Rivière 1993] Clément, A., and Rivière, A., 1993, "Tolerancing versus nominal modelling in next generation CAD/CAM system", *Proceedings, CIRP Seminar on Computer Aided Tolerancing*, ENS de Cachan, Paris, pp. 97 – 113.

[Clément et al. 1994] Clément A., Rivière A., Temmerman M., 1994, *Cotation tridimensionelle des systèmes mécaniques, théorie & pratique*, PYC Edition, Yvry–Sur–Seine Cedex (ISBN 2–85330–132–X), in French (English version: *Three dimensional tolerancing of mechanical systems*, Addison Wesley Editions).

[Controllab 1998] Controllab, 1998, http://www.rt.el.utwente.nl/20sim/clp.htm

[Dijk 1994] Dijk J. van, (1994), *On the role of bond graph causality in modeling mechatronic systems*, PhD thesis, University of Twente, Enschede (NL).

[Funabashi et al. 1978] Funabashi H., Ogawa K., Horie M., 1978, A Dynamic analysis of mechanisms with clearances, *Bulletin of the JSME*, Vol. 21, no 161, 1652–1659.

[Funabashi et al. 1980] Funabashi H., Ogawa K., Horie M., Iida H., 1980, A Dynamic analysis of the plane crank– and rocker mechanisms with clearances, *Bulletin of the JSME*, Vol. 23, no 177, 446–452.

[Miedema & Mansour 1976] Miedema B., Mansour W.M., 1976, Mechanical joints with clearance: a three mode model, *ASME J.of Eng. for Industry*, Vol. 98, 1319–1323

[Salomons 1995] Salomons, O.W., 1995, "*Computer support in the design of mechanical products: constraint specification and satisfaction in feature based design for manufacturing*", Ph.D. Thesis, University of Twente, Enschede (NL).

[Salomons et al. 1995] Salomons, O.W., Jonge Poerink, H.J., Slooten, F. van, Houten, F.J.A.M. van, and Kals, H.J.J., 1995, "A computer aided tolerancing tool based on kinematic analogies", *Proceedings, CIRP seminar on computer aided tolerancing*, Tokyo, pp. 53–72 (and to be published by Chapman & Hall)

[Salomons 1996a] Salomons, O.W., Jonge Poerink, H.J.,Haalboom, F.J, Slooten, F. van, Houten, F.J.A.M. van, Kals H.J.J. 1996a, A Computer Aided Tolerancing Tool I: Tolerance Specification, *Computers in Industry*, Vol. 31, No.2, 1996, 161–174.

[Salomons 1996b] Salomons, O.W., Haalboom, F.J, Jonge Poerink,H.J., Slooten, F. van, Houten, F.J.A.M. van, Kals H.J.J., 1996b, A Computer Aided Tolerancing Tool II: Tolerance Analysis, *Computers in Industry*, Vol. 31, No.2 1996, 175–186.

[Townsend & Mansour 1975] Townsend M.A., Mansour W.M., 1975, A pendulating model for mechanisms with clearances in the revolutes, *ASME J. for Engineering for Industry*, Vol. 97, 354–358.

[Zijlstra 1997] Zijlstra J., 1997, Dynamic tolerance analysis using FROOM & 20–sim, MSc thesis, PT–616, University of Twente, Enschede

[Zwaag 1998 a] Zwaag J. van der, 1998, Towards dynamic tolerance analysis using bond graphs, MSc thesis, PT–620, University of Twente, Enschede

[Zwaag 1998b] Zwaag J. van der, 1998, Spelingsmodellen met discrete toestanden, internal report, University of Twente, Enschede

Tolerance evaluation system of manufactured parts based on soft gaging

Fumiki Tanaka and Takeshi Kishinami
Graduate School of Engineering, Hokkaido University
Kita-13, Nishi-8, Kita-ku, Sapporo, Hokkaido, 060-8628, JAPAN
tanaka@coin.eng.hokudai.ac.jp, kishinami@coin.eng.hokudai.ac.jp

Abstract: This report deals with soft gaging as an inspection technique for manufactured parts. First, we propose a modeling of soft gaging for inspection of tolerances on features with or without a relationship between the feature and the datum. The soft gaging model that conforms to the geometric tolerance standards is represented using ISO 10303-11 EXPRESS-G notation. We also propose mathematical definitions of soft gages for such tolerances and the procedure for the evaluation of such tolerances of machined parts based on a CMM measured point set in this paper. Second, we develop a tolerance evaluation system of machined parts based on the proposed soft gaging and a fitting procedure based on a small displacement screw method considering spatial constraints, which enables the fitting of the soft gage into a measured point set of the machined parts and thus deriving a deviation. Finally, we also show that the proposed method is very effective in its application to certain inspection examples.
Keywords: Coordinate metrology, Computational metrology, Geometric tolerance evaluation

1. INTRODUCTION

Gage thinking underlies some of the main principles and definitions in geometric tolerancing, and hard gages have often been the conformance technology of choice when cost has to be considered. On the other hand, a three-dimensional measurement device such as a coordinate measuring machine (CMM) samples the surface of a measured object and generates a measured point set from that surface. After the measured point set is collected by CMM, we can evaluate geometric variations or verify tolerance conformance by an independent numerical analysis.

One of the major attractions of sampled coordinate metrology, as implemented currently with CMM, is the promise of soft gaging, that is, the emulation of hard gages and the implementation of criteria not physically gage-able (e.g. most LMC tolerances) thought programmed data collection and reduction. To date, various research works have reported on soft gaging for only specific tolerance verifications, such as form tolerance [Choi et al., 1998] and position tolerance [Etesami, 1991]. The first step of the procedure for all research related to soft gaging was to establish a mathematical definition of geometric tolerances. However, these mathematical definitions are applicable for only specific tolerances, and not suitable for other tolerances. Additionally there were few mathematical definitions for the tolerances related to the datum. On the other hand, other mathematical definitions of tolerances which were proposed by Requicha [Requicha, 1983], or specified in ASME Y14.5.1-1994, were for communication of the meaning of tolerances among wider applications such as CAD and CAM. However, these mathematical models have no capability for fitting between a measured point set and soft gage, and it is difficult to construct a soft gage model directly from these mathematical models. Therefore, we must construct another mathematical model of tolerance for soft gaging.

The second step in soft gaging is to make a soft gage model based on a proposed mathematical model. In many cases, the soft gage model is the model of geometric deviation based on the tolerance model directly, or the functional gage in the MMC cases. However, there is no explicit relationship between the soft gage model and tolerance model, and the soft gage model can not be constructed from the tolerance model automatically. Therefore, it is necessary to represent the explicit relationship between the soft gage model and the tolerance model.

The last step in soft gaging is the fitting procedure between the measured point set and the gage model. The minimum zone including all measured points should be derived in order to calculate a deviation because the tolerance zone is based on a minimum zone method in the standard. This means that it is necessary to propose a fitting algorithm preserving the spatial relationship between the datum and the gage.

To solve these problems we propose the soft gage model as follows: First, we propose an extended tolerance model with a mathematical representation of the tolerance zone. Second, we derive a soft gage model from the extended tolerance model. Third, we develop a tolerance evaluation system of machined parts with the proposed soft gage model and a fitting procedure based on the small displacement screw method [Tanaka et al., 1996] considering spatial constraints, which enables the fitting of the soft gage into the measured point set of the machined parts and thus deriving a deviation. Finally, we also show that the proposed method is very effective by applying it to an inspection example.

2. BASIC IDEA FOR SOFT GAGING MODEL

In the ASME Y14.5.1M-1995, the reference (or true position) is the center of the tolerance zone, and the real feature from which the measured point set is obtained conforms with the tolerance specification if the distance from it to the real feature is smaller than the allowance (the tolerance value). This means that the boundary of the zone is defined implicitly. For example, the mathematical definition of cylindricity is given by equation 1 (see figure 1).

$$\left\| l \times (P - A) \right| - r \right\| \le \frac{t}{2} \tag{1}$$

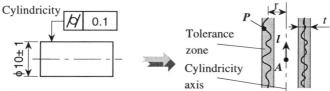

P : A position vector of any points in the tolerance zone
l : A direction vector of the tolerance zone
A : A position vector locating a point on the spine
r : A radial distance from the spine to the center of the circularity zone
t : tolerance value

Figure 1; *Cylindricity definition in the ANSI Y 14.5.1M-1994.*

However, it is difficult to consider the shape of the tolerance zone in this definition. On the other hand, the tolerance zone definition specified by ISO 10303-47 has the explicit representation of the boundary of the tolerance zone. Figure 2 shows the EXPRESS-G notation of the tolerance zone definition of the ISO 10303-47. Moreover, according to the boundary representation of the tolerance zone, not only the deviation of the real feature but also the distribution of the measured point set can be calculated from the inclusion zone which can include all measured points. Consequently, we adopt the boundary representation of the tolerance zone in the tolerance model for constructing the soft gage model. The conformance testing of the tolerance can be checked from the inclusion relationship between the tolerance zone and the inclusion zone which can include all measured points (measured point inclusion zone).

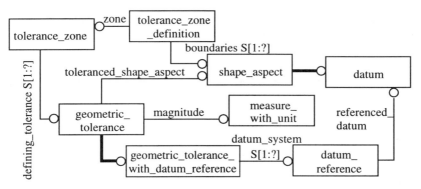

Figure 2; *Tolerance zone definition in the ISO 10303-47*

Second, we consider the representation scheme of the relationship between the tolerance model and soft gage model. The relationships between the tolerance and the tolerance zone, between zone boundary elements, and between datum and features are represented in the tolerance model as shown in figure 2. These relationships should be represented in the measured point set inclusion zone for deriving the deviation of the real feature. This means that the measured point set inclusion zone inherits the property of the tolerance zone. That is, the soft gage model consists of the measured point set inclusion zone that is a subtype of the tolerance zone as shown in figure 3. This is the basic idea for constructing the soft gage model.

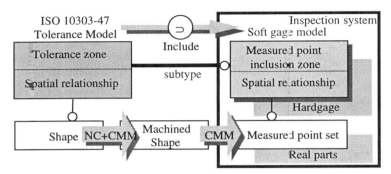

Figure 3; *Basic idea of the soft gage model*

3. TOLERANCE MODEL

Figure 4 shows the EXPRESS-G notation of the extended tolerance model that is based on the ISO 10303-47 tolerance model. The major entities of the extended tolerance model are the tolerance_zone of the geometric_tolerance, the tolelance_zone_definition which defines the tolerance_zone by its boundary, and the shape_aspect which is the toleranced feature, datum, or tolerance zone boundary. The shape_aspect has the geometric shape (plane or cylinder) with the location A and axis l vector.

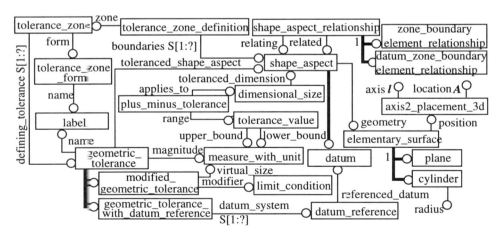

Figure 4; *EXPRESS-G notation of tolerance model*

The mathematical model of the tolerance zone consists of the zone boundary element relationship, and datum and zone boundary element relationship as shown in table I. In this table, C_D is the direction vector of the datum.

Table I; Mathematical model of tolerance zone

zone form	tolerance	zone boundary element relationship	datum and zone boundary element relationship
parallel plane	Form	parallel $l_1 = l_2$ boundaries distance $(A_2 - A_1) \cdot l = $ const.	
parallel plane	Orientation		$l_i \cdot C_D = $ const.
parallel plane	Position		$l_i \cdot C_D = $ const. $A_i \cdot C_D = $ const.
coaxial cylinder	Form	coaxial $l_1 = l_2$ $A_1 = A_2$ boundaries distance $r_2 - r_1 = $ const.	
coaxial cylinder	Orientation		$l_i \cdot C_D = $ const.
coaxial cylinder	Position		$l_i \cdot C_D = $ const. $A_i \cdot C_D = $ const.
coaxial cylinder	Runout		$l_i \cdot C_D = $ const.

4. SOFT GAGE MODEL

The soft gage model is a subtype of the extended tolerance model. Figure 5 shows the EXPRESS-G notation of the proposed soft gage model. The major entities of the soft gaging model are the measure_point_inclusion_zone which includes all measured points, the measure_point_inclusion_zone_definition which defines the measured point inclusion zone by its boundary, the datum_bounadry_element_relationship and boundary_element_relationship.

Before proposing a mathematical model of the soft gage, we explain the small displacement screw method (SDS), the fitting method between the model and the measured point set. According to the SDS, the normal distance (e_i) between the measured point and boundary element of the gage after fitting is given by equation 2 as shown in figure 6.

$$e_i = \overrightarrow{M_{thi}M_i} \cdot n - D_A \cdot n_i + (AM_i \times n_i) \cdot R_A \tag{2}$$

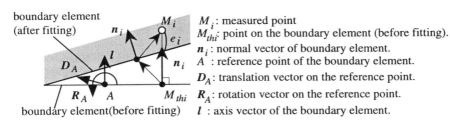

boundary element (after fitting)

boundary element(before fitting)

M_i: measured point
M_{thi}: point on the boundary element (before fitting).
n_i: normal vector of boundary element.
A : reference point of the boundary element.
D_A: translation vector on the reference point.
R_A: rotation vector on the reference point.
l : axis vector of the boundary element.

Figure 6; Small displacement screw method

298

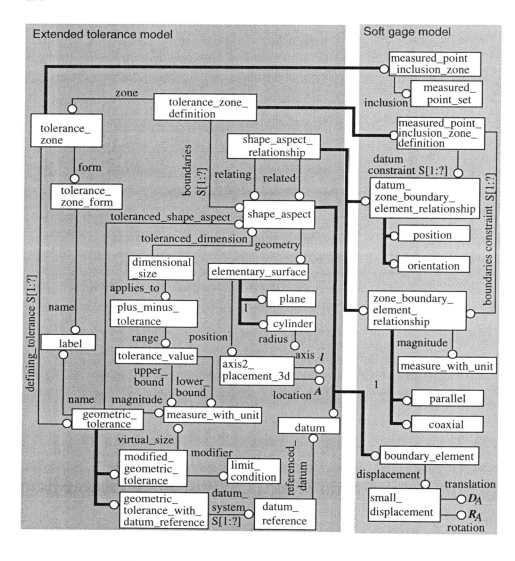

Figure 5; EXPRESS-G notation of soft gage model

The mathematical model of the soft gage consists of the zone boundary element relationship, datum and zone boundary element relationship, and measured point inclusion zone definition as shown in table II. In table II, M_{pt} is the point on the zone boundary element after fitting.

Table II; *Mathematical model of soft gage model*

zone form	tolerance	zone boundary element relationship	datum and zone boundary element relationship	measured point inclusion zone definition
parallel plane	Form	parallel $R_{A1} = R_{A2}$ $D_{A1} = d_1 l_1$ $D_{A2} = d_2 l_2$ boundaries distance $\left\{\left(A_2+D_{A2}\right)-\left(A_1+D_{A1}\right)\right\}\cdot l$		boundary1 definition $\overrightarrow{M_{thi}M}_i \cdot n_i'$ $-\overrightarrow{M_{thi}M}_{opt}\cdot n_i' \geq 0$ boundary2 definition $\overrightarrow{M_{thi}M}_i \cdot n_i'$ $-\overrightarrow{M_{thi}M}_{opt}\cdot n_i' \leq 0$ $n_i'=n_i$
	Orientation		axial constraint $\left(R_A \times l\right)\cdot C_D = 0$	
	Position		Plane locational constraint $\left(R_{A1}\times l_1\right)\cdot C_D = 0$ $\left(R_{A2}\times l_2\right)\cdot C_D = 0$ $\dfrac{\left(D_{A1}+D_{A2}\right)}{2}\cdot C_D=0$	small displacement $\overrightarrow{M_{thi}M}_{opt}\cdot n_i =$ $D_A \cdot n_i+\left(\overrightarrow{AM_i}\times n_i\right)\cdot R_A$ plane $\overrightarrow{M_{thi}M}_i \cdot n_i = \overrightarrow{AM}_i \cdot l$ $n_i=l$
coaxial cylinder	Form	coaxial $R_{A1} = R_{A2}$ $D_{A1} = D_{A2}$ boundaries distance $r_2 - r_1$		boundary1 definition $\overrightarrow{M_{thi}M}_i \cdot n_i'$ $-\overrightarrow{M_{thi}M}_{opt}\cdot n_i' \geq 0$ boundary2 definition $\overrightarrow{M_{thi}M}_i \cdot n_i'$ $-\overrightarrow{M_{thi}M}_{opt}\cdot n_i' \leq 0$ small displacement $\overrightarrow{M_{thi}M}_{opt}\cdot n_i =$ $D_A \cdot n_i+\left(\overrightarrow{AM_i}\times n_i\right)\cdot R_A$ cylinder $\overrightarrow{M_{thi}M}_i \cdot n_i=\left|\overrightarrow{AM}_i\times l\right|-r$ $n_i=\dfrac{\overrightarrow{AM}_i-\left(\overrightarrow{AM}_i\cdot l\right) l}{\left|\overrightarrow{AM}_i-\left(\overrightarrow{AM}_i\cdot l\right) l\right|}$
	Orientation		axial constraint $\left(R_A \times l\right)\cdot C_D = 0$	
	Position		Cylindrical locational constraint $\left(R_A \times l\right)\cdot C_D = 0$ $D_A \cdot C_D = 0$	
	Runout			

5. TOLERANCE EVALUATION SYSTEM BASED ON SOFT GAGE MODEL

The tolerance evaluation system consists of two functions as shown in figure 7. One is the soft gage modeling function and another is the fitting function. The soft gage modeling function translates the tolerance model data into the soft gage model data

300

automatically in consideration of the extended tolerance model and the soft gage model. The fitting function fits the soft gage model to the measured point set using linear programming based on the small displacement screw method, and outputs the fitting results.

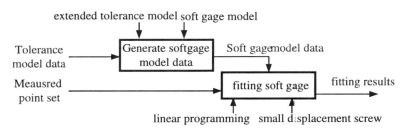

Figure 7; *Tolerance evaluation system based on soft gaging model*

6. EXAMPLE

Figure 8 shows the example of the tolerance model for the coaxiality tolerance on the maximum material condition and tolerance zone definition of that tolerance specification.

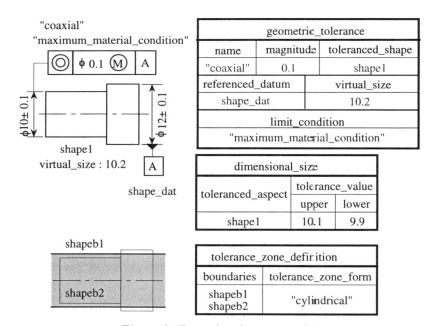

Figure 8; *Example tolerance model*

The soft gage modeling function translates the tolerance model data as shown in figure 8 into the soft gage model data as shown in figure 9. The fitting function fits the soft gage model to the measured point data set which conforms to the specified tolerance as shown in figure 10(a) and outputs the fitting results as shown in figure 10(b). From these figures, we can see that the result is correct and the proposed method is very effective.

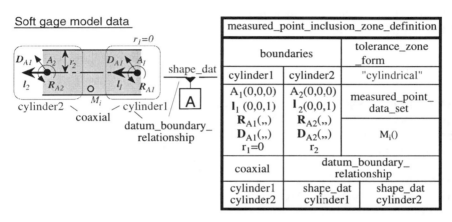

measured_point_inclusion_zone_definition		
boundaries		tolerance_zone _form
cylinder1	cylinder2	"cylindrical"
$A_1(0,0,0)$	$A_2(0,0,0)$	measured_point_ data_set
$l_1(0,0,1)$	$l_2(0,0,1)$	
$R_{A1}(,,)$	$R_{A2}(,,)$	$M_i()$
$D_{A1}(,,)$	$D_{A2}(,,)$	
$r_1=0$	r_2	
coaxial	datum_boundary_ relationship	
cylinder1 cylinder2	shape_dat cylinder1	shape_dat cylinder2

Figure 9; An example of a soft gage model

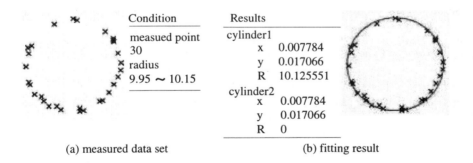

Condition	Results
measued point 30	cylinder1
	x 0.007784
radius	y 0.017066
9.95 ~ 10.15	R 10.125551
	cylinder2
	x 0.007784
	y 0.017066
	R 0

(a) measured data set　　　　　　　　　　(b) fitting result

Figure 10; Evaluation results of the example

7. CONCLUSIONS

The conclusions of this research are as follows:
1) We propose the basic idea of defining the tolerances and of representing a method of the relationship between the tolerances and soft gage model.

2) We propose the extended tolerance model and the soft gage model in the EXPRESS-G notation. We also propose the mathematical representation of the tolerance zone and the measured point set inclusion zone.

3) We develop a tolerance evaluation system of machined parts based on the proposed soft gaging and a fitting procedure based on a small displacement screw method considering spatial constraints, which enables the fitting of the soft gage into a measured point set of the machined parts and thus deriving a deviation.

4) We show that the proposed method is very effective by applying it to an inspection example.

REFERENCES

[ASME, 1994] ASME, Mathematical Definition of Dimensioning and Tolerancing Principles ASME Y 14.5.1M-1994,New York, 1994.

[Choi and Kurfess, 1998] W. Choi and T. R Kurfess, Uncertainty of extreme fit evaluation for three-dimensional measurement data analysis. *CAD* Vol. 30 No 7, 1998, p 549-557.

[Elmaraghy et. al., 1990] W.H.Elmaraghy et. al., Determination of Actual Geometric Deviations Using Coordinate Measuring Machine Data, *Manufacturing Review* Vol. 3. No.1, 1990, p 32-39.

[Etesami ,1991] F. Etesami, Position Tolerance verification Using Simulated Gaging, *The International Journal of Robotics Research* Vol. 10 No. 4, 1991, p358-370.

[Fortin and Chatelain., 1990] C.Fortin and J.F.Chatelain, A soft gaging approach for complex cases including datum shift analysis of geometrical tolerances, *Computer-Aided Tolerancing, Proceedings of the 4th CIRP Design Seminar*, 1996, p 298-311.

[ISO 10303-47, 1994] ISO, Industrial automation systems and integration-- product data representation and exchange-- Integrated generic resources: Shape variation tolerances ISO 10303-47-1997, Geneve, 1997.

[Kurfess and Banks, 1995] T. R. Kurfess and D.L. Banks, Statistical verification of conformance to geometric tolerance, *CAD* Vol. 27 No.5, 1995, p353-361.

[Requicha, 1983] A.A.G. Requicha, Toward a Theory of Geometric Tolerancing, The *International Journal of Robotics Research* Vol. 2 No.4, 1983, p45-60.

[Tanaka et. al., 1990] F.Tanaka et. al., Inspection method for geometrical tolerances using coordinate measuring machine, *Computer-Aided Tolerancing, Proceedings of the 4th CIRP Design Seminar*, 1996, p 298-311.

[Voelcker, 1993] H. Voelcker, A Current Perspective On Tolerancing and Metrology, *Manufacturing Review* Vol.6 No.4, 1993, p258-268.

AN EXPERT SYSTEM FOR ASSESSING FLATNESS WITH A CMM

Dr. Sunpasit Limnararat
Department of Industrial Engineering
King Mongkut's Institute of Technology, Ladkrabang
Bangkok 10520
Thailand

Dr. Leonard E. Farmer
School of Mechanical and Manufacturing Engineering
The University of New South Wales
Sydney NSW 2052
Australia
Email: L.Farmer@unsw.edu.au

ABSTRACT: A knowledge based system is described for controlling the inspection process of a CMM. The objectives of the knowledge base inspection system is to provide accurate results quickly with a minimum number of inspection points and minimum intervention by an operator. Results of simulated CMM inspection trials for a flatness specification are described and these show good potential for realising the above objectives, particularly where CMMs are integrated into continuous and batch manufacturing systems.
Keywords: Dimensioning, Tolerancing, Co-ordinate Measuring Machines, Knowledge Based Systems, Flatness

1.   INTRODUCTION

Co-ordinate Measuring Machines (CMMs) are now used extensively for assessing the conformance of workpiece features to dimensional specifications [Coyne, 1994]. Most CMMs have full time operators and a typical sequence of tasks performed by an operator for assessing a single attribute of a workpiece is illustrated in Figure 1. This task sequence assumes that a variety of workpiece types are being produced in batches and inspected by the CMM for single attributes, such as the flatness of a surface. Further, the inspection operation is continued until a degree of confidence is reached to decide whether the workpiece is to be accepted or rejected. Hence, the inclusion of the feedback loop for inspecting additional points on the workpiece.

303

304

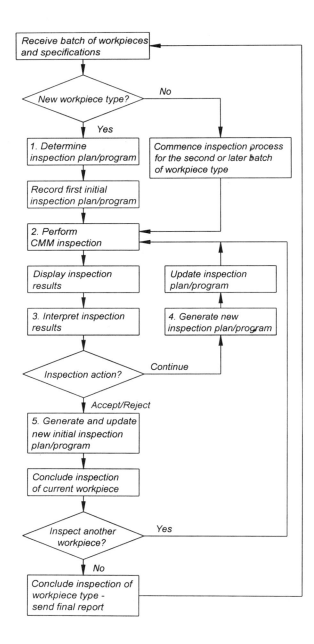

Figure 1; *Typical task sequence for operating a CMM*

Each task in this CMM inspection routine requires the operator to make decisions. First, the operator has to determine an appropriate inspection plan/program for the attribute of the workpiece being measured. The number of inspection points required and their position on the workpiece have to be chosen carefully to ensure sufficiently accurate accept/reject decisions can be reached for the minimum inspection time. This can be a critical issue where routine inspection is performed on mass produced component parts or where inspection times are long and, hence, costly.

Second, the involvement of operators with routine machine operation has been greatly reduced with the continuing improvement of the software supplied with CMMs. Nevertheless, it is usual for operators to be required to load workpieces onto CMMs, start the inspection program and then unload the inspected workpiece from CMMs.

Finally, the interpretation of inspection results can also be a difficult task requiring the application of considerable skill and knowledge from the operator. This is particularly evident in border line cases where the operator has to decide whether to accept, reject or take further inspection points with the CMM's probe (the operator having to select the number of additional points and their location on the workpiece). This task requires not only the interpretation of quantitative inspection results, such as those for a flatness specification, but also the justification of the accept/reject/continue inspecting decision.

From the above, it can be seen that the operator of a CMM requires a high level of knowledge and skill to effectively plan and operate the CMM as well as the necessary knowledge and skills to interpret a wide range of inspection results. The performance of operators improves considerably with experience and when they have the ability to apply "common sense" logic. This knowledge/skill/experience profile of CMM operators has lead to the proposal of investigating the feasibility of using knowledge base or Expert Systems to alleviate the decision task load on CMM operators and at the same time improve the overall productivity of the process.

2. EXPERT SYSTEM - OVERVIEW

The two most common features that occur in workpieces are flat planes and cylinders. Therefore, the case of assessing a flatness specification on a machined surface of a workpiece was chosen to develop the concept of an Expert System for assisting the operation of a CMM. The proposed Expert System is illustrated in Figure 2, the Expert Flatness Inspection Module interfaces with a CAD Drawing Module, the CMM and Supporting Software Module and an Inspection Results Module.

The operation of the Expert System will now be discussed in the context of the following example. The workpiece to be assessed for deviations from flatness is shown in Figure 3 with nominal dimensions and flatness specification. Superimposed on the surface of the workpiece is an assumed feature that results in deviations from flatness. The feature appears as a hemispherical indentation in the surface of depth 0.04 ±0.005

with position dimensions of 350 ±42 and 250 ±42 in the x and y directions respectively. The variations in the x, y and z directions are random and uniformly distributed.

2.1 CAD Drawing Module

The CAD Drawing Module provides a CAD file of an accurately scaled two dimensional profile of the workpiece surface that is to be inspected, such as shown in Figure 3. Any holes or surface discontinuities are to be included in the drawing as well as the flatness specification for the surface. The purpose of this module is to provide data that will allow a set of inspection data points to be generated for the workpiece surface in the Expert Flatness Inspection Module.

2.2 Expert Flatness Inspection Module

Central to the Expert System is the Expert Flatness Inspection Module. This module comprises five blocks or components. The first is the System Management Block and this performs the functions of managing the information flow between the modules and blocks of the system as well as providing the interfacing and the sequencing of operations between the modules and blocks of the system.

The second block is the Data Points Generator/Data Points Selector Block and this performs the functions of determining the locations for the points for the inspection process. Four significant tasks are carried out in this block. The first is the generation of a set of feasible inspection points for the inspection of the surface of the workpiece. This process requires access to the CAD drawing of the workpiece surface and the application of the Quadtree Meshing Method [Yerry et al., 1983] [Potyondy et al., 1995] to the surface. A mesh with the required density is produced on the surface and the mid

Figure 2; The proposed Expert System for assessing surface flatness specifications

point of the each mesh element is stored in the system's database as a feasible inspection point. This process is illustrated in Figure 4 where the surface has four primary

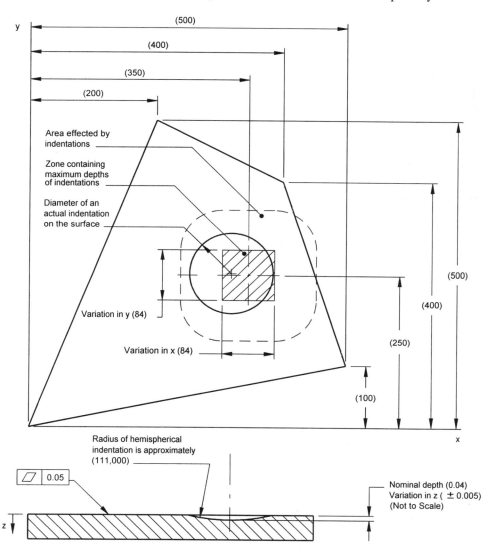

Figure 3; *Sample workpiece with flatness specification and details of an assumed feature that results in deviations from flatness.*

elements. Points 1,2,3 and 4 are inspection points on the primary mesh layer, 31, 32, 33, 34, 41, 42, 43 and 44 are inspection points on the secondary mesh layer for the primary element 3 and 4 and 311, 312, 313 and 314 are inspection points on the tertiary layer for secondary layer element 31.

The second task performed by the block is to select the initial set of inspection datapoints for the first workpiece. These are usually selected by the operator and have the status of being mandatory. They are positioned to give a good representation of the surface as well as being located in areas where there is likely to be deviations in flatness. Examples are indicated by the 8 diamond surrounded points in Figure 4. That is, points 11, 22, 23, 24, 31, 32, 33 and 44. The System Management Block transmits this data to the CMM and Supporting Software Module where these 8 points are used to commence inspection on the first workpiece.

The third task occurs when an *Accept/Reject* decision cannot be reached from the information provided by the already inspected points and it is necessary to *Continue Inspection*. That is, supplementary inspection points are nominated from rules in the

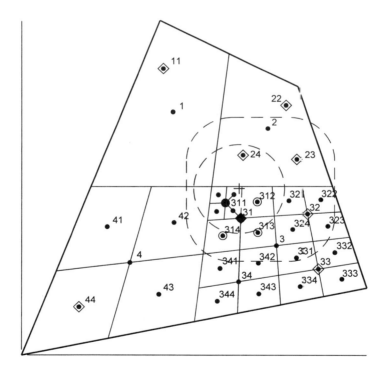

Figure 4; *Sample workpiece profile with mesh superimposed on surface.*

Decision Maker Block for the next cycle of inspection by the CMM. The co-ordinates of the nominated supplementary points are determined in the Data Points Generator/Data Points Selector Block and the System Management Block transmits this data to the CMM and Supporting Software Module. This process is illustrated in Figure 4 where point 31 would have given the greatest deviation from flatness for the first 8 inspection points. The Decision Maker Block, would have used the Quadtree Search method, supplementary points for the second cycle of inspection by the CMM. If a third cycle of inspection was required from the CMM, the supplementary points would be 3111, 3112, described by [Limnararat, 1998], to suggest points 311, 312, 313 and 314 to be the 3113 and 3114. The path followed by the Quadtree Search method is shown converging on the point of maximum deviation from flatness in Figure 4.

When an *Accept/Reject* decision is reached for a workpiece the fourth and final task performed by the Data Points Generator/Data Points Selector Block can be performed. This is to provide a new initial dataset for the next workpiece. Once again the actions of the Data Points Generator/Data Points Selector Block are determined by rules in the Decision Maker Block. That is, in general, the mandatory points would be nominated (8 for our example) plus supplementary points from the inspection of previous workpieces. The selection and inclusion of these supplementary points in the new dataset is based on the knowledge gained from inspecting previous workpieces and they are intended to increase the efficiency of the inspection process. That is, reach an *Accept/Reject* decision in less time.

The third block in the module is the Decision Maker Block. It contains the Expert System modules and decision rules that are required to interpret the Actual Flatness estimates of the surface calculated in the Surface Flatness Evaluation Block and the rules for selecting supplementary inspection points and updating the initial sets of inspection points, etc. The interpretation of the actual flatness estimate result comprises rules and strategies for;

- Deciding whether the inspection action is to be terminated (*Accept/Reject*) or continued (*Continue Inspection*),
- If the inspection process is to be terminated, do the results indicate the workpiece inspected is acceptable or unacceptable and what are the next actions to be performed by the system?
- Alternatively, if the inspecting action is to be continued, what are the next actions to be performed by the system?

The fourth block in the module is the Surface Flatness Evaluation Block. This contains an algorithm that accurately and quickly estimates the Actual Flatness of the surface from the measured x, y and z co-ordinates of the inspection points. A mandatory requirement of the algorithm is that it conforms strictly with the interpretation of the flatness specification as set down by the ISO. The main input of this module are the x, y and z co-ordinates obtained from inspection operations on the CMM. The output of this module is the estimated Actual Flatness of the surface that was calculated from the CMM input inspection data.

The fifth and last block of the module is the Database The data base is required in the Expert Flatness Inspection Module to store the past and present inspection data, the decision rules for surface inspection and the results of past and present surface flatness inspection analyses.

2.3 CMM and Supporting Software Module

The CMM and Supporting Software is the third module in the system. It comprises the CMM, its proprietary software and data input/output interfaces. The requirements of this module are:

- the CMM interface will read the x and y co-ordinates of inspection points from an external source (Expert Flatness Inspection Module),
- these points can be accepted as inspection points by programmed inspection routines that have been previously written for inspecting the workpiece on the CMM and
- the z co-ordinates of the x and y co-ordinates of the inspection points obtained by the CMM can be exported to, and read by, the Expert Flatness Inspection Module.

2.4 Inspection Results Module

The final module in the system is the Inspection Results Module. This module receives the inspection results from the Expert Flatness Inspection Module and propagates this information, as is required, in the Quality Reporting System of the manufacturing organisation. The requirement here is that a compatible interface is provided between the Expert Flatness Inspection Module and the Quality Reporting System.

3. TRIALS AND RESULTS

A set of simulated CMM inspection trials were performed with the Expert System. The workpiece for the trials was described Section 2 and is illustrated in Figure 3. The trials were simulated in the sense that the CMM was replaced by a mathematical model of the workpiece with the position and depth of the flatness deviation feature varying randomly, within the previously defined parameters, for each workpiece.

A Quadtree mesh was generated for the surface of the workpiece as shown in Figure 4 to five division levels. That is, the surface was replaced by some 1,364 feasible inspection points. The magnitudes of the size and variation in position of the flatness deviation feature were selected to approximately represent the deviations from flatness that might be expected to occur in practice for a machined plate. That is, the square area, shown in Figure 3, within which the maximum deviation from flatness would occur was approximately 5 percent of the total workpiece surface and the projected circular area resulting from a hemispherical indentation was approximately 10 percent of the total workpiece area.

The selection of the 8 mandatory initial inspection points was described earlier in Section 2. When selecting these points it was assumed that some knowledge was available concerning the likely size and nature of the flatness deviation feature. The positions of these points were chosen to be outside the square zone that will contain the maximum deviation from flatness point. However, 4 points were located within the area effected by the indentations. Only one supplementary point was carried to later workpieces, to replace any previous supplementary point. That is, for second and later workpieces size of the initial set of inspection points was 9. The supplementary point was the point resulting in the maximum flatness deviation for the previous workpiece. This model provided some limited opportunity for the system to display its learning ability from with supplementary points.

The results from trials on the first ten randomly generated workpieces are shown in Table 1. All trials were terminated by the inadequate progress rule in the Decision Maker Block. This rule requires that; if the difference between the current cycle and the previous cycle estimate of flatness is less than 10 percent of the flatness specification on the workpiece surface (0.05 for the example), or the difference is less than 0.001, the *Continue* inspection cycle is stopped and an *Accept/Reject* decision is taken on the current flatness estimate.

Table I; *Results for Expert Flatness Inspection System Trials*

Workpiece Number	Number of Inspection Cycles	Number of Inspection Points (Total)	Estimated Actual Flatness (EAF)	True Actual Flatness (AF)	Difference: DF = $\lvert$EAF - AF$\rvert$	Percent Difference: (DF/AF) x 100
1	6	28	0.0415	0.0417	0.0002	0.48
2	4	21	0.0361	0.0363	0.0002	0.55
3	4	21	0.0349	0.0351	0.0002	0.57
4	5	25	0.0386	0.0387	0.0001	0.26
5	5	25	0.0446	0.0447	0.0001	0.22
6	5	25	0.0391	0.0393	0.0002	0.51
7	4	21	0.0429	0.0431	0.0002	0.46
8	4	21	0.0435	0.0436	0.0001	0.23
9	4	21	0.0365	0.0368	0.0003	0.82
10	4	21	0.0359	0.0365	0.0006	1.64

The results from the trials show a small learning effect with the number of inspection cycles of 6, for the first workpiece, reducing to either 4 or 5 for the following workpieces. This is mainly due to the supplementary inspection point always being in the square area of maximum deviation from flatness. The variation from 4 to 5 inspection cycles is due to the position of the supplementary point relative to the maximum deviation from flatness point for the workpiece. It is expected that if the supplementary inspection points were expanded to include those for the preceding 3 or 4 workpieces then the number of inspection cycles would be reduced on average. A second significant aspect of the results is the consistently small differences between the theoretical actual flatness values for the workpieces and the values estimated by inspection with the expert system. These differences are less than 0.001mm, a value generally outside the accuracy of most CMMs.

4. CONCLUDING REMARKS

A model of an Expert System for assessing flatness specifications with CMMs has been described and shown to have potential for:

- producing consistently accurate estimates of the actual flatness of workpiece surfaces,
- reduces the dependency on, and the decision making workload of, CMM operators and
- is applicable to irregular shaped surfaces containing holes and discontinuities.

Whilst the Expert System, in its current form, is probably slower than a system where the operator nominates a number of points to be inspected by the CMM program in a single inspection cycle, improvements in the decision making rules could improve its efficiency. Its performance relative to other single cycle fixed point nomination systems would also improve with inspection tasks that include greater opportunities for learning than the current example. That is, where there are predictable trends in the size and position of a flatness deviation feature in the surface of a workpiece.

REFERENCES

[Coyne, 1994] Coyne, B.; "*Add Value to your CMM*"; In: Quality Today, pp. 16-20.

[Limnararat, 1998] Limnararat, S.; "*An Investigation of the Feasibility of Using an Expert System for Assessing Surface Flatness with a Coordinate Measuring Machine*"; PhD Thesis, In: The University of New South Wales, Australia.

[Potyondy et al., 1995] Potyondy, D.O.; Wawrzynek, P.A.; Ingraffea, A.R.; "*An Algorithm to Generate Quadrilateral or Triangular Element Surface Meshes in Arbitrary Domains with Applications to Crack Propagation*"; In: International Journal for Numerical Methods in Engineering, Vol 38, pp. 2677-2701.

[Yerry et al., 1983] Yerry, M.A.; Shephard, M.; "*A modified Quadtree Approach to Finite element Mesh Generation*"; In: IEEE CG&A, pp. 39-46.

Cost Tolerance Sensitivity Analysis for Concurrent Engineering Design Support

Richard J. Gerth, Ph. D.
Industrial and Manufacturing Systems Engineering - Ohio University
Athens, OH 45701-2979
gerth@ohiou.edu

Pavlos Klonaris, Dipl. Ing. and Tilo Pfeiffer, Ph. D.
Lehrstuhl für Fertigungsmeßtechnik und Qualitätsmanagement
Rheinisch Westphälische Technische Hochschule – Aachen, Germany

Abstract: This paper presents cost tolerance sensitivity analysis (CTSA), a novel method of determining which features are critical and non-critical early in the design phase when cost information is uncertain. The method utilizes minimum cost tolerancing methods combined with designed experiments to determine which features are sensitive or insensitive to uncertainties in the cost–tolerance curve estimates. The method successfully identified critical and non-critical dimensions on the case study of a windmill transmission. The results also show that both cost and tolerances must be considered simultaneously, and examination of near optimal solutions improves the results.

Keywords: Minimum Cost Tolerancing, Designed Experiments.

1. INTRODUCTION

A critical concurrent engineering interface is the product-process-inspection design interface. Tolerances present a natural language for the interface because tolerances affect product function, process capability, and measurement accuracy requirements. However, tolerance issues are not usually addressed till relatively late in the design cycle. It is a thesis of this paper, that in any product there are a few critical design aspects that make it unique, and that these design aspects are resolved first. Hence, it may be possible to determine at an earlier stage which features are likely to be critical. Once this is known, the information can be passed to process and inspection planning so that work in these areas may begin.

In the early design phases the product designer will develop and evaluate various systems. The designer will usually know the tolerance for a critical assembly performance measure needed to guarantee assembly function. The relationship of the performance measure as a function of lower level units is called the stackup function and assumed to be known or estimable. For example, in a transmission, which consists of gears, shafts, and a housing, the geometric position of a critical gear may be precisely or approximately known as a function of the other component geometries.

If the stackup relationship is known or estimable, then a tolerance analysis can be conducted to determine which feature dimensions are critical to product function. Common tolerance analysis methods include worst case tolerancing, statistical tolerancing, and Monte Carlo simulation methods [Gerth, 1996b; Zhang and Huq, 1992]. These methods determine the critical dimensions based on their variance contribution.

Alternatives to tolerance analysis methods are tolerance allocation or synthesis methods. Whereas tolerance analysis predicts the assembly tolerance given the component feature tolerances, tolerance synthesis methods determine the component feature tolerances that will satisfy a specified assembly tolerance. Tolerance synthesis methods include minimum cost tolerancing [Speckhart, 1972; Spotts, 1973; Chase, et al., 1990; Zhang and Wang, 1993], proportional scaling [Chase and Greenwood, 1988], and compensatory tolerancing [Gerth, 1996a]. Of interest here, is minimum cost tolerancing.

Minimum cost tolerancing determines the optimal feature tolerances that will result in the minimum cost assembly. The non-linear optimization problem is formulated as:

Objective Function:
$$MIN[C_{Total}] = \sum_{i=1}^{k} C_i(tol_i)$$
[1]

Constraints
$$Tol_Y \geq Cp_Y \sqrt{\left(\frac{\partial Y}{\partial X_i}\right)^2 \left(\frac{tol_i}{Cp_i}\right)^2}$$
[2]

$$LLtol_i \leq tol_i$$
[3]

$$tol_i \leq ULtol_i$$
[4]

where

C_{Total} — Total Cost

Tol_Y — Assembly tolerance

tol_i — ith component feature tolerance

$C_i(tol_i)$ — Cost function associated with tol_i

Cp_Y — Desired process capability for the assembly (yield)

Cp_i — Desired process capability for the ith feature

$\dfrac{\partial Y}{\partial X_i}$ — Partial derivative of the stackup function with respect to the ith feature

$LLtol_i$, $ULtol_i$ — Lower and upper limit on tol_i

The objective function to be minimized is simply the sum of the individual feature costs. These costs are presumed to be a function of the feature dimension tolerances. The constraints on the objective function are that

1. the assembly tolerance not be exceeded. This ensures that the assembly will function as desired. It is also a measure of assembly yield and is

thus called the yield constraint. The constraint is evaluated by computing the statistical stackup of the feature tolerances.

2. the individual component tolerances be bound between an upper and lower limit. This prevents the optimization program from driving tolerances to 0 or infinity, and thus forces a solution. The upper and lower bounds represent the range of alternative processing options under consideration, and are codetermined with process engineering.

Minimum cost tolerancing assumes that for every tolerance, there exists an appropriate cost. Unfortunately, it is difficult to develop accurate cost estimates. Hence minimum cost tolerancing tends to provide overly precise tolerances given the imprecise nature of the cost estimates, and thus has not found wide spread acceptance. In the early design stages, however, it is understood that the numbers are not exact, but merely estimates. Thus, accuracy is less of an issue. What is required is an accuracy level capable of aiding in design decisions. Hence, a method is needed to determine which factors have the greatest impact on product function and part cost given the uncertainty in the cost estimates.

2. COST TOLERANCE SENSITIVITY ANALYSIS

CTSA utilizes minimum cost tolerancing concepts. Currently the method is limited to a single performance measure. CTSA involves determining the stackup equation, estimating the cost tolerance relationships, and conducting a cost tolerance sensitivity experiment and analysis.

First, the assembly performance measure and its tolerance stack are determined or estimated. This includes the functional assembly tolerance, the features in the stack, and the partial derivatives for the stackup function. There are a variety of methods of determining the stackup equation and the interested reader is referred to Gerth, 1996b for a brief synopsis of these methods.

Next, cost tolerance (CT) curves are estimated for each feature in the stack. The cost tolerance relationship typically takes one of three forms, discrete, piecewise continuous, or continuous (see Figure 1). The shape of the curve has an impact on the type of optimization algorithms that can be used. A variety of cost tolerance curves and solution methods, including dynamic programming [Enrick, 1985], Lagrange multiplier [Spotts, 1973], 0-1 programming [Ostwald and Huang, 1977], non-linear programming [Zhang and Wang, 1993], branch and bound [Lee, et al., 1993], etc. have been applied to this problem. For a summary of the various CT curves and optimization methods, the interested reader is referred to Gerth, 1996b.

Usually, minimum cost optimization methods require a single CT curve. CTSA, however requires 3 curves, namely a nominal curve, an upper estimate and a lower estimate curve. The nominal curve represents the most likely CT relationship. However, since this is likely to be only an estimate, especially in the early design phases, the analyst must estimate a likely upper and lower bound on these estimates. In essence these upper and lower bounds define the region within which the true CT relationship is likely to be.

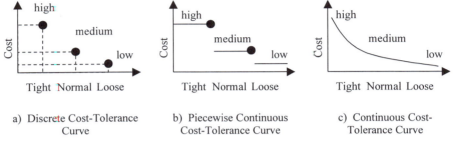

a) Discrete Cost-Tolerance
Curve

b) Piecewise Continuous
Cost-Tolerance Curve

c) Continuous Cost-
Tolerance Curve

Figure 1. Discrete, Piecewise Continuous, and Continuous Cost Tolerance Curves.

At this point, a cautionary note is appropriate. Costs are not only a function of the feature tolerances, but also a function of the feature nominal dimension. This aspect has not been addressed in this paper, and the CT curves presented do not capture this information. When equipment purchases are being considered, other factors such as flexibility, variety of features that can be processed, multiple fixturing, etc. become an issue. The current method does not take these issues into account and assumes that each feature is made on a single machine.

Once the CT curves have been estimated it is possible to conduct the cost tolerance sensitivity analysis. A series of minimum-cost tolerancing solutions are generated based on the upper and lower cost tolerance curves. The decision which CT curve to use for a particular feature is determined according to an experimental design paradigm. Since the purpose of CTSA is to determine which features from many are critical or non-critical, a screening design (resolution III fractional factorial design) is recommended. The independent variables are the CT curves where the upper and lower estimates represent the two levels. The dependent variable is the distance of a particular component tolerance from its upper or lower limit. In essence, the cost-tolerance space is explored in a designed experimental fashion.

For every experiment, the same stackup equation is evaluated and solved for its minimum cost solution assuming different cost tolerance curves. Depending on the particular set of CT curves used, the optimization program will determine a particular set of tolerances to be the optimal minimum cost solution. As a result, some of the feature tolerances will be driven to their tight tolerance limit and others to their loose tolerance limit. A different experimental run involving a different set of CT curves will presumably result in a different set of optimal tolerances. Again, certain feature tolerances will be driven to their tight or loss tolerance limits. However, certain feature tolerances may have been driven to the same tolerance limits in both experiments.

If a feature tolerance is always driven to its lower bound in every experiment, it is a critical feature, because regardless which cost estimate is used, the optimization procedure results in a tight tolerance. This information can be passed on to process and inspection planning, for they will need to ensure the process capability and acceptability of the features. Conversely, if a feature tolerance is always driven to its upper tolerance limit regardless of the CT curve, the feature is non-critical and the cheapest manufacturing method may be employed. This information can also be passed onto process design, as they can begin process planning on these features without too much concern about process capability. All other features will vacillate between the upper and lower bounds depending on the specific cost tolerance curve used. These features are sensitive to the cost estimates and additional tradeoff studies will be necessary to determine their specific tolerance and impact on product performance and cost.

3. CASE STUDY - WINDMILL TRANSMISSION

The scenario studied is a fictitious windmill transmission. The authors, however, worked with a German transmission manufacturer, which preferred to remain anonymous. The company's primary business is industrial transmissions, but they also build 150 kW windmill transmissions. The company provided first stage drawings, some of the cost tolerance information, the stackup function, and general validation of the results.

3.1 The stackup function

A major customer criterion, as determined from a QFD analysis was noise. Transmission noise is directly a function of how well the gear teeth mesh together. The proposed design was a two stage planetary system, where the first stage sun is allowed to float among the planets. This has the advantage that the gear forces will cause the sun to find a natural center among the planets. Thus, a major performance measure is the amount of movement the sun has relative to the planets. An excessive amount of movement would result in excessive noise. At the company's suggestion, only the 1st stage was modeled to simplify the case study.

The transmission consists of a drive and output housing, a drive and output shaft, a universal ring gear, 3 planets, a planet holder, and a sun (see Figure 2). Each of these components has feature dimensions with associated tolerances (see Table 1). The cooperating company supplied the drawings for the planet, planet holder, sun, universal gear, shafts, and transmission assembly with their tolerances. The company engineers accepted the general simplifications used in the development of the stackup function, which is given in equation [5].

$$Y = \frac{35}{109}\left[\left(L - \frac{\tan(20°)}{\left(1 + \frac{\tan(20°)}{2.097}\right)}(J + K)\right) - \left(A + B + (C - D) + (H - E) + F + G + I\right)\right] \quad [5]$$

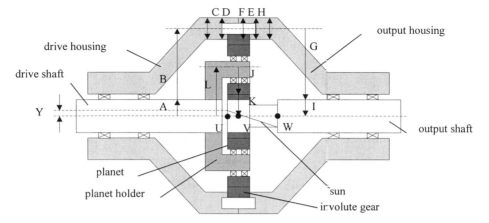

Figure 2. Sketch of Single Stage Planetary Gear Transmission.

Table 1. Feature Tolerances and their Partial Derivatives for the Windmill Transmission.

Component	Feature	Tolerance	Abbreviation	Value (mm)	Derivative
Drive shaft	Housing bearing	Runout	A	0.015	0.321
Drive housing	Universal Gear Pilot	Runout	B	0.05	0.321
		Mating clearance	C	+0.000 -0.044	0.321
Universal gear	Output housing pilot	Mating clearance	D	+0.070 -0.000	0.321
	Drive housing pilot	Mating clearance	E	+0.070 -0.000	0.321
		Runout	F	0.05	0.321
Output housing	Universal Gear Pilot	Runout	G	0.05	0.321
		Mating clearance	H	+0.000 -0.044	0.321
Output shaft	Housing bearing	Runout	I	0.015	0.321
Sun	Gear profile	Profile deviation	J	+/- 0.012	0.440
3 Planetary gears	Gear profile	Profile deviation	K	+/- 0.012	0.440
Planet holder	Sun to planet center	distance variation	L	+/- 0.02	0.321

3.2 The cost tolerance curves

The next step is to estimate the cost tolerance relationships. The cost tolerance (CT) information was very difficult to obtain. The company provided the process capabilities for their machinery, as well as the incremental unit costs for tightening or loosening the tolerances within particular machining centers. However, for most of the component features, with the exception of the gear tooth profile deviations, no cost differences existed for tightening or loosening tolerances with the machining center's capabilities. There were occasions where discussions with engineers knowledgeable in specific areas did lead to cost tolerance curves directly and quickly. This is interpreted as a verification of the basic approach and existence of the CT curve knowledge. In other words, there are knowledgeable individuals who can estimate with upper and lower bounds the costs and tolerance capabilities associated with alternative manufacturing strategies for particular components or component features.

The CT curves were generated from a matrix of cost estimates obtained from engineers. The engineers estimated from a reference design how much it would cost at a minimum and a maximum to increase or decrease the tolerances for each dimension. Where no information was available, the authors estimated the cost. Hence the costs represent an incremental per unit cost relative to an existing reference design. The cost estimates were used to compute continuous or piecewise continuous curves using an inverse power function. The CT curve and upper and lower tolerance limits are presented in Table 2. The tolerances are expressed in millimeters, and the costs are expressed in DM.

3.3 Sensitivity experiment and analysis

A 2^{12-8} fractional factorial design was constructed, and each experimental run consisted of solving the minimum cost tolerance problem using different cost tolerance curves. The independent variables were the upper or lower cost tolerance curve for each tolerance. The dependent variable was the difference between the solver tolerance solution value and the lower tolerance limit set as a constraint in the optimization procedure. This resulted in a 12 x 16 matrix. The assembly tolerance was set at 20 μm. The standard deviation across the 16 experiments was computed for each tolerance, and the results are shown in Table 3.

As can be seen, only 4 tolerances, C, D, E, and H show a sigma of 0. This means they were consistently driven to the same tolerance value. These are the clearance tolerances, which CT curves are flat (see Table 2). Thus, the tolerances will always be driven to their tight tolerance since there is no cost involved. The remaining tolerances have non-zero standard deviations, and thus, conclusions are difficult to make; they are all sensitive to the cost estimates made. Closer, examination of the data reveals that for 5 feature dimensions, the tolerances are all pushed to one limit or the other, with only a single value being larger or smaller than the rest (bold values in Table 3). This leads to the belief that for the most part, these tolerances will not be very sensitive to the cost estimates.

Table 2. Cost Tolerance Curve Equations

Feature	Cost Tolerance Curve Equations
Sun gear profile deviation (J)	$C_{Lower} = 14.22 - \dfrac{5.44 \cdot 10^4}{tol^{-1.85}} \quad for\, 0.006 \le tol \le 0.025$ $C_{Upper} = 40.01 - \dfrac{2.28 \cdot 10^3}{tol^{-0.93}} \quad for\, 0.006 \le tol \le 0.025$
Drive and output shaft runouts (A and I)	$C_{Lower} = 10.44 - \dfrac{1.37 \cdot 10^7}{tol^{-2.88}} \quad for\, 0.0025 \le tol \le 0.0125$ $C_{Upper} = 26.22 - \dfrac{1.59 \cdot 10^4}{tol^{-1.31}} \quad for\, 0.0025 \le tol \le 0.0125$
Drive and output housing runouts (B and G)	$C_{Lower} = \begin{cases} 90 & for \quad 0.005 \le tol < 0.02 \\ 0 & for \quad 0.020 \le tol < 0.025 \\ -280 \cdot tol + 7 & for \quad 0.025 \le tol \le 0.05 \end{cases}$ $C_{Upper} = \begin{cases} 110 & for \quad 0.005 \le tol < 0.02 \\ -1400 \cdot tol + 35 & for \quad 0.02 \le tol < 0.025 \\ 0 & for \quad 0.025 \le tol \le 0.05 \end{cases}$
Universal gear pilot runout (F)	$C_{Lower} = \begin{cases} 125 & for \quad 0.005 \le tol < 0.02 \\ 0 & for \quad 0.020 \le tol < 0.025 \\ -600 \cdot tol + 15 & for \quad 0.025 \le tol \le 0.05 \end{cases}$ $C_{Upper} = \begin{cases} 175 & for \quad 0.005 \le tol < 0.02 \\ -5.31 + \dfrac{1.33 \cdot 10^{-6}}{tol^{4.12}} & for \quad 0.020 \le tol \le 0.05 \end{cases}$
Planet gear profile deviation (K)	$C_{Lower} = 38.16 - \dfrac{6.36 \cdot 10^3}{tol^{-1.15}} \quad for\, 0.006 \le tol \le 0.025$ $C_{Upper} = 71.56 - \dfrac{1.67 \cdot 10^3}{tol^{-0.72}} \quad for\, 0.006 \le tol \le 0.025$
Planet holder(L)	$C_{Lower} = \begin{cases} 175 & for \quad 0.01 \le tol < 0.02 \\ 0 & for \quad 0.02 \le tol \le 0.10 \end{cases}$ $C_{Upper} = \begin{cases} 250 & for \quad 0.01 \le tol < 0.02 \\ 0 & for \quad 0.02 \le tol \le 0.10 \end{cases}$
All clearances (C, D, E, and H)	$C_{Lower} = C_{Upper} = 0 \quad for \quad 0.022 \le tol \le 0.0875$

Table 3. Results of the 20mm Experimental Series.

Component	Tolerance	Drive Shaft	Drive Housing	Drive Housing	Universal Drive	Universal Output	Universal	Output Housing	Output Housing	Output Shaft	Sun	Planet	Planet holder
		Runout	Runout	Pilot Clearance	Pilot Clearance	Pilot Clearance	Runout	Runout	Pilot Clearance	Runout	Profile Deviation	Profile Deviation	Center to Center Distance
experiment		A	B	C	D	E	F	G	H	I	J	K	L
1		0.0100	0.0150	0.0000	0.00000	0.00000	0.0150	0.01500	0.00000	0.00688	0.00938	0.01059	0.01000
2		**0.0089**	0.0150	0.0000	0.00000	0.00000	0.0150	0.01500	0.00000	0.00950	0.00720	0.00769	0.01000
3		0.0100	0.0150	0.0000	0.00000	0.00000	0.0150	0.01500	0.00000	0.00847	0.00682	0.00746	0.01000
4		0.0100	**0.0167**	0.0000	0.00000	0.00000	**0.0161**	**0.00000**	0.00000	0.01000	0.01900	0.01900	**0.01110**
5		0.0100	0.0150	0.0000	0.00000	0.00000	0.0150	0.01500	0.00000	0.00838	0.00703	0.00760	0.01000
6		0.0100	0.0150	0.0000	0.00000	0.00000	0.0150	0.01500	0.00000	0.00857	0.00677	0.00707	0.01000
7		0.0100	0.0150	0.0000	0.00000	0.00000	0.0150	0.01500	0.00000	0.00849	0.00676	0.00742	0.01000
8		0.0100	0.0150	0.0000	0.00000	0.00000	0.0150	0.01500	0.00000	0.00790	0.00860	0.00795	0.01000
9		0.0100	0.0150	0.0000	0.00000	0.00000	0.0150	0.01500	0.00000	0.00857	0.00663	0.00721	0.01000
10		0.0100	0.0150	0.0000	0.00000	0.00000	0.0150	0.01500	0.00000	0.00840	0.00676	0.00777	0.01000
11		0.0100	0.0150	0.0000	0.00000	0.00000	0.0150	0.01500	0.00000	0.00859	0.00656	0.00719	0.01000
12		0.0100	0.0150	0.0000	0.00000	0.00000	0.0150	0.01500	0.00000	0.00846	0.00670	0.00759	0.01000
13		0.0100	0.0150	0.0000	0.00000	0.00000	0.0150	0.01500	0.00000	0.00858	0.00665	0.00712	0.01000
14		0.0100	0.0150	0.0000	0.00000	0.00000	0.0150	0.01500	0.00000	0.00846	0.00666	0.00762	0.01000
15		0.0100	0.0150	0.0000	0.00000	0.00000	0.0150	0.01500	0.00000	0.00858	0.00677	0.00700	0.01000
16		0.0100	0.0150	0.0000	0.00000	0.00000	0.0150	0.01500	0.00000	0.00846	0.00696	0.00734	0.01000
sigma		0.0003	0.0004	0.0000	0.0000	0.0000	0.0003	0.0038	0.0000	0.0006	0.0031	0.0030	0.0003
tolerance		Loose	Tight	Tight	Tight	Tight	Tight	?	Tight	?	?	?	Tight
Conclusion		Non-crit.	Critical	Critical	Critical	Critical	Critical	Unclear	Critical	Unclear	Unclear	Unclear	Critical

The tolerance values associated with factors B, F, and L are all slightly larger than the optimal values for the other experimental runs. Thus, it would be possible to tighten these factors without violating any of the boundary conditions. Indeed doing so does not increase cost and reduces the overall assembly variation to 0.0196 μm from the limit of 0.020 μm. Thus, one can conclude that factors B, C, D, E, F, H, and L are critical dimensions that should be held to their tight tolerances. Factor A is a little tighter in experiment 2 than all other experiments. If one were to loosen it to match the other experimental values the cost drops by 12%, but the assembly tolerance increases to 20.07 μm. Since the analysis was conducted in the early design stages, the 20.07 μm is an acceptable deviation, and one can conclude that factor A is not a critical dimension and can be held at its loose tolerance.

Factor G has a single instance where the tolerance is driven from its 0.015 mm above its tight tolerance limit of 0.005 mm to its tight tolerance limit. Examination of its CT curve equation (see Table 2) shows that it is step function, with a large jump at 0.020 mm. Thus, G has a moderate effect on the tolerance stack, and is thus always driven to the jump point. In only a single instance was it forced to go to the high limit to meet the assembly yield constraint. However, because of its moderate effect and its large jump, it cannot be concluded to be a critical or non-critical factor (denoted with a question mark).

The factors I, J, and K have large sigma values indicating that their tolerances change as a function of the particular CT curve employed. Thus, they are very sensitive to the precision of the cost estimates, and their effect on the sun position variation would need to be investigated further before their criticality can be determined.

4. DISCUSSION AND CONCLUSIONS

The work presented here is a first attempt at bringing tolerance information early into the design phase. Cost tolerance sensitivity analysis is one of the first methods of its kind in that it combines minimum cost tolerancing and design of experiment concepts to deal with the inherent uncertainty of the cost estimates early in the design stages. The methodology was evaluated on a case study of a windmill transmission and shown to lead to useful results. However, there are a few additional points worth noting.

First, the factors with the smallest impact in the stackup function, namely factors J and K, were not found to be non-critical dimensions. In other words, they were not always driven to their loose tolerance limit. Conversely, the factors with a high impact, such as factor A, were not found to be critical dimensions, i. e., they were not always driven to their tight tolerance limit. Thus, one cannot determine dimension criticality based on the stackup function alone.

Similarly, one cannot determine the criticality of the factors based on the processing costs. Factors associated with expensive processes, such as factor L, were not driven to their loose limit, and factors associated with inexpensive processes, such as factor A, were not driven to their tight limit. Thus, one must conclude that the problems associated with identifying a cost effective assembly cannot be viewed purely from a product performance point of view, which would only consider a factor's effect on performance, nor from a processing point of view, which would only consider the cost of producing the dimension. Rather, both sides must be examined simultaneously to identify the tradeoff opportunities. CTSA is a method that enables the CE team to identify some of the cost effective tradeoffs early in the design.

Lastly, the factor A was deemed non-critical and factors B, D, and L critical only after examination of the individual data points. Loosening the tolerance on A and tightening it on B, D, and L on a single experimental run enabled these conclusions. These modifications represent near optimal solutions, which may also be acceptable solutions. Therefore, examination of the near solutions which either exceed the yield constraint and/or are not minimum cost solutions, may provide better insight into the factor relationships, the processing alternatives, and product performance requirements. This understanding would hopefully lead to a better insight into the tradeoffs between product performance and manufacturing cost.

The most difficult aspect of the methodology is obtaining cost tolerance information. Although there is every indication that the knowledge is available in every company, it is in the head of individuals. As such the information is neither organized, nor available from any central location. It is also usually not a simple task to identify the people with the knowledge. The cost tolerance curves which were referenced as provided by the company were provided by two individuals who did not document the information nor were able to point to any independent sources other than their own experience. Research into this area, i.e., how cost-tolerance information should be structured, obtained, organized, and disseminated is a difficult and important task.

ACKNOWLEDGEMENTS

The authors thank the Deutsche Forschungs Gemeinschaft for their funding of this research.

REFERENCES

[Chase and Greenwood, 1988] Chase, K. W. and Greenwood, W. H.; "Design issues in mechanical tolerance analysis"; *Manufacturing Review*, Vol. 1, 50-59.

[Chase, et al., 1990] Chase, K. W., Greenwood, W. H., Loosli, B. G., and Hauglund, L. F.; "Least cost tolerance allocation for mechanical assemblies with automated process selection"; *Manufacturing Review*, Vol. 3, No. 1, pp. 49-57.

[Enrick, 1985] Enrick, N. L.; *Quality, Reliability, and Process Improvement*, Industrial Press Inc.

324

[**Gerth, 1996a**] Gerth, R. J.; "Compensatory Tolerancing"; In: *Proceedings of the 5th Industrial Engineering Research Conference*, May 18-20, Minneapolis, MN, pp. 333-338.

[**Gerth, 1996b**] Gerth, R. J.; "Engineering tolerancing: a review of tolerance analysis and allocation methods"; *Engineering Design and Automation*, 2(1), 3-22.

[**Lee, et al., 1993**] Lee, W. J., Woo, T. C., and Chou, S.Y.; "Tolerance synthesis for non-linear systems based on non-linear programming"; *IIE Transactions*, 25(1): 51-61.

[**Ostwald and Huang, 1977**] Ostwald, P. F. and Huang, J.; "A method of optimal tolerance selection"; *Journal of Engineering for Industry*, 558-565.

[**Speckhart, 1972**] Speckhart, F. H.; "Calculation of tolerance based on a minimum cost approach"; *Journal of Engineering for Industry*, May, pp. 447 - 453.

[**Spotts, 1973**] Spotts, M., F.; "Allocation of tolerances to minimize cost of assembly"; *Journal of Engineering for Industry*, August, 762-764.

[**Zhang and Huq, 1992**] Zhang, H. C., and Huq, M. E.; "Tolerancing techniques: the state-of-the-art"; *International Journal of Production Research*, **30** (9), 2111-2135.

[**Zhang and Wang, 1993**] Zhang, C., and Wang, B.; "Tolerance analysis and synthesis for cam mechanisms"; *International Journal of Production Research*, 31: 1229-1245.

Motion constraints in a 2D polygonal assembly

Bing Li and Utpal Roy
Department of Mechanical, Aerospace and Manufacturing Engineering
Syracuse University
Syracuse, New York 13244
uroy@mailbox.syr.edu

Abstract: The paper analyzes the mating natures of a 2D polygonal assembly. For an assembly composed of an object part and a target part, any geometric variation will introduce extra degrees of freedom between two parts and result in some different assembly configurations. In order to investigate the assembly configuration uncertainties, translational and rotational motion constraints are generated by analyzing geometric contact relations between variant parts. Design function in terms of dimensional measurements has different values with regard to different assembly configurations. The part position corresponding to the optimal (maximum or minimum) design function values can be obtained. Relative positioning scheme based on proposed motion constraints can be used effectively to automate the tolerance analysis for an assembly.
Key Words: Motion constraints, variational geometry, relative positioning.

1. INTRODUCTION

When an assembly model is created in a CAD system, each part is positioned relative to an adjacent part or a reference datum. Relative positioning is controlled by geometric relations among parts. Changes to the size and shape of one part would have an effect on the positions of the other parts in the assembly. From tolerance analysis point of view, we need to investigate how small geometric variations of part feature (i) alter the nominal assembly configuration and (ii) affect the design function.

Relative positioning techniques have been developed by specifying some set of constraints and an objective function. The final positions of parts in the assembly can be determined by realizing the objective function using optimization methods [Mattikalli and Khosla, 1992, Turner, 1990, Sodhi and Turner, 1994, Scott and Gabriele, 1989, Mullineux, 1987, Inui and Kimura, 1990, Inui et al., 1996].

After each variant part is positioned, the assembly is reconfigured. The new values of design functions are then obtained. In order to develop an accurate positioning scheme, motion constraints between variant parts need to be carefully studied.

2. PROPOSED RELATIVE POSITIONING SCHEME

In a two-part assembly, the object part (moveable) is positioned with respect to the target part which is fixed with the reference coordinates (Figure 1*a*).

Part features can be classified as contact feature and non-contact features. For 2D objects, they are contact edges (such as *Edge a* and *Edge b*) and non-contact edges (such as *Edge e* and *Edge f*) (Figure 1*a*). The mating relations between contact features are expressed as "*against*" such as: *Edge a* of *Part 1* is *against Edge b* of *Part 2*. The *against* mating condition means that two faces are parallel each other and the distance between them is zero. All the *against* mating relations cannot be always realized simultaneously under variation conditions. Figure 1*b* shows mating uncertainties of Figure 1*a*. Design function, a dimension or a distance vector (for example, the clearance F_d between *Edge e* and *Edge f* in Figure 1*a*) will be affected by geometric variations in individual parts.

Realizing the *against* mating relations for some specific mating feature pairs has been pursued as an objective in previous researches. However, assembly configurations resulted from mating relation oriented positioning scheme may not be complete since some configurations might be omitted. It therefore necessitates a total enumeration of assembly situations in order to achieve the optimal assembly configuration.

We assume the object part is a convex polygonal object while the target part, concave. The proposed positioning scheme for 2D polygonal assemblies will be four steps: (1) *Nominal positioning*. (2) *Initial variant positioning*: the object part is positioned with one of the mating features being *against* its counterpart of the target part. (3) *Continued variant part positioning*: the object part will try to realize the *against* contact status for another mating feature pair by translation and rotation repeatedly until all the potential candidates of the mating feature pairs are positioned. (4) *Final positioning*: for each mating feature pair positioning, the design function value is calculated. After Step 3 is finished, the positioning with the optimal design function value (max. or min.) is chosen as the final assembly configuration.

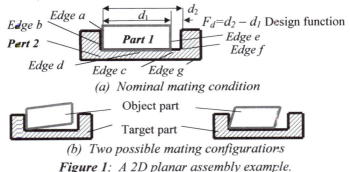

(a) *Nominal mating condition*

Object part
Target part

(b) *Two possible mating configurations*

Figure 1; *A 2D planar assembly example.*

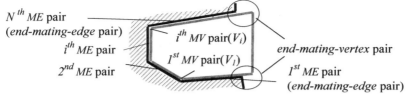

Figure 2; Mating edges and mating pairs.

Figure 3; Translational constraint classifications.

3. MOTION CONSTRAINTS

Suppose two parts mate with more than three mating-edge pairs. One of the end mating-edge pair is named as the 1st mating-edge (*ME*) pair. The adjacent mating edge pair is the 2nd *ME* pair. The naming process goes on until it reaches the other end-mating-edge pair (as the Nth *ME* pair). The mating-vertices are also numbered the same way (Figure 2).

The motion constraints are realized when mating-features of different parts contacts, among which the vertex is an active-vertex, the edge is an active-edge, and the point where the contact occurs is a contact-point. Distance D between mating-features (the active-vertex and the contact-point) is the translational constraint. Rotation angle φ derived from the distance between mating-features is the rotational constraint.

3.1 Translational constraint

Translational constraints are classified into three types: Type I(active mating-vertex pair), Type II(inactive mating-vertex pair) and Type III(end-mating-vertex pair) (Figure 3).

When the object part translates in direction T, vertex P_2 may contact edge l_1^{target} or l_2^{target} of target part. Edge l_1^{target} and l_2^{target} are adjacent edges of vertex Q_2. Vertex P_2 and Q_2 are mating-vertices. Constraint distance D is a signed distance between P_2 and contact-point Q_2^* (Figure 4).

- Draw two lines from vertex P_2, translation line l^T is in direction T and test line l^{test} connects vertex Q_2.
- Translation line l^T intersect with edge l_1^{target} or l_2^{target} depending on T. The slopes of translation line l^T and test line l^{test} need to be compared to determine the active edge.

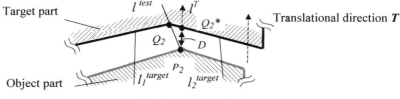

(a) Constraint distance

(b) Edge l_1^{target} as an active-edge   (c) Edge l_2^{target} as an active-edge

Figure 4; Translational constraint.

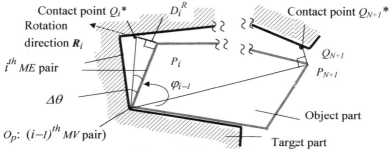

Figure 5; Pure rotation.

3.2 Rotational constraints

With small variations, two types rotational constraint angle φ are derived from translational constraint distances. They are pure rotation (object part rotate around a fixed pivot O_P) and complex rotation (object part not only rotate but also translate).

In pure rotation, the object part rotates around a fixed pivot point: contact-point O_P (Figure 5). Suppose point O_P occurs at vertex P_{i-1} of the ith mating-vertex pair, and vertex P_i is another mating-vertex in the ith mating-vertex pair of the object part. Similarly, Q_i is the mating-vertex of the target part. The rotational constraint angle for the ith ME pair, φ_i^P, can be calculated as $\varphi_i^P = \min.\ (\varphi_i,\ \varphi_{i+1},\ ...,\ \varphi_{N+1},\ \Delta\theta)$. $\Delta\theta$ is the angle between inner mating angles of the active i^{th} ME pair. $\varphi_k = \dfrac{D_k^R}{O_P P_k}$ ($k = i,\ ...,\ N+1$), where D_k^R can be calculated the same way as described in Section 3.1.

329

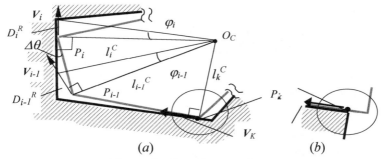

(a)   *(b)*

Figure 6; *Complex rotation.*

Unlike pure rotation, a pivot point O_C of complex rotation is generated as the instantaneous rotating center under small variation assumption. The procedure to obtain the rotational constraint angle for the ith ME pair, φ_i^C, is illustrated in Figure 6's example.

- During rotation, two i^{th} mating edges remain in contact at the *contact-point* P_i. P_i's velocity vector (V_i) is parallel to the mating edge of the target part. The instantaneous center is in a *connect line* l_i^C from point P_i perpendicular to V_i.
- *Contact-point* P_k of the object part in its active end mating edge is defined in two situations shown in Figure 6. During rotation, two end mating edges remain in contact at the point P_k whose velocity vector (V_k) is parallel to the active end mating edge of the target part. Draw another connect line (l_k^C) from point P_k perpendicular to V_k. The pivot point Q_C is defined as the intersection point of two connect-lines.
- Draw connection lines from pivot point Q_C to ($N- i$) vertices of the object part. The velocity vectors of those vertices are perpendicular to their connect lines, and their magnitudes (D^R) can be calculated the same way as described in Section 3.1.
- The rotational constraint angle φ_i^C is obtained as $\varphi_i^C = \min. (\varphi_i, \varphi_{i+1}, ..., \varphi_{N+1}, \Delta\theta)$. $\Delta\theta$ is the different angle between two inner mating angles of the active ME pair and

$$\varphi_k = \frac{D_k^R}{O_C P_k}, (k = i, ..., N+1), \text{ where } D_k^R \text{ is the velocity vector magnitude.}$$

3.3 Mating procedures

The positioning begins from the 1st ME pair and ends at the Nth ME pair (1st ME$\rightarrow 2^{nd}$ ME$\rightarrow$ ..., $\rightarrow(N-1)^{th}$ ME$\rightarrow N^{th}$ ME), which may skip some mating-edge pairs due to constraints. The positioning from the ith pair to the $(i+1)^{th}$ pair is composed of 3 steps.

Step 1 (*Pre-translation position*): The two part mate with their i^{th} ME pair (*active ME pair*) in a line-contact pattern.

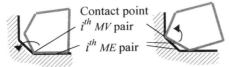

(a) post-translation   *(b) post-rotation*

Figure 7; *Type I mating examples of*
pre-rotation position.

(a) post-translation   *(b) post-rotation*

Figure 8; *Type II mating examples of*
pre-rotation position.

Step 2 (*Pre-rotation position*): The object part slides along the active *ME* pair of the target part in direction of making the $(i+1)^{th}$ *MV* pair closer by the translational constraints distance: $D_i{}^{object}$ = min. $(D_1, D_2, ..., D_j, ..., D_N)$, where D_j is the translational distance for the vertex $V_j{}^{object}$ and can be derived by the way as described in Section 3.1.

This *pre-rotation* for the ith *ME* pair, classified into two mating types, may be the result of a translation (*post-translation result*) or another rotation (*post-rotation result*). (i) If the object part will rotate around the *contact-point* where two vertices of the ith *MV* pair coincide, then this configuration is *Type I Mating* for the ith *ME* pair and the rotation is a *pure rotation* (Figure 7). (ii) If the object part will rotate around a pivot point which is an *instantaneous rotation center* from the configuration, then this configuration is *Type II Mating* for the ith *ME* pair and the rotation is a *complex rotation* (Figure 8).

Step 3 (*Post-rotation position*): The object part rotates around some pivot point under rotation constraints in form of pure rotation or complex rotation.

(1) Rotational constraints for *Type I mating*: The rotational angle φ_{Ii} = min. (φ_1, φ_2, ..., φ_j, ..., φ_N, $\Delta\theta_i$), where $\Delta\theta_i = \theta_i{}^{target} - \theta_i{}^{object}$ and φ_j is the allowable rotational angle for vertex $V_j{}^{object}$. The *post-rotation* position will be one of the two cases:

Case I: If two edges of the $(i+1)^{th}$ *ME* pair mate in a line-contact pattern, then the $(i+1)^{th}$ *ME* pair is in the *pre-translation* position, the positioning on the $(i+1)^{th}$ *ME* pair will begin from *pre-translation* position as described in the previous section (Figure 9).

Case II: If two edges of the $(i+1)^{th}$ *ME* pair do not mate in a line-contact pattern and two parts contact at two points which don't belong to the same *ME* pair (i.e., one contact-point is the pivot point of the ith *MV* pair and the other contact-point belongs to the kth *MV* pair ($k \geq i+1$), then the $(i+1)^{th}$ *ME* pair is in the *pre-rotation* position again and the object part will undergo a complex rotation (Figure 10).

(2) Rotational constraints for *Type II mating*: The complex rotational direction is determined by the rotation goal: to make two vertices of the kth *MV* pair as close as possible. k ($k \geq i+1$) is the number of mating pair which follows the current (the ith) mating pair, immediately or non-immediately. In Figure 10a, $V_k{}^{target}$ contact the kth *ME* edge of the target part, and the object part will rotate clockwise to make two vertices closer. In the other example shown in Figure 10b, $V_k{}^{target}$ contact the $(k+1)^{th}$ *ME* edge of the target part, the object part will then rotate counterclockwise to reach the goal.

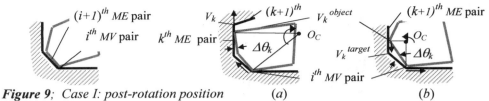

Figure 9; *Case I: post-rotation position is a pre-translation position for $(i+1)^{th}$ ME pair*

(a) (b)

Figure 10; *Case II: post-rotation position is a pre-rotation position for k^{th} ME pair*

There are two different scenarios to obtain the instantaneous rotating center. (i) If *Type II Mating* occurs as the result of *post-translation* where two parts mate with one contact-point and one pair of line-contact edges (Figure 8a), then the derivation of the pivot point follows the same procedure as described in Section 3.2. (ii) If *Type II Mating* occurs as the result of *post-rotation* where two parts mate with two contact-points, the pivot point is then the intersected point between two connect-lines (Figure 10a, b).

After the pivot point is generated, the rotational constraint angle $\varphi_{II\,k}$ (for the kth ME pair) can be obtained as $\varphi_{II\,k}$ = min. (φ_1, φ_2, ... , φ_j ,... , φ_N, $\Delta\theta_k$), where $\Delta\theta_k$ is the angle between two edges of the kth ME pair (Figure 10) and φ_k is the allowable rotational angle for vertex V_k^{object} which can be calculated in the way as described in Section 3.2.

After rotation, the post-rotation position of the object part will be either pre-translation position or pre-rotation position for the mth MV pair ($m \geq k + 1$).

4.   CASE STUDY

The positioning scheme we have discussed in previous sections can also be applied to cases where one or two end-edge (or side-edge) pairs do not mate in nominal situations. As an example, we consider a simple assembly: a rectangular object sitting in a slot. (This is a cross section projection of a 3D slide-guideway assembly.)   In this example, there are only two mating-edge pairs, and the clearance is the design function. The design function value corresponding to different positionings can be obtained by an analytical method.

From perspective of tolerance analysis, we are more interested in the maximum value of the design function (clearance) since it controls the functionality of the assembly. The clearance is formed by two non-contact features, one being in the object part, the other in the target part.   The maximum clearance leads to largest separation of these two non-contact features, i.e., the shortest distance between the non-contact feature of the object part (*Edge2*) and point V_1, the intersection vertex of *Edge0* and *Edge1* of the target part (Figure 12).

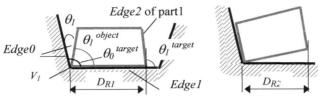

$$(a) \; \textit{Pre-rotation position} \qquad (b) \; \textit{Post-rotation position}$$

Figure 11; *RAD for type I Mating*

This newly introduced distance is termed as the Real Acting Distance (*RAD*) D_R. *RAD* is a projected distance on *Edge1* of the target part and measured from V_1 to the farthest vertex of *Edge2* (end-point) of the object part (Figure 11). *RAD* reflects the approximity of the object part and the target part in some directions.

The following terms are introduced. $\theta_1 = \theta_0{}^{target} - \theta_1{}^{object}$ is named as *mating-angle* where $\theta_0{}^{target}$ and $\theta_1{}^{object}$ are the angles between *Edge0* and *Edge1* of the target part and the object part respectively. $\theta_1{}^{target}$ is the angle between *Edge2* and *Edge1* of the object part. Suppose the design function is the distance between *Edge0* of the target part and *Edge2* of the object part, our goal is to find the configuration of the assembly which leads to the minimum design function value.

Suppose two parts first mates with their *Edge1*s in the line-contact pattern, then the object part slides along its *Edge1* pair to reach *pre-rotation* position. There are only two mating types for this L-shape assembly, i.e., *Type I mating* and *Type II mating*.

4.1 Type I mating: when mating-angle $\theta_1 > 0$

With small variation of $\theta_1{}^{target}$, we noticed $D_{R1} \approx D_{R2}$. Therefore, the final position can be either *pre-rotation* position or *post-rotation* position (Figure 11).

4.2 Type II mating: when mating angle $\theta_1 < 0$

Case I ($\theta_1{}^{target} < \pi/2$): Vertex *B* is the *end-point* and $D^B{}_R$ is the *RAD*. The complex rotation will be realized with two steps (Figure 13*a,b,g*).

Step1 (pure rotation): The object part rotates (clockwise in this example) around vertex *A* to make its *Edge0* (h_1) parallel to *Edge0* of the target part (line a_1a_2). With the new position of h_1 (line b_1b_2), the distance between line a_1a_2 and b_1b_2 is r. $r = h_1\sin\theta_1 \approx h_1\theta_1$. Vertex B is now located in its new position by distance $\Delta_B \approx h_2\theta_1 \, (\rightarrow)$.

Step 2 (translation): The object part then translates (left in this example) to make *Edge0* (h_1) collinear with line a_1a_2. The translation distance is

$$\Delta_h = \frac{r}{\sin\theta_0{}^{target}} = \frac{h_1\sin\theta_1}{\sin\theta_0{}^{target}} \approx \frac{h_1\theta_1}{1} = h_1\theta_1 \; (\leftarrow)$$

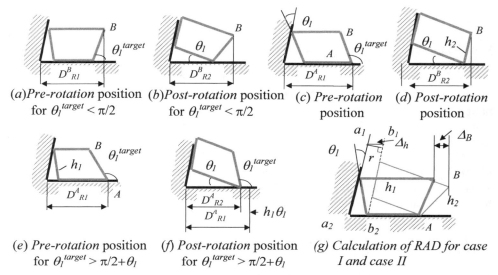

(a)Pre-rotation position   *(b)Post-rotation* position   *(c) Pre-rotation*   *(d) Post-rotation*
for $\theta_l^{target} < \pi/2$ for $\theta_l^{target} < \pi/2$ position   position

(e) Pre-rotation position   *(f) Post-rotation* position   *(g) Calculation of RAD for case*
for $\theta_l^{target} > \pi/2 + \theta_l$ for $\theta_l^{target} > \pi/2 + \theta_l$ *I and case II*

Figure 12; *RAD for Type II mating.*

The *RAD* of the post-rotation position is $D^B_{R2} = D^B_{R1} + \Delta_B - \Delta_h$. If the length of h_l is longer than that of h_2 after variation, then the pre-rotation position will be the final position since it has shorter *RAD*. Otherwise, the post-rotation position will be the final position. In preliminary analysis, the *RAD* remains unchanged in both pre-rotation and post-rotation positions (i.e., $D^A_{R1} \approx D^B_{R2}$) under small variation assumption ($h_l \approx h_2$).

Case II ($\pi/2 < \theta_l^{target} < \pi/2 + \theta_l$): Vertex A is the end-point in a pre-rotation position and vertex B is the end-point in a post-rotation position ($D^B_{R2} = D^B_{R1} + \Delta_B - \Delta_h$). The *RAD* is the shorter one of D^B_{R2} and D^A_{R1}. In preliminary analysis, the post-rotation position will be the final position (because $\Delta_B \approx \Delta_h$ and D^B_{R1} is shorter than D^A_{R1}.) (Figure 12c,d,g).

Case III (when $\theta_l^{target} > \pi/2 + \theta_l$): Vertex A is the end-point and the *RAD* of the post-rotation is reduced from D^A_{R1} to D^A_{R2} by $D^A_{R1} - D^A_{R2} = \Delta_h = h_l \theta_l$ where h_l is the length of *Edge0* of the object part (Figure 13e,f,g).

4.3 Tolerance analysis

From Section 4.2, we can derive the relationships necessary to perform tolerance analysis for assembly using Monte Carlo simulations. A series of random numbers is first generated and then input into the assembly system. The dimensional design function is measured as the system output. Statistic results can be obtained based on large amount of samples. For detailed description, see [Roy and Li, 1999].

5. CONCLUSION

A method of calculating and representing positions of 2D polygonal parts in assemblies has been presented. Translational and rotational constraints are developed due to the extra degree of freedom caused by the shape and size variation of parts. By computing translational and rotational constraints, the space of allowed motion parameters for each mating-edge pair are obtained and object part transformation is realized. Assembly configuration uncertainties caused by part variations are clarified by realizing some objective function.

One important use of this work toward a CAD system is in the solution of tolerancing problems. To produce tolerance measurements, the assembly configurations are simulated based on part variations (using either worst case or statistical model) and the variation range of measurements are obtained.

REFERENCES

[Inui and Kimura, 1991] Inui, M. and Kimura, F.; "Algebraic reasoning of position uncertainties of parts in an assembly"; *1991 ACM Symposium: Solid Modeling Foundations and CAD/CAM Applications*, pp. 419-428; Austin, TX, USA; 5-7, June.

[Inui et al., 1996] Inui, M., Miura, M. and Kimura, F.; "Positioning conditions of parts with tolerances in an assembly"; In *Proceedings of the 1996 IEEE International Conference on Robotics and Automation*, pp. 2202-2207; Minneapolis, MN, USA.

[Mattikalli and Khosla, 1992] Mattikalli, R. S. and Khosla, P. K.; "Motion constraints from contact geometry: Representation and analysis"; In *Proceedings of 1992 IEEE International Conference on Robotics and Automation*, pp. 2178-2185; Nice, France.

[Mullineux, 1987] Mullineux, G.; "Optimization scheme for assembling components"; *Computer-Aided Design*, 19(1), pp. 35-40.

[Roy and Li, 1999] Roy, U. and Li, B.; "3D variational polyhedral assembly configuration"; In *Proceedings of the 6th CIRP International Seminar on Computer Aided Tolerancing*; Enschede, the Netherlands.

[Scott and Gabriele, 1989] Scott, R.T. and Gabriele, G.A.; "Computer aided tolerance analysis of parts and assemblies"; In *Proceedings of ASME 15th Design Automation Conference*, v1, pp. 29-36; Montreal, Quebec, Canada; Sept. 17-21.

[Sodhi and Turner, 1994] Sodhi, R. and Turner, J.U.; "Relative positioning of variational part models for design analysis"; *Computer-Aided Design*, 26(5), pp. 366-378.

[Turner, 1990] Turner, J.U.; "Relative positioning of parts in assemblies using mathematical programming"; *Computer-Aided Design*, 22(7), pp. 394-400.

3D variational polyhedral assembly configuration

Utpal Roy and Bing Li
Department of Mechanical, Aerospace and Manufacturing Engineering
Syracuse University
Syracuse, New York 13244
uroy@mailbox.syr.edu

Abstract: Parts with geometric (size and shape) variations will generate assembly configuration uncertainties. To perform tolerance analysis, the real positions of variant parts and variant assembly configuration need to be investigated. In this paper, a relative positioning scheme is proposed to determine the optimal configuration of variant parts in an assembly. A method of calculating and representing positions of 3D polyhedral parts in assembly has been presented. By computing translational and rotational constraints, the configuration space for allowed motion parameters for each mating pair is obtained and the transformation of the object part is realized. Assembly configuration uncertainties caused by part variations are clarified by realizing the objective function. A 3D example shows the proposed positioning scheme on tolerance analysis of assemblies.
Key Words: Relative positioning, variational geometry, tolerance analysis.

1. INTRODUCTION

The positions of parts in an assembly are determined by geometric constraints imposed on the degree of freedom between mating parts. When tolerances are specified for the parts, changes to the size and shape of one part would have an impact on the positions of the other parts in the assembly because the part variations generate some extra degrees of freedom. The process of positioning a part relative to another part in an assembly is termed as relative positioning, which is used extensively in tolerance analysis.

In order to conduct such a tolerance analysis, Monte Carlo simulation is usually used, which includes the following procedures: (i) generate variational models of variant parts according to their tolerance specifications, (ii) analyze the mating natures between variant parts, and (iii) evaluate the functionality of different sets of part variations (generated randomly) by performing numerical calculations. A method is presented in this paper to analyze part position uncertainties between two assembled parts (by analyzing the translational and rotational constraints between them), i.e., procedure (ii) and (iii) of above mentioned Monte Carlo simulation.

2. LITERATURE REVIEW ON RELATIVE POSITIONING

All relative positioning techniques can be classified into two broad categories in terms of constraint conditions. Explicit technique applies exact constraints to parts without taking into consideration of any part variations [Lee and Andrews, 1985, Rocheleau and Lee, 1987] and is limited in application to the positioning of nominal parts in assemblies. Implicit technique specifies some set of constraints and an objective function rather than applies exact constraints to parts. Final positions of the parts in the assembly can be determined by realizing the objective function. The implicit technique can be applied to both nominal and variant situations [Turner, 1990, Sodhi and Turner, 1994, Scott and Gabriele, 1989, Mullineux, 1987, Inui and Kimura, 1990, Inui et al, 1996].

The assembly is reconfigured after variant parts are positioned. The new values of design functions are then obtained. Since all positioning schemes are determined by their objective functions, and different objective functions give rise to different assembly configurations and different design function values, design function is only expressed in terms of objective function. In existing approaches only mating relations are used to formulate the objective functions. Hence, those design functions are functions of mating relations, and those positioning strategies are mating relation oriented.

3. 3D POLYHEDRON POSITIONING

A 3D object has six degrees of freedom (*DOF*), three translational and three rotational. For a polyhedral assembly composed of an object part and a target part, the object part is positioned relative to the target part in three steps by controlling the six DOFs. Step 1. The primary constraint requires that two surfaces (or three points) of two parts come into contact. Step 2. The secondary constraint requires that a straight line (or two points) of one part come into contact with a surface of the other part. Step 3. The tertiary constraint requires that one point on one part comes into contact with a plane of the other part.

Tolerance analysis for assemblies studies how small geometric variations (i) alter the nominal assembly and (ii) affect the design function. In this research, a slider-guideway assembly, which is a simplification of the real slider-crank mechanism (Figure 1), is studied as an example of 3D polyhedral positioning. The slider part (object part) is allowed to move back and forth along the Z-axis in the guideway part (target part) which is fixed with the global coordinate systems. In simulation, the object part is located in different positions along the Z-axis with its bottom and side faces mating their counterpart faces (i.e., plane P_1 and P_2) in the target part. The Z-position of the object part is constrained by a virtual plane P_3 ($z=z_0$).

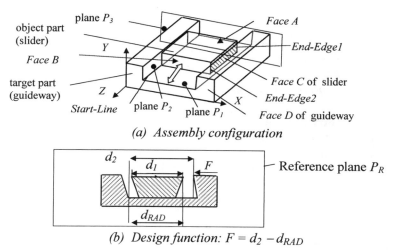

(a) Assembly configuration

(b) Design function: $F = d_2 - d_{RAD}$

Figure 1; *Slider-Guideway assembly.*

The clearance between side faces of the slider and the guideway is of most concern and is therefore defined as the design function (Figure 1b). Using the Real Acting Distance (*RAD*) concept introduced in [Li and Roy, 1999], we define the objective function as minimizing the *RAD* — d_{RAD}. In Figure 1b, d_1 is the external dimension of the object part and d_2 is the internal dimension of the target part.

Plane P_3 ($z=z_0$) is the tertiary plane that locates the object part's Z-position. Since z_0 is randomly generated in some intervals in Z-axis, the tertiary constraints can be omitted in the Monte Carlo simulation. The primary and secondary constraint will take planes P_1 and P_2 as constraint planes. Plane P_1 is first considered as the primary plane, and plane P_2 as the secondary plane. We will see later on why this sequence needs to be reversed.

3.1 Primary constraint requirement

The two parts mate first with one of their mating faces in the face-face contact pattern. In Figure 2, N_1^{target} is the normal vector of the target part's primary mating face (plane P_1), N_1^{object} is the normal vector of the object part's primary mating face (bottom face), α_1 is the separating angle between two primary mating faces, $R_1^{rotation}$ is the rotation vector, i.e., the intersection line of two primary faces,

$$\alpha_1 = cos^{-1} (N_1^{target} \bullet -N_1^{object}) \tag{1}$$

$$R_1^{rotation} = -N_1^{object} \times N_1^{target} \tag{2}$$

The primary constraint requirement is realized by rotating the object part about the axis $R_1^{rotation}$ through an angle α_1.

Figure 2; *Primary constraint requirement.*

3.2 Secondary constraint requirement

The object part then transforms to contact the target part with their secondary mating faces (side faces). The constraint for the secondary mating consists of orientation and translation constraints. The primary constraint must be maintained during the secondary contact, i.e., the object part rotates about some axis parallel to N_1^{target}, and the secondary translation is realized by a translation normal to N_1^{target}.

Because of geometric variations, two parts cannot contact each other in a face-face pattern for the secondary constraint. It is likely that they mate in an edge-face pattern with one edge of a plane contact with the surface of the other plane (Figure 3). The contacting edge and face are termed as an active secondary edge and face (Figure 4a).

Viewing in direction normal to the secondary face (*Face F_a*) of one part, if the whole secondary edge (*Edge E_b*) of the other part falls into *Face F_a* (Figure 4a), then *Face F_a* is the active secondary face and *Edge E_b* is the active secondary edge. If *Edge E_b* falls partially into *Face F_a*, then the status of the active mating feature are ambiguous, i.e., the active edge can belong to either the object part or the target part (so as to the active face) (Figure 4b,c).

This ambiguity can be clarified by a different mating sequence. The side face of the slider and the side face of the guideway (plane P_2) mate first as primary mating, the bottom face of the object part and the slot-face of the target part (plane P_1) mate as the secondary mating. Thus, the active secondary edge always belongs to the object part (slider) and the active secondary face belongs to the target part (guideway).

3.3 The orientation constraint

The orientation constraint is completed by rotating the object part about N_1^{target} to make its active secondary edge parallel to the active secondary face of the target part. Two cases will occur: bottom edge-face mating and top edge-face mating.

In Figure 6, N_1^{target} is the normal of the target part's primary mating face, N_2^{object} is the normal of the object part's secondary mating face, and $N_2^{targe'}$ is the normal of the target part's secondary mating face. The angle between N_2^{object} and N_1^{target} is θ^{object}, the angle between N_2^{target} and N_1^{target} is θ^{target}, and $\theta_1 = \theta^{target} - \theta^{object}$ is termed as *mating angle*.

$$\theta^{object} = cos^{-1} (N_2^{object} \bullet N_1^{target}) \tag{3}$$
$$\theta^{target} = cos^{-1} (N_1^{target} \bullet N_2^{target}) \tag{4}$$

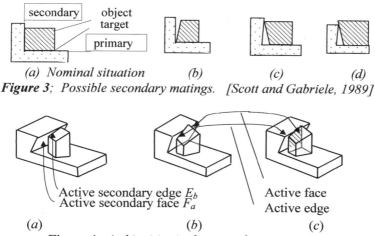

(a) Nominal situation (b) (c) (d)

Figure 3; *Possible secondary matings. [Scott and Gabriele, 1989]*

Active secondary edge E_b
Active secondary face F_a

Active face
Active edge

(a) (b) (c)

Figure 4; *Ambiguities in the secondary constraint.*

(1) Bottom edge-face secondary mating. If $\theta_1<0$, then bottom edge-face secondary mating occurs, i.e., the active bottom edge on the object part mates with the active secondary face on the target part (Figure 5). The orientation angle α_2 is the angle between two bottom edges (Figure 5b).

$$\alpha_2 = cos^{-1} (V_2^{object} \bullet V_2^{target}) \tag{5}$$

The orientation constraint is realized by rotating the object part about some axis R_2 that is parallel to N_1^{target} through angle α_2. For convenience, the rotation axis R_2 can be directed passing through one of the end vertices of the bottom edge on the object part.

(2) Top edge-face secondary mating. If $\theta_1>0$, then top edge-face secondary mating occurs, i.e., the active top edge on the object part mates with the active secondary face on the target part (Figure 6). The secondary orientation constraint requires that the object part rotate by angle α_2 so that the top edge vector V_2^{object} is parallel to the active secondary face on the target part. In other words, vector $V_2^{object}*$, which is the projected vector of V_2^{object} on the primary face, is parallel to the bottom edge vector on the target part, V_2^{target} (Figure 6a). The rotation angle α_2 can be calculated by:

$$\alpha_2 = cos^{-1} (V_2^{object}* \bullet V_2^{target}) \tag{6}$$

where
$$V_2^{object}* = V_2^{object} \bullet cos\beta_2 \tag{7}$$

and
$$\beta_2 = \pi/2 - cos^{-1} (V_2^{object} \bullet N_1^{target}) \tag{8}$$

The orientation constraint can be realized in a similar way as the case of bottom edge-face situation. The object part is rotated about axis R_2, which passes one of the end vertices of the top edge on the object part and is in the same direction as N_1^{target}.

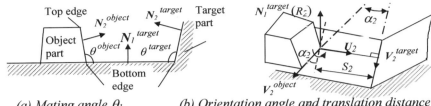

(a) Mating angle θ_1   *(b) Orientation angle and translation distance*

Figure 5; *Bottom edge-face secondary mating.*

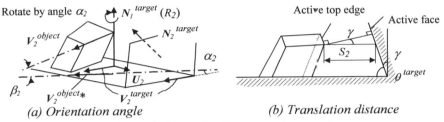

(a) Orientation angle   *(b) Translation distance*

Figure 6; *Top edge-face secondary mating.*

3.4 The translation constraint

There are infinite possible directions for the object part to slide on the primary mating plane to accomplish its secondary mating. The translation direction, U_2, proposed here is perpendicular to both N_1^{target} and V_2^{target},

$$U_2 = N_1^{target} \times V_2^{target} \qquad (9)$$

After the secondary orientation, the secondary translation is realized by moving the object part along U_2 by distance S_2. S_2 is calculated differently for bottom and top edge-face mating situations. For the bottom edge-face mating situation, S_2 is the distance between these two bottom edges (see Figure 5b). For top edge-face mating situation (Figure 6b),

$$S_2 = S_d/\cos\gamma \qquad (10)$$

where S_d is the distance between the active top edge on the object part and the active face on the target part, the angle γ is obtained by

$$\gamma = \pi/2 - \theta^{target} \qquad (11)$$

4. VARIATION ANALYSIS OF A 2D CASE

A 2D assembly example is investigated in order to provide a necessary assistance to the 3D problem. From section 4. (Case study) of companion paper [Li and Roy, 1999], the relationships necessary to perform tolerance analysis for 2D assembly using Monte Carlo simulations are obtained. A series of random numbers is first generated and then input into the assembly system. The dimensional design function is measured as the system output. Statistical results can be obtained based on large amount of samples.

In plus-minus tolerance analysis, only the size of parts will vary while geometric shape remains the same. In Figure 7a, with d_1^0, d_2^0, a, b, c and d are all positive, the design function is calculated as a simple arithmetic summation,

$$d_1 = d_1^0 + \Delta d_1, \quad \Delta d_1 \in [-a, b], \quad d_2 = d_2^0 + \Delta d_2, \quad \Delta d_2 \in [-c, d] \qquad (12)$$

In geometric tolerance analysis, both size and shape of parts will vary (Figure 7b). Assume the dimensions are measured as the distance between two farthest vertices for the *external dimension* (d_1 in Figure 7a,b) and two closest vertices for the *internal dimension* (d_2 in Figure 7a,b). The object parts and the target parts of different variation strategies in Figure 7 will have the same length (i.e., d_1 and d_2) along X-axis. What happens after the parts are assembled?

If the objective function aims to minimize the distance of *RAD*, then the final position of the object part will produce larger clearance than the conventional tolerance analysis. It is because the proposed relative positioning scheme takes advantage of the inclined edges and eliminates the air space between the assembled parts (Figure 7b). In section 5, this phenomenon will be justified with numerical results.

5.   FINAL *RAD* AND DESIGN FUNCTION

The *RAD* can be obtained with the following two issues being addressed: (i) how to measure the *RAD* and the dimensional design function F_d? and (ii) will different order of the primary and secondary constraints affect *RAD* and F_d?

The *RAD* is measured from the start-point to the end-point in a 2D assembly where the start-point is the intersection vertex of two edges of the target part. For a 3D assembly, there will be a start-line, which is the intersection edge of two faces of the target part. The *RAD* and F_d are measured in a reference plane (P_R) whose normal is parallel to the start-line. When the assembly is projected into plane P_R, the *RAD* and F_d will be calculated from two 2D assemblies (Figure 1b), i.e., one 2D assembly being the cross-section of the 3D assembly where *Face A* resides (RAD_A), and the other being the cross-section where *Face B* resides (RAD_B). *Face A* and *Face B* are two end faces of the slider part in Z-axis. Because of linear variation for the polyhedral assembly, the extreme values (i.e., max. and min.) of *RAD* and F_d will occur on these two cross-sections. In 3D assembly, however, unlike the case in 2D assembly where the min. *RAD* means the max. *F*, the min. *RAD* in one cross-section may not lead to the F_d in this cross-section being the max. for the assembly because there are two deciding *RAD*-F_d pairs.

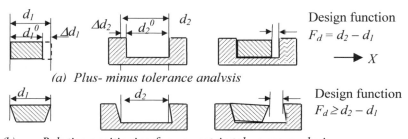

(a) Plus- minus tolerance analysis

(b)   Relative positioning for geometric tolerance analysis
Figure 7; Comparison of two different tolerance analysis strategies.

For each of the two cross-sections in the reference plane P_R, the RAD and F_d can be obtained the way we discussed in section 4 of [Li and Roy, 1999]. For RAD, the start-line becomes a start-point, and the end-point will be one of the two projected vertices of the end-edges on the object part (Figure 1a). Also in plane P_R, F_d is measured as the distance between two closest vertices of two edges. One edge is the projected end-edge of the slider part, i.e., *End-Edge1* or *End-Edge2*. The other edge is the intersected edge between *FaceD* and the cross-section plane where *FaceA* or *FaceB* resides. See Figure 7b.

For the second issue, different mating sequence will generate different assembly configurations, and therefore generate different RAD and F_d. See section 4 of [Li and Roy, 1999] for detailed description.

5.1 Geometric variation and arithmetic variation
The dimensions and tolerance specifications are assigned to the slider part and the slot-guideway part (Figure 8). Each variant face is represented by planar equation, which is not discussed here in detail.

For each simulation in the plus-minus stack-up routine, the variation of the toleranced faces of each part will generate the same amount of *dimensional measurements* (i.e., the internal and external dimensions) as their geometric variant counterparts (Figure 7). For example, the dimension d_1 of Figure 7a measures the same length as the dimension d_1 of Figure 7b. The clearance of this scheme can be calculated by the simple arithmetic summation of two variables as shown in Figure 7a.

5.2 Numerical implementation
The numerical experiments on the proposed relative positioning scheme (Model-2) and the conventional plus-minus stack-up scheme (Model-1) for the slider-guideway assembly are implemented on PC platform using C++ language. Monte Carlo simulation is conducted 1000 times for two groups. The simulation results are analyzed with MS Excel spreadsheet. Table 1 shows the numerical test results for the two groups.

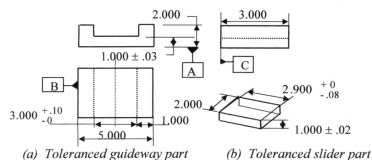

(a) Toleranced guideway part   (b) Toleranced slider part

Figure 8; Test part with tolerance specifications.

Table 1 Numerical test results of the example

	GROUP #1			GROUP #2		
	Model-1	Model-2	difference	Model-1	Model-2	Difference
Mean	0.1422	0.1686	18.53%	0.1369	0.1600	16.92%
Median	0.1420	0.1638	15.36%	0.1352	0.1568	15.98%
Standard Deviation	0.0167	0.0231	38.88%	0.0165	0.0241	46.10%
Range	0.0712	0.1078	51.43%	0.0608	0.1316	116.48%
Minimum	0.1036	0.1340	29.37%	0.1096	0.1152	5.15%
Maximum	0.1748	0.2418	38.36%	0.1704	0.2469	44.87%
Count	1000	1000		1000	1000	

5.3 Result analysis

It is observed in Table 1 that (i) the mean, minimum and maximum values of clearance (F_d) of Model-2 are greater than those values of Model-1, and (ii) the standard deviation and the variation range of F_d of Model-2 are also greater than those values of Model-1.

The first observation shows that geometric variation leads to larger clearance (F_d) when the relative positioning scheme is applied than the conventional plus-minus tolerancing scheme. The second observation shows that geometric variation also results in more scattered distribution of the clearance than the conventional plus-minus tolerancing scheme. It means that the geometric positioning scheme is more sensitive than the plus-minus scheme.

It is expected that the above discovery can be utilized in at least two areas of assembly design analysis. (1) In assembly planing, the actual clearance between assembled parts may be lager than the estimated value obtained by using plus-minus tolerancing scheme. Applying relative positioning scheme will lead to more precise results and reduce the production cost by avoiding tighter than necessary tolerance specifications. (2) In mechanism analysis, the clearance affects the performance of the assembly functionality. Since the variation range of design function becomes larger when relative positioning scheme is applied, it is recommended to specify a more stringent tolerance specification on the moving parts in order to improve quality.

6. CONCLUSION

A relative positioning scheme is proposed to determine the optimal configuration of variant parts in an assembly. Assembly configuration uncertainties caused by part variations are clarified by realizing the objective function. This technique is developed to deal with 3D polyhedral objects in an assembly. A mathematical programming approach is formulated and implemented on a slider guideway assembly example.

One important use of this work is in solving tolerancing problems for assemblies. To produce tolerance measurements, we simulate the assembly configurations based on part variations (using either worst case or statistical model) and the variation range of measurements are obtained. We believe the proposed relative positioning scheme is an important step towards automated tolerance analysis for assemblies.

REFERENCES

[Inui and Kimura, 1991] Inui, M. and Kimura, F.; "Algebraic reasoning of position uncertainties of parts in an assembly"; *1991 ACM Symposium: Solid Modeling Foundations and CAD/CAM Applications*, pp. 419-428; Austin, TX, USA; 5-7, June.

[Inui et al., 1996] Inui, M., Miura, M. and Kimura, F.; "Positioning conditions of parts with tolerances in an assembly"; In *Proceedings of the 1996 IEEE International Conference on Robotics and Automation,* pp. 2202-2207; Minneapolis, MN, USA.

[Lee and Andrews, 1985] Lee, K. and Andrews, G.; "Inference of the position of components in an assembly: Part 2"; *Computer-Aided Design*, 17(1), pp. 20-24.

[Li and Roy, 1999] Li, B. and Roy, U.; "Motion constraints in a 2D polygonal assembly"; In *Proceedings of the 6th CIRP International Seminar on Computer Aided Tolerancing*; Enschede, the Netherlands.

[Mullineux, 1987] Mullineux, G.; "Optimization scheme for assembling components"; *Computer-Aided Design,* 19(1), pp. 35-40.

[Rocheleau and Lee, 1987] Rocheleau, D.N. and Lee, K.; "System for interactive assembly modeling"; *Computer-Aided Design,* 19(2), pp. 65-72.

[Scott and Gabriele, 1989] Scott, R.T. and Gabriele, G.A.; "Computer aided tolerance analysis of parts and assemblies"; In *Proceedings of ASME 15th Design Automation Conference*, v1, pp. 29-36; Montreal, Quebec, Canada; Sept. 17-21.

[Sodhi and Turner, 1994] Sodhi, R. and Turner, J.U.; "Relative positioning of variational part models for design analysis"; *Computer-Aided Design,* 26(5), pp. 366-378.

[Turner, 1990] Turner, J.U.; "Relative positioning of parts in assemblies using mathematical programming"; *Computer-Aided Design,* 22(7), pp. 394-400.

A Development of Statistical Tolerancing System for Optical Products

Toyoharu SASAKI, Masahiko SHINKAI, Kohichiro HIGASHIYAMA
Nikon Corp., Technical System Department
1-6-3 Nishi-Ohi Shinagawa-Ku Tokyo, 140-8601 Japan
Fumiki TANAKA and Takeshi KISHINAMI
Hokkaido Univ., Graduate School of Engineering
Kita-13 Nishi-8 Kita-Ku Sapporo, 060-8628 Japan.
sasaki @ nikongw.nikon.co.jp

Abstract: This study is the development of a statistical tolerancing system for optical products based on computer simulations. This system can quantitatively evaluate optical performance, productivity and tolerance sensitivity in mass production at the design stage. By using this system, the designer is able to determine the more suitable set of tolerances for a given optical product. This paper reports the concept of proposal system, assembly model of SLR lens, realized computer software and it's reliability.
Keywords: tolerance, optics, assembly model, statistical method, computer simulation.

1. INTRODUCTION

Determination of tolerances is important during product manufacturing in order to maintain low cost while achieving a consistent level of product performance. However, with optical products, precise with quantitative prediction of tolerance sensitivities, optical performance, and productivity in a mass production environment have been very difficult in the design stage. Therefore, difficulties, unanticipated in the design stage, the troubles which can not be expected in design stage, have occurred frequently once into the production stage. This has resulted in the prolongation of the development period, and an increase in manufacturing costs for new products. The cause of these difficulties has been optical tolerance designs that have neglected or inadequately addressed errors occurring in manufacturing and assembly in the mass production stage.

The purpose of this study was the development of a statistical tolerancing system for optical products based on computer simulations of the manufacturing process with the goal of increasing productivity (called Virtual PT System, PT: Production Try). This system can quantitatively evaluate optical performance, productivity and tolerance sensitivity in mass production at the design stage. By using this system, the designer is able to determine the more suitable set of tolerances for a given optical product by taking into

345

346

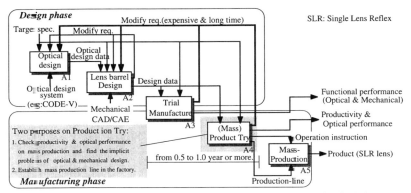

Figure 1; IDEF0 of conventional development process for SLR lens

account manufacturing/assembling errors occurring in mass production.

Firstly, this paper reports the concept of the Virtual PT System for optical products like as SLR lens. Next, the development of the both mass-production simulation system and complex tolerance sensitivity analyzing system is reported. Finally, the reliability results based on this mass-production simulation and the confirmation of the effect using the sensitivity analysis result by the mass-production simulation is reported.

2. CONVENTIONAL DEVELOPMENT PROCESS OF SLR LENSES

Figure 1 shows the IDEF0 of a conventional product development process for a SLR lens. The conventional product development process consists of an optical design (A1), a lens barrel design (A2) and a trial manufacturing (A3). Designed performance and proper functioning of the products are confirmed in the trial manufacturing process (A4). If problems occur, they are feedback into the optical design (A1) or the lens barrel design (A2). After this confirmation-modification cycle (usually repeated a few times) is completed in the design phase, a production try process (A4) is done. The purpose of this process is to confirm the production line design, and to detect and solve any number of various problems arising during mass production.

When optical performance problems occur during the production try process (A4), there are two ways to solve the problem. The first requires that the machining and assembly accuracy of each station in the production try process be improved in order to meet the required tolerances. The second requires a re-design of the product in order to lower the required tolerances. However, both ways require the expenditure of a great deal of time and cost. So far, the influence of individual tolerances on the total optical performance of the product is unknown. This, then, requires that each tolerance be strictly adhered to.

We propose the third way to determine the required tolerance of each part from the manufacturing viewpoint. If we can realize a more suitable tolerance design, taking into

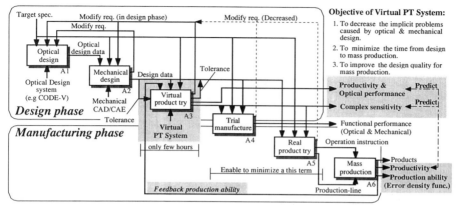

Figure 2; *IDEF0 of new development process for SLR lens by Virtual PT System*

consideration the manufacturing capability of the workshop, and the optical performance in mass production and productivity, preventing the occurrence of problems in the mass production process. Then product development can be conducted efficiently without an increase in time and cost.

3. VIRTUAL PT SYSTEM

3.1 Virtual PT System

As method to solve the above-mentioned problem, We have developed the "Virtual PT System" detailed in this study (Figure 3 shows this IDEF0). Figure 2 shows the improved product development process after the application of this system. In a real production try process, various kinds of problems which unpredicted in the design stage are occur. Solving these problems requires a great deal of time and expense since the problems must be identified, and the fixes verified through a small-scale mass production run. Therefore, we believe that the proper approach to this problem is to model the production try process (A4) with a numerical computer simulation. If such a simulation can be achieved then quantitative evaluation of productivity and tolerances will be possible in the design stage, and the designer will be able to determine the most suitable set of tolerances given the manufacturing/assembling error of the workshop, and the mechanics of mass production.

The Virtual PT System proposed by this study consists of the following three systems. Activity A31 simulates in a computer the mass production process of an optical product. In particular, it is an important point that, in addition to optical tolerances, this simulation also treats the lens barrel structure and the lens barrel component tolerances directly in the same way. Activity A32 is the system that analyzes the tolerance sensitivity for optical performance. Analysis has done by using a Multivariate Analysis Method on the data set output from the mass production simulation (A31). In this sensitivity analysis, the sensi-

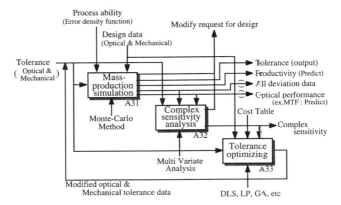

Figure 3; *IDEF0 of Virtual PT System for SLR lens*

tivity of two tolerances, changed simultaneously, (called complex tolerance sensitivity) must be precisely determined. Activity A33 is system to decide automatically the most suitable tolerances based on the analysis from the mass production simulation results, the complex sensitivity analysis results and the manufacturing cost table.

By building this system, we can realize a virtual production try process in the computer. It then becomes possible to precisely predict, in the design stage, problems, productivity and optical performance in the mass production stage. In addition, by using this Virtual PT System in the optics and lens barrel design, it becomes possible to prevent problems from occurring in the real mass production process.

4. REALIZATION OF COMPUTER SOFTWARE

4.1 Mass production simulation system

Figure 4 shows the IDEF0 of the mass production simulation system A31 (shown in Figure 3). The mass production simulation system consists of five activities (A311 ~ A315: it meaning software module).

"A311" is the module that generates tolerance data (text file). In this data file, 1) the total sample number, 2) the error distribution functions, 3) the tolerance data and 4) the command strings to the optical design system are described as input data for the mass production simulation. This tolerance data describes information on the decentration occurring due to mechanical structure and tolerance of the lens barrel components as well as the independent optical tolerances.

"A312" is the module that generates error data (deviation from nominal value) according to error distribution functions designated for each tolerance and based on the input tolerance data. The output of this module is a command batch file for optical design system, which is then combined with the generated error data and the command strings for

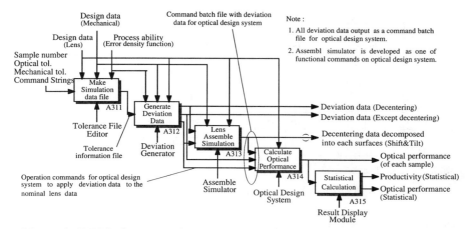

Figure 4; IDEF0 of mass production simulation for SLR lens on Virtual PT System

optical performance calculation.

"A313" is a module that simulates the assembly of the lens barrel and the optical components. According to the description given in the command batch file, this simulator adds the quantity of decentration designated (derived from CAD/CAE/FEM) to each lens surface of the lens data set. In the final, lens barrel assembling simulation, the module calculates the decentration amount (tilt & shift) for each lens surface from a theoretical optical axis and once again outputs to a command batch file.

"A314" is the optical design system (e.g. CODE-V). This activity calculates the optical performance (e.g. MTF, etc.) according to the commands described in the batch file. All the calculated optical performance data is automatically output to file.

"A315" executes a statistical calculation and a validation calculation on optical performance and productivity for all estimation points derived from the optical performance data output from module A314. The final module generates a graphic output and an accompanying listing according to the procedure listed below. All of the calculations mentioned above are done automatically with the exception of the generation of the tolerance data file. And Table1 shows currently available tolerances on this system.

4.2 Assembly model on Virtual PT System for SLR lens

Figure 5 shows EXPRESS-G of the assembly model for SLR lens using this system. When we demand the optical performance of the manufactured product by numerical calculation, both the manufacturing error of the optical component and the assembly error caused by the error (or tolerance) of mechanical parts are required. The optical parts are held by the lens barrel parts to the base surface (or base coordinate). Therefore, the actual position and orientation of each lens surface on the base coordinate are obtained by tracing "contact relationship" from the base surface (usually mount surface) to the lens sur-

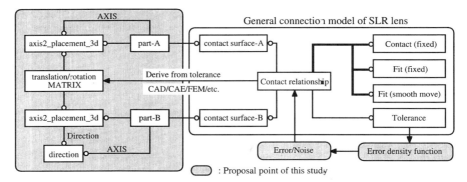

Figure 5; EXPRESS-G of general lens barrel assembly model for SLR lens

faces. Then, the "contact relationship" of the contacting surface between the two parts must be expressed directly as the right side EXPRESS-G shown in Figure 5.

By the way, from viewpoint of the calculation for the optical performance, this structure is able to replace the another structure that is the relationship of coordinate transformation by the transformation matrix (called decentering transformation matrix). In this model (left side EXPRESS-G in Figure 5), The coordinate transformation is equivalent to the decentering caused by the error of the contacting point. Currently, the data for making the decentering transformation matrix is obtained as both shift and tilt (with the center of rotation) value for the each contact relationship using CAD/CAE/FEM/etc. systems. The decetnering transformation matrix has generated from this shift and tilt values in the lens assemble simulation module (A313).

4.3 Complex sensitivity analysis system

Figure 6 shows the IDEF0 of the complex sensitivity analysis system A32 (shown in Figure 3). The output from the mass production simulation system (A31) is a data pair consisting of the used error data and the calculated optical performance based on simulation results. This data is very convenient data for obtaining optical performance tolerance sensitivities. By analyzing the relationship between these data sets using some mathematical method, we should be able to easily obtain the tolerance sensitivities.

In this system, the polynomial equation (1) is used to describe the relationship between the complex tolerance sensitivity, the error and the optical performance [Rimmer, 1978] [Koch, 1978] [Hilbert, 1979].

$$ \mathrm{MTF} = \sum_i^n a_i p_i + \sum_i^n b_i p_i^2 + \sum_i^n \sum_{i<j}^n c_{ij} p_i p_j + d \tag{1} $$

Where, MTF is Modulation Transfer Function, n is total number of tolerances, p is deviation value and a, b, c, d are coefficient must be solved. This equation, then, gives us the ability to quantify complex relationships between two tolerances .

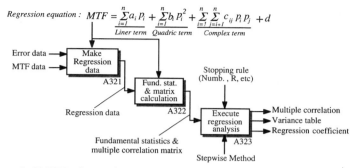

Regression equation : $MTF = \sum_{i=1}^{n} a_i P_i + \sum_{i=1}^{n} b_i P_i^2 + \sum_{i=1}^{n} \sum_{j=i+1}^{n} c_{ij} P_i P_j + d$

Figure 6; *IDEF0 of complex sensitivity analysis system on Virtual PT System*

As a result of the examination of various approaches, we selected the "Stepwise Multiple Regression Method", one of several multivariate analysis methods. There were a number of reasons for selecting this method: 1) it can execute stable analysis calculations on an enormous number of parameters, 2) in the real manufacturing stage, there is no need to know which parameters have low tolerance sensitivity for optical performance.

5. EXECUTE EXAMPLE AND RELIABILITY INSPECTION RESULT

5.1 An example of mass production simulation

Figure 7 shows an executed example of a mass production simulation of a SLR lens. If the errors corresponding to each tolerance can be considered to be a random phenomenon dependent on an error density function, then, from the central limit theorem, we can assume that the distribution of the optical performance (MTF) has a Gaussian distribution. A Gaussian distribution is completely determined by its average and standard deviation. The MTF is plotted on the vertical scale with the image height (estimation point) plotted on the horizontal axis (see left side illustration). The MTF curves for the nominal value, the average value, and the -1σ, -2σ and -3σ values, corresponding to a productivity of 50.0%, 84.1%, 97.7% and 99.9% are all plotted on the same graph. This illustrates, in an easily understandable manner, the correlation between product performance and productivity in mass production.

In the example results shown in Figure 7, the simulation result graph (right side) has indicates that the difference in optical performance between the nominal value and the 99.9% value will be large for image heights from 10.8 mm to 15.1 mm. This means that if the lower limit value of the product standard for the MTF at 15.1mm is 0.2, many substandard products (100.0%-84.1%=15.1%) will be generated by this set of tolerances. In this way, the designer can quantitatively evaluate the productivity and adequacy of the tolerances based on the simulation result. By adjusting the tolerances and re-executing the mass production simulation, the designer can easily find the most suitable set of toler-

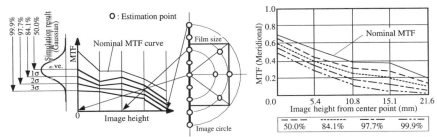

Figure 7; *Display method of mass product simulation result (Productivity graph)*

ances and prevent problems from occurring in the mass production phase without the time and expense of a trial production run.

5.2 Reliability of mass production simulation

Figure 8 show the two results. The left side graph is the relationship graph between measured MTF vs. calculated MTF from measured error value. This result shows that correlation between two values is almost good. And the right side graph shows the productivity that compares the measured value with the value predicted by the simulation. This graph shows the productivity comparison for the Meridional MTF and Sagital MTF for image heights of 0.0 mm and 15.1 mm. The measured productivity is shown as a continuous line and represents the 95% confidence level. The predicted productivity is shown as a dashed line. Figure 8 shows that the value predicted by this system roughly corresponds to the 95% confidence level. Even if there are tolerances that cannot yet be treated by this (e.g. surface deformation, etc.), we can understand and confirm the reliability of the simulation results of this system sufficiently.

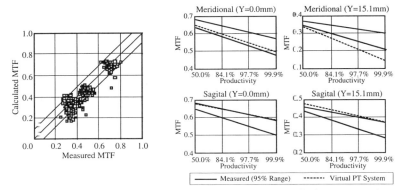

Figure 8; *Reliability check of mass product simulation result (by measured result)*

5.3 An example of complex sensitivity analysis

Figure 9 shows an example of the complex sensitivity analysis result that has been calculated by the Stepwise Multiple Regression method. The output listing shown in Figure 9 is the top of 6 parameter of a regression analysis result that was specified to search for 15 most sensitive parameters.

A multiple correlation coefficient (%), called the contribution rate, represents the rate at which the variables selected can account for the variation in system performance. The partial regression coefficient (bi) is the regression coefficient of the assumed polynomial equation. For the purpose of this analysis, we only want to know the sensitivity for the tolerances that does not depend tolerance type, unit, width and etc. The standard partial regression coefficient (std.bi) is a regression coefficient that is normalized by the standard deviation of all parameters ($\sigma=1$). This value indicates the correlation of performance variation for a given tolerance width regardless of tolerance type. If there is not a dependency on other variables, this value will be the same as the simple correlation coefficient (Rxy). Then, the relationship between two tolerances is provided in this regression equation as well.

To confirm the effects and reliability of the complex sensitivity analysis result, we performed the experiment by simulation. The experiment was executed with the method as follow. At the first, mass production simulation and complex sensibility analysis were executed by initial tolerance. And then, top five-tolerance width was corrected by 1/2, and

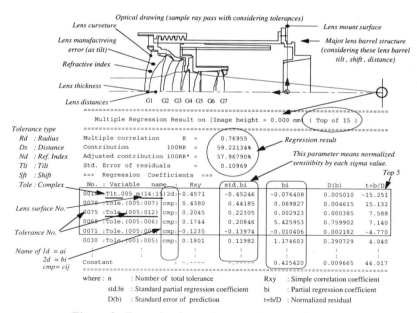

Figure 9; Example of complex sensitivity analysis result

mass production simulation was done again. The mass production simulation result was well improved as had predicted from the complex sensitivity analysis result. Consequently the reliability of this system was confirmed sufficiently.

6. CONCLUSION

1) We proposed the "Virtual PT System" concept with the assembly model of SLR lens as an approach to improve product development process of SLR lenses.
2) Based on this proposal, "Mass Production Simulation System" and "Complex Sensitivity Analysis System" are realized as the computer software.
3) The reliability of the mass production simulation was confirmed by comparing them with measured results of the manufacturing process. And the reliability of the complex sensitivity analysis was also confirmed sufficiently by the mass production simulation.

REFERENCES

[Rimmer, 1978] Matthew P. Rimmer; "A tolerancing procedure based on modulation transfer function (MTF)"; In: *SPIE Vol.147 Computer-Aided Optical Design*, pp.66-70
[Koch, 1978] Donald G. Koch; "A statistical approach to lens tolerancing"; In: *SPIE Vol.147 Computer-Aided Optical Design,* pp.71-81
[Hilbert, 1979] Rovert S. Hilbert; "Semi-automatic modulation transfer function (MTF) tolerancing"; In: *SPIE Vol.193 Optical System Engineering*, pp.34-43
[Force,1996] D.Forse; "Statistical Tolerancing for Optics", In: *SPIE Vol.2775 Proceeding of SPIE*, pp18-27
[Drake,1997] Pail Drake; "Auto-Tolerancing on Optical Systems (Six Sigm Approach to Designing Optical Systems)" , In: *SPIE Vol.3130 Proceeding of SPIE,*pp136-147
[Sasaki et al., 1998a] Toyoharu Sasaki, Masahiko Shinkai, Kcichro Higashiyama, Fumiki Tanaka, Takeshi Kishinami; "Development of Statistical Tolerancing System for Optical Product"; In: *SPIE Vol.3482 Proceeding of IODC 1998*, pp.528-537
[Sasaki et al., 1998b] Toyoharu Sasaki, Masahiko Shinkai, Kcichro Higashiyama, Fumiki Tanaka, Takeshi Kishinami; "Development of Statistical Tolerancing System for Optical Products"; In: *Jounal of the Japan Society for Precision Engineering Vol.64 No.7*, pp.1090-1095

Modelling-Simulation-Testing of Compliant Assemblies

Eric Sellem[1], Charles André de Hillerin[1], André Clement[1], Alain Rivière[2]
[1] *DASSAULT-SYSTEMES - 9, quai Marcel Dassault - BP 310 -*
92156 Suresnes Cedex - France
[2] *GRIIEM-LISMMA Laboratory of CESTI-ISMCM*
93407 Saint-Ouen cedex, France
eric_sellem@ds-fr.com

Abstract : Current statistical tolerance analysis of assemblies rely heavily on Monte Carlo simulation. The available software tools using this technique model the assembly of rigid parts, by only considering the kinematic laws. [Sellem et al., 1998] proposed a linear mechanical approach taking deformation into account in the computation of tolerances for welded, riveted, bolted or glued assemblies of compliant parts, avoiding in this way the expensive Monte Carlo simulation. This paper presents some improvements in the modelling of the assembly process, and describes a sensitivity analysis approach to identify the key characteristics of the assembly. The accuracy of this method is evaluated by a comparison with measurements performed on an actual assembly of four complex parts.
Keywords : statistical analysis, assembly process, compliant part, key characteristics, validation of the approach.

1. INTRODUCTION

With the new generation of integrated CAD-CAE applications, there is an increasing demand for SW tools that can simulate industrial processes such as part assembly by welding, gluing, riveting or bolting.

In industries such as automobile and aerospace, controlling the sheet-metal assembly process is of critical importance. Due to the propagation of part defects or of process imperfections, it becomes necessary to be able to predict variations in the final shape of an assembly resulting from known error distributions at the pre-assembly level. This is true both in the early phases of the project, at the part and process tolerancing level, and in later phases such as inspection and measurement tool calibration.

For welded flexible parts, there is no efficient exhaustive computational aid on tolerancing and metrology available on the market, although some suggestions for the implementation of such tools have already been made.

355

[Chang et al., 1997] propose a methodology to model the assembly process of compliant parts describing the PCFR and PCMR cycles. [Merkley et al., 1997] develops a linear method of solving certain contact problems of the mating surfaces by using super-elements. [Hu et al., 1996], [Hu et al., 1997] provide a method based on the concept of influence coefficients, by using a Monte Carlo simulation to generate the source of variation.

This paper will present an alternative FEM-based approach. More specifically, we present a program using as input the nominal CAD geometries of the different parts and the distributions of variabilities both at the geometry and at the process level, which predicts the distribution of shape variability at any number specified inspection points in the final assembly.

2. FASTENING PROCESS

2.1. Description

The assembly process for two sheet-metal parts can be described as follows :

1) each part is placed on statistically determined "isostatic" supports (at points termed « positioning points »),
2) each part is brought in conformity with additional supports (at points termed « additional points »),
3) the parts are brought in contact by the robot (at points termed « fastening points »),
4) the parts are welded by attaching together the couples of corresponding points,
5) the robots are released,
6) the assembly is released (aside from one arbitrary set of "isostatic" supports).

The simulation should model the assembly process as closely as possible. Therefore, the following chapters describe some concepts to consider.

2.2. Different Kinds of Positioning

The first step corresponds to the positioning of each part to assemble. One method consists of locating each part relative to an absolute reference frame by using the fixtures (see **Figure 1**). Another method consists of positioning a part relative to another one by using the so called "Hole to Hole" method (see **Figure 2**). In the field of the Aircraft Industries, [Marguet et al., 1997] gives a good description of these two methods by making a comparison.

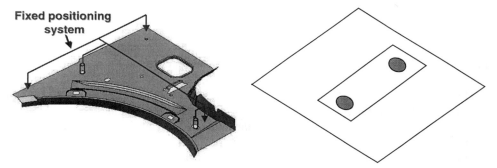

Figure 1: *assembly on tooling*   **Figure 2:** *assembly "hole to hole"*

2.3. Different Ways of Bringing the Parts into Contact

This is usually achieved :

- either by applying equal and opposite loads on each side of the assembly (case of a light / flexible robot, see **Figure 3**). In this case, if the robot is considered as perfect, parts are assembled in the deformed configuration.
- or by applying a load which forces the parts in the nominal configuration (case of a massive / rigid robot, see **Figure 4**; note that in this case fastening occurs at a fixed location relative to an absolute reference frame). In this case, parts are assembled in the nominal configuration.

Figure 3: *opposite forces*   **Figure 4:** *One force*

2.4. Different Ways of Measuring the Assembly

As opposed to rigid parts, which are measured while in a free, unconstrained state, flexible parts are constrained prior to inspection (that means additional supports are used). This constrained state corresponds to the one which will be used for the positioning of this assembly for the next assembly process.

3. HYBRID MODEL THIN SHELL/BEAM

In contrast with the automotive industry, where sheet-metal parts are stiffened through stamping prior to the assembly process (thus requiring exclusively thin shell FEA models), most aerospace industry assemblies are stiffened by riveting strings and fasteners to sheet-metal parts (see **Figure 5**). A study has shown that the use of beam elements instead of thin shell elements to mesh the string can decrease the number of degrees of freedom by 90 percent. Moreover the beam element formulation is more appropriate to describe the physical behaviour of the strings.

This suggests that an industrial tool should have the capability to model the strings by beam elements in order to decrease considerably the CPU time needed to perform the simulation.

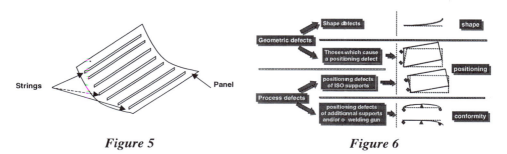

Figure 5   *Figure 6*

3.1. The 3 types of variation

The method enables the simulation of a fastening assembly process by taking into account (see **Figure 6**):

7) the **positioning variabilities** : these variations represent either a geometric defect of the part localised at the positioning points or a positioning defect of the "isostatic" supports (3-2-1). Such a variability generates a rigid body motion of the corresponding part,

8) the **conformity variabilities :** these are due to the positioning defect of the robot which fastens the parts, or possibly to the positioning defect of the additional support (3-N),

9) the **shape variabilities** : these correspond to all profile defects (except those related to rigid body motions).

4. THEORETICAL BASIS REVIEW

The simulation of the process consists of splitting the computation in two distinct phases.

In the first phase, the program performs computations corresponding to « unit displacements » at the various points. This gives rise to a number of solution vectors equal to the number of simulations. These vectors grouped together form three matrices of influence coefficients $[w]_{RB}^{P}$, $[w]_{C}^{A,F}$, $[w]_{S}^{S}$ corresponding to three types of variability (each calculated vector is the effect of one defect on the assembly).

The linear behaviour assumption enables us to avoid the expensive use of the Monte-Carlo simulation. Therefore, the approach is based on the finite elements method (FEM) which models the deformation effect of given displacements and loads applied to linear elastic structures. Each sheet-metal part is considered as an elastic structure.

In the second phase, these matrices are linearly combined with any set of actual input distributions at the relevant points, to produce the final configuration.

The mean deviation of one point of the assembly in one direction is then given by :

$$\mu_{M,\bar{n}} = \{n\}^{T} \left([w_M]_{RB}^{P} \{\mu_P\}_{RB} + [w_M]_{C}^{A,F} \{\mu_{A,F}\}_{C} + [w_M]_{S}^{S} \{\mu_S\} \right)$$

Similarly, the variance of the deviation of this same point is given by an appropriate tensor product involving the direction in which the deviation should be calculated, the matrices of influence coefficients and the full covariance matrix[1] of the input variabilities.

5. VALIDATION OF THE METHOD

A tool has been implemented in CATIA as a prototype in order to validate this approach. The example corresponds to an actual assembly composed of four complex stamped sheet-metal parts. The assembly is part of a Body-In-White car (see **Figure 7**). From the CAD model each part has been meshed so that for each point (ISO support, Additional supports, Fastening points, Geometric points, …), required to specify the process and the assembly, a node has been created. The Geometric points correspond to the points where the measurement has been performed before the assembly on each part relative to its reference frame.

---

[1] Note that the covariance matrix is not assumed to be diagonal.

360

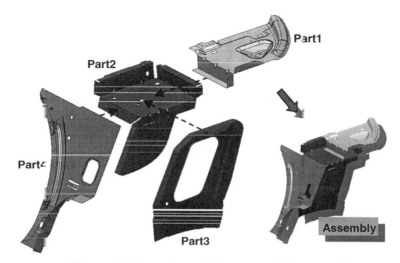

Figure 7: *illustration of the parts and the assembly*

In this first comparison, only the shape variability was simulated. In fact, no information about the positioning and conformity variabilities was provided. So the first step was to calculate the matrix of influence coefficients $[W]_G$.

a) Mean

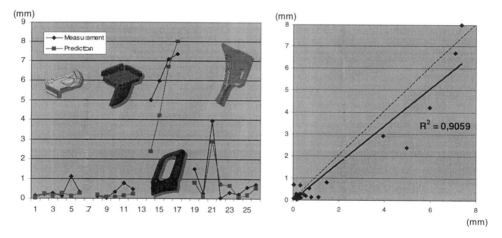

Figure 8: *Mean comparison between measurement and simulation*

The predicted means exhibit the same behaviour as the measured ones with values of the same order. The difference obtained at the inspection points of the third part seems to reveal a positioning variability during the measurement. Although this last defect cannot be evaluated with respect to the measurement data, a strong correlation is still obtained with a correlation coefficient of 0,90.

b) Range

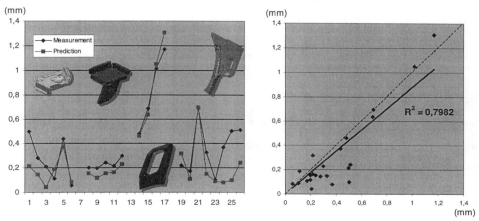

Figure 9: *Range comparison between measurement and simulation*

Here the predicted ranges exhibit a behaviour similar to the one obtained by measurement. The correlation coefficient is equal to 0,80.

The inspection made on the assembly has been performed in a constrained state. The corresponding state provided by the computation is represented in $\mu + 3\sigma$ by the following picture:

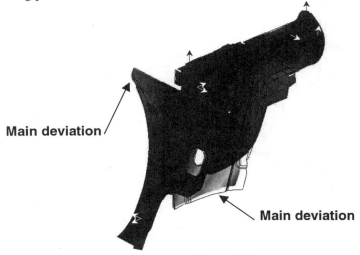

Figure 10

This first case study provides encouraging results relative to the given input. In fact, even though only shape variability measurements were provided, this simulation gives the same global behaviour for the mean and the range on the inspection points deviations. Moreover, this comparison has been carried out on an actual assembly of complex parts. A second case study with more data should confirm this statement.

6. SENSITIVITY ANALYSIS

Let $\sigma^2_{M,\bar{n}}$ denote the variance of one point M in the direction n due to all input variabilities and $\sigma^2_{j_{M,\bar{n}}}$ the variance due to the variability of point j. The influence $s_{Mj,\bar{n}}$ of the jth defect relative to the influence of all defects at one point M is given by the ratio :

$$ s_{jM,\bar{n}} = \frac{\sigma^2_{j_{M,\bar{n}}}}{\sigma^2_{M,\bar{n}}} $$

7. ROOT CAUSES IDENTIFICATION

The previous example shows that this tool gives a strong correlation between measurement and prediction. It would be interesting to identify the root cause of variations at a given inspection point in the assembly. The sensitivity analysis is able to provide such insight. **Figure 11** shows the root causes of the deviation of one point of the green part (Part3) whereas **Figure 12** shows the root causes of the deviation of one inspection point of the red part (Part4).

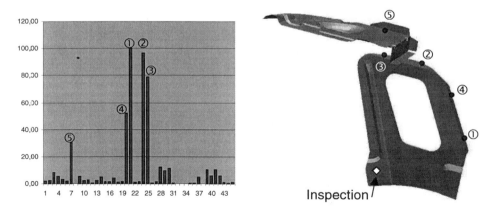

Figure 11: First sensitivity analysis to a given inspection point

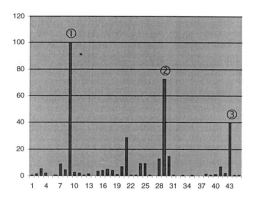
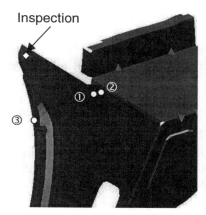

Figure 12: Second sensitivity analysis to another inspection point

Note that only few of the initially provided inspection points turn out to be influential on the results. This suggest that the method could be used to propose a reduced (and more significant) set of inspection points.

Moreover, this analysis can provide the key characteristics of the assembly and the next simulation using this assembly as a part could use only this reduced set of input points. The CPU time required would decrease.

8. CONCLUSION

This paper has described a model based on the computation of three matrices of influence coefficients for the simulation of the assembly process for compliant non-ideal assemblies. By allowing the user to choose between various ways of positioning and fastening the parts, the tool provides the required flexibility for a more realistic and finer statistical analysis of the process. The integration of alternative ways to mesh each part (shell/beam elements) considerably decreases the computation time. Finally the first comparison made between prediction and measurements confirms the interest of such an approach.

9. NOMENCLATURE

μ, σ Mean and standard deviation of a statistical distribution.

$\{D_A\}_B^C$ D is the type of data, B the type of variability analysed. C is the set of points where unit displacements are applied. A is the set of points where D is given. If A is not mentioned, D is given at all nodes.

$\{w\}$ Nodal displacement vector.

Types of variability

RB   Rigid-body-related variability (isostatic positioning)
C   Process-related variability (conformity defect)
S   Geometry-related variability (Shape defect)

Types of point sets

P   « Isostatic » (Positioning) points/supports
A   Additional points/supports
F   Fastening points (welding spot, riveting, bolted or gluing)
S   « Shape » points (used to describe the shape defects)

10. REFERENCES

[Chang et al, 1997] Chang M.; and Gossard D., « Modelling the assembly of compliant, non-ideal parts », In: *Computer-Aided Design*, 1997, Vol. 29, N°10, 701-708.

[Hu et al., 1996] Hu S. J., Long Y., « Tolerance Analysis for Deformable Sheet Metal Assembly », In: *ASME Journal of Mechanical Design*, Vol. 118 n°1, April 1996.

[Hu et al., 1997] Hu S. J., and Long Y., « Variation Simulation for Compliant Workpiece Assembly », In: *ADCATS '97*.

[Marguet et al.,] Marguet B.; Mathieu L.; « Tolerancing Problems for Aircraft Industries », In: *the 5th CIRP International Seminar on Computer Aided Tolerancing*, Toronto.

[Merkley et al., *1996*] Merkley K. G.; Chase K. W., Perry E., « An Introduction to Tolerance of Flexible Assemblies », In: *MSC 1996 World Users' Conference Proceeding*, www.Nafems.org.

[Sellem et al., 1998] Sellem E.; Rivière A. « Tolerance analysis of deformable assemblies », In: *Proceedings of DETC98, 1998 ASME Engineering Technical Conference, Atlanta*.

Aircraft Assembly Analysis Method Taking Into Account Part Geometric Variations

Benoit MARGUET
PhD student
Département Industrialisation et Productique,
Centre Commun de Recherche
AEROSPATIALE, Suresnes, France
benoit.marguet@siege .aerospatiale.fr

Luc MATHIEU
Associate professor
Laboratoire Universitaire de Recherche en
Production Automatisée (LURPA),
ENS de Cachan, Cachan - France,
mathieu@lurpa.ens-cachan.fr

Abstract :

In order to be more competitive, aerospace industry has to develop new design, manufacturing and assembly methods. To insure that all the aircraft's components, from elementary part to main sectors will fit properly together and that all functions will be realized is an important task for product producibility. We propose in this article a new assembly analysis method based on assembly requirements and geometric variations. This methodology analyses and optimizes the assembly sequence to find the best product configuration. Sequence selections are based on assembly graph analysis, key characteristic and assembly sensibility.

Keywords : Tolerance Analysis - Functional Tolerancing - Assembly Analysis Method - Industrial applications and CAT systems - Aerospace Industry.

1. INTRODUCTION

Tolerance Analysis is a critical step to design and build a product such as an aircraft. Whitout or with an ineffectual tolerance study on a sub-component can lead to some huge problems during the assembly process. These problems will introduce additional reworking time and product costs which are not compatible with today's aircraft industry requirements. It is even possible that the product design may have to be subsequently changed because of unforeseen tolerance problems not detected prior to actual assembly took place. Of course in this case, costs to the business will be tremendous.

To identify and analysis the impact of tolerances on a product, as soon as possible in the product life, is, therefore a guarantee for the designer that the chosen assembly process is reliable. But tolerance analysis is not easy to perform, especially for complex aerospace assemblies. So the aid of Computer is called for. Today Computer Aided Tolerance Software is readily available [Salomons 1997] but even if these tools provide good results they have not been widely used especially at aerospace industries. We believe that this situation can be explained by a lack of methods towards tolerancing problems. What are the functions of the product , how do we flow down these "Key" Product Functions through into its detail parts ? How to optimize the assembly process in order to reduce the tolerance impacts on these functions ? etc...

Today, all these questions do have not clear answers, and unfortunately without methods in place to deal with these questions, computer tools are not usable.

The goal of this paper is to present a method for the design analysis of assemblies based on geometrical and dimension variations. This method identifies at the earliest stage, the most attractive assembly sequences for the product in order to manage geometrical and dimension variations on its Key Product Functions. Our method can be used during preliminary design where there is still time to make product changes without incurring extra costs. Finally this method has been developed for aircraft assemblies, however it can also be used for any others mechanical assemblies or products.

This article is in two parts. The first part explains how the method is able to recognize poor assembly sequences. The second shows how this method can assist in selecting the optimal assembly sequence and build process.

2. MACROSCOPIC ASSEMBLY ANALYSES

Let us begin by describing a very simple example. Imagine that we are able to define the assembly process of an aircraft floor as depicted in figure 1. We shall assume that we are in the preliminary design phase and that all the parts of this assembly are rigid. The floor is composed of 9 parts : 3 Floor Beams, 4 Inter-Costal and 2 Seat Tracks. We have defined the constraints of the assembly sequence as follows : two parts have to be located on a fixture, other parts are self-locating through hole to hole assembly features. These constraints are fictional, but include two important aircraft assembly techniques : part location by fixturing and part to part assembly.

In our example, we shall assume that a functional analysis has been already completed. During this analysis, functions have been cascaded from the aircraft's upper level into the floor structure itself [Marguet 97]. All functions have been identified for the floor assembly and risk analysis has allowed us to select the "Key" Functions. The floor possesses a main Key Assembly Function which is the alignment between two floors during its assembly join-up. This function is related to a Key Characteristic with a satisfactory level : the alignment of two floors is related to the dimension between the two seat tracks which has a tolerance value : 10+/-0,1.

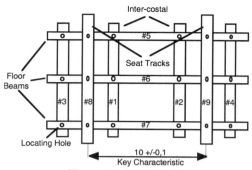

Figure 1: *Case Study*

The goal of our method is to identify the best assembly sequences and build operations in order to achieve the Key Characteristic (KC) within its given tolerance. The method includes geometrical and dimension variations on the parts, requirements on the assembly sequences and build processes.

2.1 Assembly Sequence

If we take a build aircraft and then disassemble it, step by step, we can identify its assembly sequence in reverse. It may be that more one possible assembly sequence is found. All these sequences, in consequence, results in a certain product quality, and not all of the sequences may be able to support the achieve Key Characteristics. Which of these sequences are most practical and which of these sequences will provide the best product quality are generally two pertinent questions.

In order to answer to these questions, we begin by representing all the feasible assembly sequences in a graph. Every box in the graph represents a feasible assembly state and every path from the top to bottom of the graph represents a feasible assembly sequence [Baldwin 91]. For example, all assembly sequences of an assembly with 4 assembly states and without any assembly constraints are represented on figure 2. Within this graph we have 36 paths which travel from the top to bottom giving 36 assembly sequences.

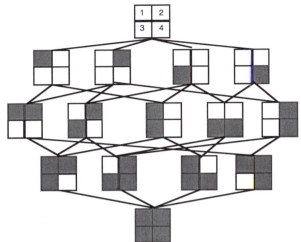

Figure 2 : *Assembly Sequences*

One solution to find the optimal sequence would be to use Computer Aided Tolerance Software. But because of the excessive number of assembly sequences this option may not be realistic within a practical period of time.

In order to reduce the number of sequences to a manageable number, we propose a new analysis method able to reject all inappropriate and unworkable assembly sequences quickly. This method is based on an Assembly Oriented Graph.

2.2 Assembly Oriented Graph

An Assembly Oriented Graph (OAG) is a graph able to represent all kinds of mechanical assemblies. Each node of the graph represents a component and each oriented link represents a constrained position going from the master component to its slave. A component is either a part or a sub-assembly (black node) either a fixturing tool (crossed node).

The Assembly Oriented Graph and the assembly sequences are linked. If component A locates component B then A has to be assembled before B. Each Assembly Oriented Graph is then associated to a family of assembly sequences.

The Assembly Oriented Graph has to respect a few rules. These rules are :

- There are no loops in the AOG. A loop would infer a self-positioning of a component, which is not possible.
- A node with only output links is called a "base component". There is always at least one base component in a graph.
- There is at least one way going from a base component to another component in the graph.
- A node with only input links is called a "final component". There is always at least one final component in a graph.

Figure 3 shows two Assembly Oriented Graphs for the floor assembly. These two graphs represent two different assembly sequences. In the first graph (graph n°1), we begin assembling by locating 2 intercostals (parts n°5 and n°7) on the fixture. The second graph begins by locating two floor beams (parts n°1and n°2) on the fixture (graph n°2).

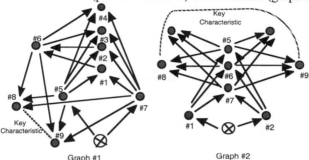

Graph #1   Graph #2

Figure 3 : *Assembly Oriented Graphs*

Each feasible assembly sequences are now represented by a single Oriented Assembly Graph. We may now begin to analyze the Oriented Assembly Graphs in order to discover the most promising assembly sequences and remove the unpromising ones.

2.3 Position Links

Each oriented link on the Assembly Oriented Graph represents a constraint location for the slave component related to the master component. The position of any component in the graph related to the base component is included by all the oriented paths going from the base component to the studied component.

If a Key Characteristic is achieved by the positioning of two components, this KC can be represented on the Oriented Assembly Graph. Figure 3 shows our KC of the Floor Assembly. The relative position of the two components achieving the KC, is given by a set of oriented paths for each component as described in the previous paragraph. These oriented paths are called Delivery Chains by Cunningham as the chains show how Key Characteristics are delivered [Cunningham 97]. In our example the dimension between the two Seat Tracks is achieved by the position of components n°8 and n°9. Each of these components is located related to the base component by 6 oriented paths (figure 4).

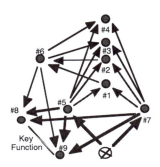

Figure 4 : *Oriented Paths*

We now classify each Oriented Assembly Graph based on an evaluation of the sets of Oriented Paths required to achieve Key Characteristics.

This classification is made through a risk analysis developed by the authors. This risk analysis takes into account the length of the minimal Delivery Chain for each Key Characteristic. A risk number is computed for each KC and each Oriented Assembly Graph. Summation of all the risk numbers, for all the KC in an Oriented Assembly Graph gives a Final Assembly Risk Number. Graphs with the highest assembly risk number are those that are more likely to provide trouble for assembly. For our floor assembly, the length of the minimal Delivery Chain for the alignment Key Characteristic is 2 for graph n°1 and 3 for graph n°2. Risk Analysis concludes that graph n°1 is better than graph n°2.

Based only on Oriented Paths and Risk Analysis we are then able to dismiss quickly all the unprofitable Oriented Assembly Graphs. The second next step in our method is to identify the optimal solution from the remanding graphs.

3. MICROSCOPIC ASSEMBLY ANALYSES

In order to identify our optimal solution, we need to look in more detail how components are related in the assembly and which surfaces of the components participate to the assembly link. To represent the importance of surfaces in our method we introduce the Oriented Contact Graph.

3.1 Oriented Contact Graph

We transform the Oriented Assembly Graph into an Oriented Contact Graph (OCG). Each node of this graph represents a component's surface. All surfaces of the same component are encompassed within a circle. An oriented link in the graph represents the assembly link between two surfaces travelling from the master component to its slave. An extract of an Oriented Contact Graph for the Floor Assembly is shown figure 5.

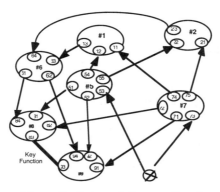

Figure 5 : *Oriented Contact Graph*

Based on the type of surface, it is possible to identify the sort of the assembly link to match. Based on the assembly link type, it is possible to study how the geometric and dimensional variations are transferred through its assembly link.

3.2 Allowed Displacements Analyses

Each assembly link, between 2 surfaces, is composed by one or several location parameters. Each parameter constrains one translation or one rotation position of the slave component relative to its master component. The type of assembly link is given by the type of 2 linked surfaces. If we employ 7 invariant classes of surface, Professor Clement has found 28 different configurations between these surfaces. Each of these configurations can be expressed by 7 types of assembly links.

Unfortunately, a typical component in an assembly has multiple assembly links with other components in the assembly. Not all of these assembly links transfer position constraints and it is essential to distinguish the ones that do from others that are over-constrained. Whitney calls an assembly link that establishes a component's position as "mates" while an assembly link which is over-constrained, he calls "contacts" [Mantripragada 1997]. To identify automatically which parameters of the link are "mates" and which ones are "contact", we perform an Allowed Displacement Analysis for each component. This analysis is based on the work of Ballot and Bourdet [Ballot 1997].

The identification of the parameter types, by an Allowed Displacement Analysis, is related to the sequence of the assembly links. Each assembly sequence selected from the previous chapter can be related to one or a few of assembly link sequences. These assembly link sequences are called "build process".

Let us consider an example in figure 6. Two plates are assembled together by 4 assembly links. Each assembly link is composed of two hole surfaces. The assembly sequence is part A before part B which mean part A locates part B. Based on Professor Clement's work, we know that each assembly link allows us to fix two rotations and two translations for part B.

We can now perform an Allowed Displacement Analysis for each build process in order to identify all the "mates" and the "contacts" for our example. For instance, if we study the build process : 1-2-3-4 then all parameters of the assembly links n°3 and n°4 and the two rotation parameters of assembly link n°2 are "contacts". If now we consider the build process 4-1-3-2, then all the parameters of the assembly links n°3 and n°2 and the two rotations parameters of assembly link 1 are "contacts".

From a tolerancing point of view, the difference between "contact" and "mate" is important. "Mate" offers the position in one direction or in one rotation of those components. So every geometrical and dimension variations of the surfaces in relation to these assembly links have a direct impact on the position variation for that component. If this component is associated with a Key Characteristic, variation can damage the product. In order to reduce this problem, small variations on surfaces relating to mates are required.

Contact Parameters have no impact on the component's position. They are mainly used to fasten and support component once it is located. So geometrical and dimensional variations on the surfaces relating to these assembly links have no impact on the component position and the quality of its function. Larger variations are permissible on these surfaces.

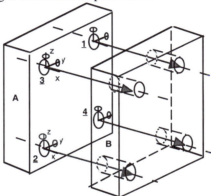

Figure 6 : *Example*

If now we represent the Oriented Contact Graph with only the "mates" we obtained an Oriented Position Graph. Whitney calls this graph with only component nodes "Datum Flow Chain". Two Oriented Position Graphs associated to the Oriented Assembly Graph n°1 is shown by the figure 7.

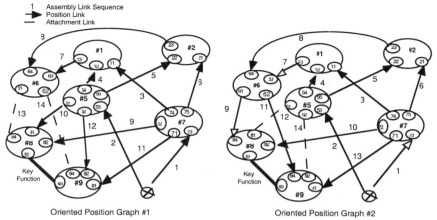

Oriented Position Graph #1   Oriented Position Graph #2

Figure 7 : *Oriented Position Graphs*

The Oriented Position Graphs show how components are geometrically related each other. All oriented paths going from the base component to a Key Characteristics become the 3D Tolerancing Chains for KC. All necessary data is now present in order to perform a proper tolerance analysis so as to select the optimal solution for our build process.

3.3 Tolerance Analysis
Based on the Oriented Position Graphs we can now perform a two step tolerance analysis.

The objective of the first step is to identify one or a handful of the best Oriented Position Graphs. In each case, we study only the variation sensitivity of all surfaces relating to mates on the Oriented Position Graphs. The preferred solutions are Graphs where variations have the least impact on Key Characteristics.

The objective for the second step of the tolerance analysis is to validate those solutions identified in the first step. Variations on all surfaces are now simulated with accurate values delivered from process capabilities. The solution can be validated if all the Key Characteristics are respected and if all components can be assembled without interference between contacts. This tolerance analysis in two steps can be performed using a Computer aided tolerancing software available on the market.

If unfortunately the validation process fails, two options are still possible. From a manufacturing process point of view, it is possible to decrease tolerance values by modifying the manufacturing and the assembly processes. From a design point of view it is possible to enhance the satisfactory level of the KC or to change the design concept to improve its performance. Cost comparisons and negotiations between design and manufacturing specialists is necessary to choose the solution which offers greatest benefit for the business.

4. CONCLUSION

Geometrical and dimensional variations of aircraft's component have to be taking into account as soon as possible in the product life cycle in order to insure product qualities. Impacts of these variations on product's qualities are function of course to the variation's values but either to the assembly sequence and to the product's build process. Select the optimal build process for an aircraft assembly and validate this process as soon as the preliminary design allows to avoid most of the problems during the assembly phase and saved a lot of money and time.

This paper has presented a new method to analyze aircraft assemblies based on geometrical and dimensional variations. The method is composed of two steps. The first step of the method is to reject all the poor assembly configurations in order to focus attention only on those most likely to provide the best quality of product. Assembly Oriented Graphs and Risk Analysis are used in order to organize Graphs into a hierarchy. The second step identifies the optimal configuration based on Oriented Position Graphs and Tolerancing Chains. This method has been applied for few real industrial cases with great benefits and we hope to widespread use for most of new aircraft assemblies.

REFERENCES

[Salomons 1997]   Salomons O.W; Van Houten F; Kals H; "Current Status of CAT Systems"; *CIRP Seminar on Tolerancing,* Toronto 1997

[Baldwin 1991]   Baldwin D.F; Whitney D; "An Integrated Computer Aid for Generating and Evaluating Assembly Sequences for Mechanical Products"; *IEEE Transactions on Robotics and Automation*, Vol. 7, pp. 78-94, 1991.

[Cunningham 1998]   Cunningham T; "Chains of function delivery : A role for product architecture in Concept Design", *PhD Thesis*, Massachusetts Institute of Technology, 1998

[Mantripragada 1998]   Mantripagada R; Whitney D.E; "The Datum Flow Chain: a Systematic Approach to Assembly Design and Modeling", *Research in Engineering Design*, October, 1998

[Marguet 1997]   Marguet B; Mathieu L; "Tolerancing Problems for Aircraft Industries", *CIRP Seminar on Tolerancing*, Toronto 1997

[Ballot 1997]   Ballot E; Bourdet P; "Computation Method for the Consequences of Geometric Errors in Mechanisms", *CIRP Seminar on Tolerancing*, Toronto 1997

[Clement 1994]   Clément A; Rivière A; "Tolerancing versus nominal modeling in next generation CAD/CAM system", *CIRP Seminar on Tolerancing,* Paris, 1993

Integrated Tolerance Information System

Dipl.-Ing. Markus Wittmann
DFG-Projekt "Qualitätsmerkmal Toleranzen"
Institute for Information Systems (IWi)
University of Saarland
Prof. Dr. Dr. h. c. A.-W. Scheer
Altenkesseler Straße 17, Geb. D2
66 115 Saarbrücken, Germany
wittmann@iwi.uni-sb.de

Abstract: The paper presents a concept for creating a tolerance information system which on the one hand calculates time and costs resulting from each tolerance and on the other hand serves as a decision support system for all business processes. Thus the concept combines the technical with the economical view to the tolerances. The purpose of the tolerance information system is to collect and manage relevant tolerance information from all business activities within the company and to support the design engineering by providing information about costs, time and feasibility of a given tolerance in a feedback loop. The focus is on the technical and management basis needed for the adjustment of all business processes, which are influenced by tolerances, to the strategic factors time cost and quality. It also tries to integrate and coordinate management and technical goals during the process of product development regarding tolerances as a quality feature.

Keywords: information system, quality feature, tolerance costs, feedback loop, business processes

1 INTRODUCTION

Engineers and manufacturers as well as quality control assistants are having to face the problem that technical products cannot be accurately manufactured according to the desired values. Systematic and/or random error is more or less unavoidable, depending on the technical expenses. However, this refers more to technical products manufactured in big series or mass production.

The designer's reaction to such errors was to tolerance all measures. This means that limit values will be directly or indirectly predefined and the actual values should vary within these imits.

Many design engineers allocate tolerances according to the principle: "as large as possible and as small as necessary" and additionally in cases of uncertainty which are more often "better smaller than larger " [Kirschling, 1988]. It leads to so called afraid-tolerances which cause increased manufacturing costs and are therefore not economically viable.

The purpose of the research project "Quality measure Tolerances" is to develop a concept for extensive consideration of tolerance information and its influence on all company operations. The technical and economical basis for the alignment of all company operations which are influenced by tolerances to the strategic factors of performance for the company will be systematically researched. The results will be presented in a company wide tolerance information system.

2   STRATEGIC FACTORS OF PERFORMANCE

The strategic factors are those which determine the success of a strategic company unit [Pieper, 199]. Theoretically they are the reason for the positive or negative development of a business enterprise and answer the question which criteria have substantial influence of the potential for success of strategic business fields. The strategic factors are measurable and controllable through indices [Fischer, 1993] [Krüger et al,. 1990].

A reference number represents the fundamentals of a productive enterprise. Reference numbers serve as a measure for analysis of internal results as well as for comparison of the performance of different companies. They not only result from manufacturing but also from other business operations. Reference numbers are relative numbers representing a combination of different substantial factors. With their help it is possible to get more precise information about the company structure or cost structures, analyze these structures and improve them. Companies working in the same field and with identical manufacturing structures should be compared.

The specific factors of a successful performance regarding the business manufacturing enterprises considered in this project (see Fig. 1) could be measured through the reference numbers quality, time and cost. However, these reference numbers are also influenced by the design engineer who determines the tolerances. Figure 1 shows the relationship between tolerance values, reference numbers and factors of a successful performance.

Providing tolerance information during the design process and also other value added manufacturing operations makes it possible to increase effectiveness that could be measured by the reference numbers time, cost and quality.

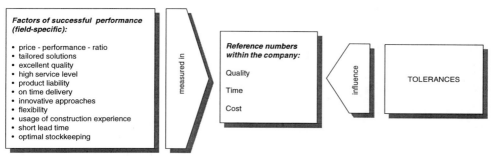

Figure 1; *Interrelation between tolerances – reference numbers – factors of successful performance*

3   INTEGRATION CONCEPT

The tolerance information system connects the tolerance data from CAD with the experimental database from the manufacturing and capacity information from the production planing and control system (PPS – system). This information, aggregated by the system, is available for the design engineer.

The following part introduces the components of the integration concept.

3.1 CAD-Linking

The data necessary for the tolerance information system from the design process is realized in the form of a so called extended tolerance feature. A feature is defined as a syntax of a form element (more specifically a geometrical element) whose semantic is the technical relevance [Baer, 1998]. For the quality measure tolerance it means that the tolerance information regarding a geometrical element is added as a semantic to the corresponding "form feature".

The classical feature-based approach considers only the relationship between function and tolerancing with its corresponding geometrical data, however, by this approach some further information is being implied in a feature. Figure 2 presents the different viewpoints regarding tolerances of a form feature that are being researched in this project.

According to the picture, the extended tolerance feature consists of a form-feature and the viewpoints of the design-engineering (corresponds to the technical function), manufacturing, measuring and management. These views are assembled in a database which is accessible for the tolerance information system.

378

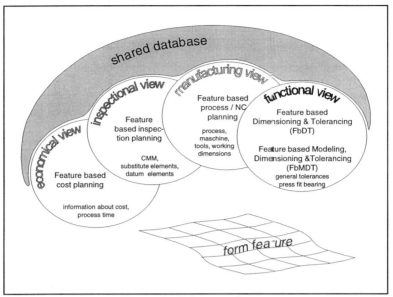

Figure 2; Tolerance feature

3.2 Experimental Database

In order to relieve the production planning when selecting parameters it is necessary to collect empirical data. One possibility to do so would be to save all the experimental data in a database as long as it is easily and quickly accessible.

This experimental database allows the derivation of process parameter needed for manufacturing on the basis of geometrical additional information and also from tolerance data. [Bley et al.,1998].

On the one hand process parameters like feeding, cutting speed, way of clamping, etc. are recorded in this database, but on the other hand there is also management data as machine hourly rate and running time. Additionally there is information about the achieved accuracy of measure, form and position. Data is collected through experiments with standard workpieces which are manufactured with different machines' parameters and finally checked accuracy to gauge. (see Figure 3).

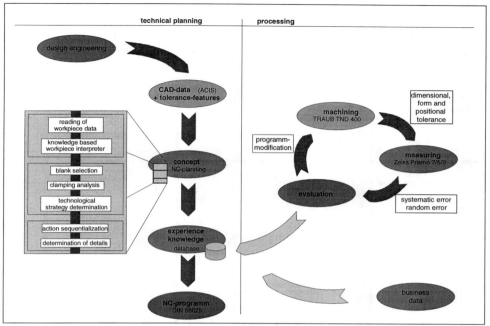

Figure 3; *Experimental database*

In an experimental circuit of machining – measuring – evaluation the technological and economical knowledge will be collected and prepared. Cost and data structures concerning the manufacturing process could be also stored in a database. It is important to separate between random error and systematic error in the step of evaluation. The modification of the NC programs based on the won results closes the experimental circuit. The experimental results in the SQL database are now evaluated and summarized. Using the methods of the statistical experimental planning guarantees that the number of required experiments can be kept small.

With the help of queries it is possible to generate special information for the work planning from the database and to support the planning when taking decisions. It is possible that the database answers questions like "Which process parameters must be chosen in order to maintain a pre-given tolerance?" The database could not only support a real work planning manager but also virtual ones. It may be that some programs for work planning operate with the information from the database and automatically generate NC-programs, for example. Queries from the database like "How much does a certain tolerance cost?" which are needed for the cost calculations are also the property of the system. This represents the interface to the developed tolerance information system.

380

3.3 Tolerance Information System

The tolerance information system first of all serves as a decision support tool for the designer during the design process. The system works like a feedback loop integrated in the value added chain. The design engineer receives feedback from the information system and according to it he can adjust his design. The system makes the tolerance relevant information transparent and is available during the whole product development and formation process.

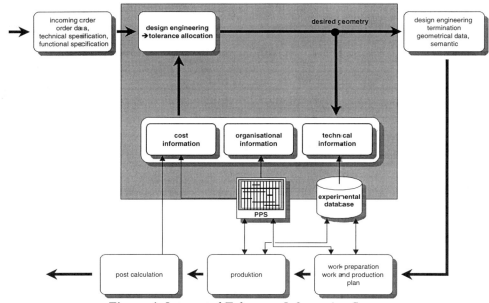

Figure 4; *Integrated Tolerance Information System*

Figure 4 presents the whole concept of the tolerance information system. The grey area shows the feedback loop. Those areas in white are the value added units, a traditional product development process.

The approach and functioning of the tolerance information system will be interpreted in the following 5 steps. This explanation uses also Fig. 4 as well as the data model of the tolerance information system (Fig. 5).

Figure 6 shows a screenshot of the user's scenario by visualizing a possible tolerance information system

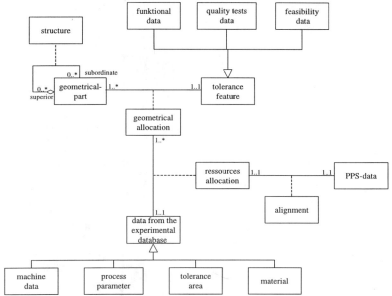

Figure 5; *Integrated Tolerance Information System*

1. The design engineer inserts a feature (e. x. borehole, arbour offset etc.) in his CAD drawing and allocates the corresponding tolerance.
2. When necessary or in cases of uncertainty he/she can chose a certain form element and receive some more information.
3. A query from the database will meet the data from the chosen tolerance feature. The result is that information is made available as to whether the chosen form element together with the allocated tolerance could be manufactured with the machinery tools saved in the experimental database (technical feasibility). If the query is positive the machinery information and the process parameters (cutting speed, running time, etc.) are available.
4. Which machines are available as well as their process costs would be checked through a interface to the PPS-system (organizational feasibility). Together with the running time it would be possible to determine a cost reference number concerning the chosen tolerance.
5. The given additional information (technical feasibility, costs, organizational feasibility) is available for the design engineer compressed in his CAD environment and on the basis of this information, the design engineer is able to change his drawing and respectively the allocated tolerances.

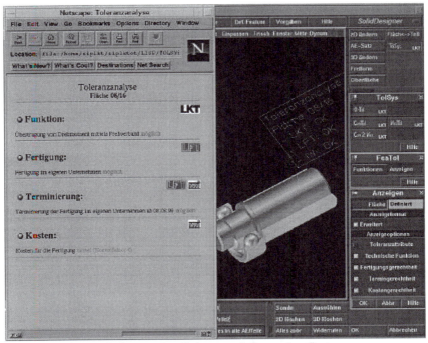

Figure 6; *Screenshot Tolerance Information System*

The data model of the tolerance information system (see Picture 5) shows the links between the single integrated components of the tolerance information system – CAD – system, the experimental database and the PPS – system. The interface between the systems are still at concept level. It is reasonable to use here standard exchange formats in order to realize the tolerance information system in different IT landscapes.

The screenshot of the user scenario (Fig. 6) clarifies how it is possible to provide information for the design engineer. A information system of a html-basis is shown in this screenshot so that the information can be obtained from the CAD system and it is not necessary for the design engineer to leave his working area.

OUTLOOK

The described functions of the tolerance information system in this paper concentrate mainly on supporting the designer during product development stage. Acquiring tolerance information for work planning, manufacturing and other value added processes is also possible using the data structure of the integrated tolerance information system described above. The exact linking of these areas and the use of tolerance information are the subject of further research within this project.

REFERENCES

[Baer,1998]   Bär, T.: Einsatz der Feature-Technologie für die Integration von FE-Berechnungen in den frühen Phasen des Konstruktionsprozesses, Dissertation des Lehrstuhls für Konstruktionstechnik/CAD, Universität des Saarlandes, Saarbrücken, 1998

[Bley et al., 1998]   Bley, H.; Oltermann, R.: Determination of technological data during NC Programming based on CAD-integrated tolerances, Lehrstuhl für Fertigungstechnik/CAM, Universität des Saarlandes, Saarbrücken, 1998

[Fischer,1993]Fischer, T.: Kostenmanagement strategischer Erfolgsfaktoren, Verlag Vahlen, 1993

[Kirschling, 1988]   Kirschling, G.: Qualitätssicherung und Toleranzen, Springer Verlag, 1988

[Krüger et al., 1990] Krüger, W.; Schwarz, G.: Konzeptionelle Analyse und praktische Bestimmung von Erfolgsfaktoren und Erfolgspotentialen, in: Bleicher, K.; Gomez, P. (Hrsg.):Zukunftsperspektiven der Organisation, Festschrift zum 65. Geburtstag von R. Staerkle, Bern, 1990

[Pieper, 1991] Pieper, R. (Hrsg.): Lexikon Management, Gabler Verlag, Wiesbaden, 1991

Integrated Tolerance Management

by

Spencer Graves and Søren Bisgaard

Center for Quality and Productivity Improvement,
University of Wisconsin-Madison and
Productive Systems Engineering,
751 Emerson Ct., San José, CA 95126 USA,
sgraves@prodyse.com,
and Institute for Technology Management,
University of St. Gallen, Switzerland
soren.bisgaard@unisg.ch

Abstract: Integrated tolerance management is a systems approach for achieving global consistency of tolerances throughout an organization. The purpose is to facilitate harmonization of tolerances throughout an organization to achieve high quality functionality at low cost. A tolerance management system should consist of three elements: planning, control and improvement. The planning function aims at developing tolerancing policies, standard practices, organizational responsibilities, and action plans for implementation of new initiatives. The control function includes besides the inspection role also the collection and storage of data required for inspection, process control, monitoring and improvement. Because tolerancing involves a compromise between what is needed and what is possible, the control function plays a critical role in delivering the data on which such compromises should be made. Data from all sources should be organized to help identify opportunities for improvement in specific products as well as applicable procedures and associated business plans. The cycle of planning, control and improvement continues by incorporating ideas for improved procedures and plans for the future.

Keywords: Tolerances, manufacturing economy, manufacturing management, process control.

1. INTRODUCTION

Tolerances play a vital role in transforming product ideas into physical products that can be manufactured economically, satisfying customers' requirements. Although widely neglected, especially in our educational institutions and by many design engineers, a rational, consistent company-wide system of tolerances can make an important contribution to the production of high quality, low cost products. Most of the current literature, including our own contributions, focuses on technical details of tolerancing a single product or assembly. To be sure, much needs to be done at that level before we have what we would consider a rational approach to tolerancing.

However, another important aspect of tolerancing is how to develop a company-wide system for the coordination and management of the tolerancing function. The focus of this article is to outline the basic elements of an integrated tolerancing management system for coordinating tolerancing policies and practices throughout an organization and beyond. Because this is a much neglected area of research, with only few previous contributions, we hope to stimulate a discussion of this issue.

2. THE TOLERANCE MANAGEMENT SYSTEM

In theory, tolerances are how design engineers communicate to manufacturing what they want in terms of precision of manufactured components. In practice this communication takes (or should take) the form of a dialog rather than a one-way dictate from the designer to the manufacturing department. How much variability can be allowed without impairing the function of a product is most often not known. Design engineers, at least when they are honest, will admit that they in most cases are guided in their allocation of tolerances mostly by intuition, past practices and process capability data. Tolerances are rarely based on hard facts about what is objectively needed as a minimum for the product to function.

Design engineers in their request for precision on parts will most often need to yield to the realities of manufacturing and economics. As a general rule the tighter the tolerances the more costly the manufacturing. Thus in reality most tolerances are obtained by negotiation and compromise between what the designer would like and what manufacturing can deliver economically. Such negotiations would be much facilitated if product designers would label tolerances as critical and non-critical for the function, as on any given product there are usually many more dimensions than what can reasonably and economically be dealt with. I f negotiations could be structured such that most time was spend on critical dimensions and non-critical dimensions could be dealt with more by policy, much time could be saved and an overall better solution obtained.

Negotiations about compromises require close cooperation between many organizational departments such as Design, Manufacturing, Quality Control, Purchasing, Sales and Service. Moreover, as [Shewhart, 1939, pp. 44-45] pointed out and as illustrated in his diagram shown in Figure 1, tolerancing is not a one-shot affair where tolerances are specified, and products are manufactured and checked. Instead, it is an ongoing iterative process that is similar to the scientific process of knowledge generation. In fact in explaining Figure 1 he said that "specification, production, and inspection correspond

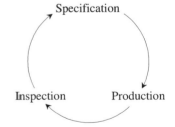

Figure 1; Shewhart's Iterative Scientific Learning Process

respectively to making a hypothesis, carrying out an experiment, and testing the hypothesis. The three steps constitute a *dynamic* scientific process of acquiring knowledge." [Emphasis added.] In most organizations we are familiar with, the tolerancing negotiations and learning processes are rather haphazard and proceed in an ad hoc fashion. It would be beneficial for an organization to formalize and coordinate this cross-functional negotiation process and to clarify the tasks, practices and responsibilities. For this, we suggest a two-layer committee framework, one at a strategic and one at a tactical level, as explained below. With the modern trend towards a global marketplace and virtual organizations, this would seem even more important.

As indicated above, tolerances are, and should be, determined based on negotiations between many relevant departments representing various interests. As shown in Figure 2, manufacturing and engineering/design are the key players. They represent in the negotiations the two extremes of what is necessary and what is economically possible. However, the negotiations, whether at a strategic or tactical level, should also include representatives from Sales, Service and Purchasing. To coordinate and moderate the tolerancing process the Quality Department should play a key role. In fact in many organizations it would be beneficial to have the Quality Department chair the tolerancing board or committee. That department in some sense is a neutral partner between Design and Manufacturing and can act as a representative for the customer.

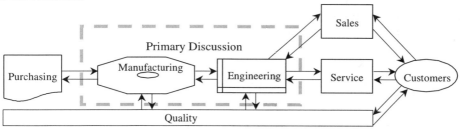

Figure 2; *Information Flow for Tolerancing*

To coordinate the tolerancing efforts and facilitate the negotiation process, we propose a two-layer organizational structure. At a strategic level we propose a company-wide *Tolerancing* policy coordination *Board*. It should be engaged in long-term planning and policy making. To shield the Tolerancing Board from excessive detail, we envision tactical level *Tolerancing Committees* coordinating the routine tolerancing process of setting, evaluating, checking and revising tolerances on a product family basis (Figure 3). The strategic Tolerancing Board's tasks and responsibilities should be to:

- Establish general policies for company-wide tolerancing
- Define authority for setting, changing and making exceptions for tolerances
- Define and maintain the company data base for process capability data and past process and product performance
- Establish and support a tolerancing research agenda
- Maintain and update the company (possibly web-based) tolerancing manual
- Mediate large scale, cross product-family tolerancing conflicts

- Negotiate with upper management for the necessary budget line items for maintaining a company-wide tolerancing management system

The Tolerancing Board should be company wide and consist of members from all organizational units. It should be chaired by a representative from the Quality department. This board should only consider long-term, high level issues. It should set policies and coordinate company-wide harmonization. It should not be concerned with everyday issues.

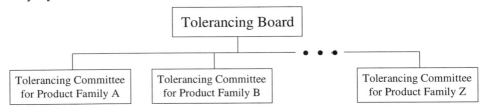

Figure 3; *Organizing for Tolerancing*

The daily tasks of setting tolerances based on the general policies laid down by the company-wide Tolerancing Board are managed by product family related Tolerancing Committees. These committees' responsibilities should be to:

- Promote the general company-wide tolerancing policies in the particular sub units
- Review new product tolerances as part of preparation for manufacturing
- Compare existing products' field performance (returns, complaints, customer service calls) with manufacturing defect rates to identify potential tolerancing issues
- Issue tolerancing guidelines specific to the product families to be approved by the company-wide Tolerancing Board and to be incorporated into the tolerancing manual (Section 6 below)
- Initiate discussions and negotiations with other business sub units when cross product family issues arise
- Police violations of company tolerancing policies and standards
- Mediate conflicts within product families

To be effective, the product family specific Tolerancing Committees should consist of the head of the design department developing the product, a Manufacturing department representative, a Quality Control department representative (who should chair the committee), a Service department representative and someone from Sales. When a specific product is being discussed, either initially or whenever tolerances are being revised, the specific designers responsible for the product or sub assembly under discussion should be asked to join the committee.

The strategic Tolerancing Board and the product family Tolerancing Committees should organize their activities around the Shewhart Plan, Control, Improvement cycle [Shewhart, 1939]. This cycle was presented as four steps (Plan, Do, Check, Act) by [Deming, 1986, p. 88] to Japanese executives in 1950, and has been popularly known as the Deming cycle ever since. It has also been used by [Juran, 1986] in his "Trilogy" to explain Quality activities by analogy with Accounting and other managerial activities. In Table 1, we extend this analogy to tolerancing. The three primary components of this management model, plan, control, improvement, are discussed in the next three sections.

Function	Finance	Quality	Tolerancing
Managed by	Finance Committee	Quality Council	Tolerancing Board
Plan	Budgeting	Quality planning	Setting specifications
Control	Cost, expense control	Quality control	Feedback from MRB, etc.
Improvement	Cost reduction, profit improvement	Quality improvement	Improving tolerancing procedures
Documentation	Accounting procedures	Quality documentation (ISO 9000)	Tolerancing handbook

Table 1; Tolerancing and the Juran Trilogy

3. TOLERANCE PLANNING

Typically, original specifications for new products are set at the end of the new product process, either by design or manufacturing engineers. When problems arise in production, specifications may be modified; changes are typically made by production engineers, possibly after consulting the engineers who originally set the specifications.

This traditional approach has often failed. We believe that the committee structure outlined above would work better and would formalize and facilitate the necessary negotiations. Many parties in the organization should be involved in tolerancing decisions, and the process should not be left to happen haphazardly. First and foremost, the Tolerancing Board and Committees need advocates for customers. While Engineering may have some interface with customers, most organizational contacts with customers are through Marketing, Sales, and Service. As a basis for negotiation for a new product, it would seem reasonable that the Engineering department present a proposal to the Tolerancing Committee outlining suggested tolerances for critical as well as non-critical characteristics. Manufacturing, as well as Quality and Service then review this proposal, preferably armed with process capability data. Later when product tolerances are revised or when problems arise, Customer Services or Customer Returns or Quality Departments may make valuable contributions based on field data. Hence these and the other groups depicted in Figure 2 should be involved in the tolerancing negotiations and thus be represented on the Tolerancing Board and Committees.

Figure 4 depicts activities relating to tolerancing relative to the product development cycle. Everything begins in the strategic planning process. The strategic plan describes which new products will be developed to serve which customers, and how much money and time will be invested in these projects. The more promising initial product concepts become full product development projects. Particularly difficult questions regarding core technological competencies of an organization may get separate attention under the heading of technology development. At various stages throughout product development, design engineers may be concerned about tolerances because they need to know if the existing design concept can be produced within the targeted per unit budget. As a new product design nears completion, more decisions are made on tolerances.

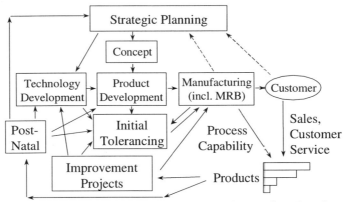

Figure 4; *Timing Tolerancing Activities Relative to the Product Development Process*

For low volume products, tolerances may be set using a "worst case" approach. However, for high volume products, even very small savings per unit can add up quickly. In such cases, it can be cost effective to establish procedures whereby data relevant to estimating process capabilities can be shared between Manufacturing, Incoming Inspection, vendors, and Technology Development. These data can then be used to make more informed decisions regarding tolerancing. For higher volume production, statistical tolerancing is as a rule much more economical. However, assumptions should be carefully checked [Graves and Bisgaard, 1997; Bisgaard and Graves, 1997].

In volume production, some percentage of new products shipped to customers lead to customer service calls and returns. Some products come back because customers just changed their minds. Other products were dead on arrival; some Customer Return Departments receive no information from the customer regarding the reason for return. The most difficult problems to understand are "No Trouble Found". Some warranty problems may stem from tolerances that are looser than they should be, while others arise from components outside specifications. It is generally difficult to determine the root cause of a problem just by looking at a customer complaint or a returned product. It is therefore wise to produce a Pareto chart of product problems (preferably of forecasted problems [Graves, 1997]), as depicted in the lower right hand corner of Figure 4, for use in selecting products for in-depth study. Some of the projects that study the selected problems will uncover and correct tolerancing problems. When that happens, there is a problem with either tolerancing procedures or practices. If the organization is to avoid repeating these problems, appropriate changes must be made either in policies, training, or the way people are managed. [Juran, 1993] discussed a related problem that contributed to serious financial difficulties for a major international company in the late 1970s: Known problems were replicated in new products while their competitors were modifying designs to eliminate those problems. The final activity in Figure 4 is a "post-natal" review of a new product development project. Such reviews are conducted by leading engineering companies as part of their process of continually improving of their product development process [Graves et al., 1991; Zangwill, 1993]. The role of tolerancing activities in product development is one issue to discuss in such a review.

4. CONTROL

Ongoing control and inspection of parts and processes are key to delivering assembled quality products to the customer. Such inspection must necessarily be based on data. As noted in Table 1, there are today substantial traditions for using accounting data to control costs and expenses against budgets (plans). Similarly there is now an increasing trend towards using data to control quality of products and processes against plans. The concept of Integrated Tolerance Management provides a vision of organizational and electronic procedures that support the collection and sharing of data relevant to tolerancing with the people who set the tolerances. In particular, in volume production, there may be substantial opportunities for cost reductions through effective statistical tolerancing. However, as noted by [Graves and Bisgaard, 1997] and [Bisgaard and Graves, 1997], these opportunities can be easily misunderstood unless process capability data are used to evaluate the appropriateness of essential assumptions. Through appropriate collection and use of process capability data, the relative validity of the assumptions of statistical tolerancing can be evaluated and used to support production with lower cost materials and production technology when they will not affect product quality. In some cases, designed experiments may be needed to evaluate and estimate the functional model needed for tolerancing [Bisgaard, 1997].

Control of tolerancing practices should include standards for documenting tolerancing decisions and standards for sizes and grades of material and parts used routinely in production, as well as labeling dimensions as critical and non-critical. Standards for documentation can make it easier for someone to review a tolerancing decision years later to determine if it should be modified or copied in a related product. Documentation standards should include a system for recording and retaining the history of what tolerancing decisions were made and why. If someone recommends changing a particular specification, it could be helpful to know the original rationale for the initial tolerancing decisions and for any previous revisions. This kind of documentation would be both easier to collect and more valuable if it were retained in an electronic format that could be searched in a variety of ways.

Compliance with standards for tolerancing decisions and documentation can in many cases be facilitated by an appropriate Computer-Aided Tolerancing system. This system can provide access to data relevant to tolerancing decisions including process capability data and relevant tolerancing history for that and similar applications. It can also recommend specific solutions following certain rules and can highlight gaps in available data. The user should have the option of attaching additional documentation that seems required to explain tolerancing decisions for future reference.

The organizational routine must also provide opportunities for explaining the importance of data collection to the people whose data are used in tolerancing. Without feedback of this nature, required data may not be collected or shared or may include so many errors that it becomes worse than useless.

5. IMPROVEMENT

Tolerancing problems are a daily phenomenon in production. A little reflection and practical experience in manufacturing will show that their economic impact is often dramatic. Tolerancing problems often mean the difference between profitable production and serious losses. Thus it is important to ask how can tolerancing problems be reduced and preferably eliminated? Progress can be built on a program of improvement projects, carefully selected using as one input a Pareto chart, as discussed with Figure 4. Each such project team should be tasked to consider the possible contribution of tolerancing to these problems. When the project team determines that tolerances appear to have contributed to the problem, an ad hoc task force should be established to study the tolerancing procedures used to determine if established procedures were followed and, if they were, should they be changed. If they were not followed, the question turns to why they were not followed and what should be done in the future to elevate the importance of following the procedures. Reports from these analyses should be made to the relevant Tolerancing Committee, described in Section 2.

Similarly, evaluation of defect rates in manufacturing in comparison with tolerances should be routinely done 6-12 months after new product introduction, as part of a "post-natal" evaluation on a new product, as mentioned with Figure 4. Problems found should be reflected either in improvements in the tolerancing manual (described next) or in procedures for training and managing the people and the organization to improve compliance.

6. DOCUMENTATION: THE COMPANY TOLERANCING MANUAL

Integrated tolerance management must be institutionalized based on documentation of procedures and organizational structures [Gilson, 1951; Juran and Gryna, 1993]. This documentation should include company-wide tolerancing policies as well as record the results of decisions regarding which departments will provide what data to which other departments, in what formats, who will use that data to set tolerances, and how the data will be used. This documentation may be organized as a company Tolerancing Manual. In this age of intranets and ISO 9000, the only official forms of this manual may be electronic, with all access to the manual being achieved by computer terminal. Related issues that are already covered in other company policies and procedures may be treated in the Tolerancing Manual at least partially with a reference to the other information. For example, the Tolerancing Manual may include references to Design for Manufacturability and its implications for tolerancing.

The Tolerancing Manual would also document previous decisions regarding which parts of the organization provide data and which make initial tolerancing decisions. Procedures for revising tolerances and for change order management must also be discussed, along with standards for documentation.

7. DISCUSSION AND CONCLUSION

Tolerancing is often a stubborn detailed technical problem (i.e. of no interest to upper management) that can derail a large operation. The American Revolution was ultimately won in part by the mobility of the French Light Artillery, which was instrumental in the defeat of the British at the Battle of Yorktown, the decisive battle of that war. The success of the French Light Artillery in that case was due in part to the "uniformity principle", advocated by General Gribeauval who distributed standard gauges throughout France to ensure that a ball made in one part of France would fit in a cannon made elsewhere [Hounshell, 1984].

Tolerancing is important but unglamorous. The economic impact of tolerances are not well understood by upper management, and most design engineers have at best only a very limited understanding of and interest in the technical and economic aspects of the problem. Even the literature reflects little on how to organize for company-wide harmonization of tolerances and how to set up structures for organizational learning around tolerancing. However, in volume production, appropriate tolerancing can make the difference between profit and loss. Manufacturers who produce in substantial quantities would be wise to look at tolerancing as a potential source for a competitive advantage. A good start would be to formally establish organizational structures and committees (as outlined in this article), responsible for coordinating the tolerancing process; this is something few if any companies have done in the past.

Significant improvements in tolerancing practices require engineers in one part of the organization to utilize data collected in other parts. In a traditional system a design or manufacturing engineer may on rare occasions access that data. This will usually have little impact on how tolerances are set and enforced over all. Integrated Tolerance Management, as described in this article, will for many organizations require substantial improvements in how data are shared throughout the organization, and how those data are used for tolerancing. This kind of change will require a certain level of commitment and involvement by management, as well as a formalized organizational framework.

ACKNOWLEDGEMENTS

This research was supported by grant number DMI 9500140 from the National Science Foundation of the US government. The presentation was improved by comments from John Wesner of Lucent Technologies and a Reports Committee of the Center for Quality and Productivity Improvement at the University of Wisconsin-Madison.

REFERENCES

[Bisgaard, 1997] Bisgaard, S.; "Designing Experiments for Tolerancing Assembled Products", *Technometrics*, Vol. 39, No. 2, pp. 142-152

[Bisgaard and Graves, 1997] Bisgaard, S., and Graves, S. B.; "A Negative Process Capability Index from Assembling Good Components? A Problem in Statistical Tolerancing", *Quality Engineering*, 10(2), pp. 409-414

[Deming, 1986] Deming, W. E.; *Out of the Crisis* (Cambridge, MA: MIT Center for Advanced Engineering Study)

[Gilson, 1951] Gilson, J.; *A New Approach to Eengineering Tolerances* (London, Machinery Pub. Co.)

[Graves, 1997] Graves, S. B.; "How to Reduce Costs Using a Tolerance Analysis Formula Tailored to Your Organization", Report 157, Center for Quality and Productivity Improvement, University of Wisconsin-Madison; www.engr.wisc.edu/centers/ cqpi

[Graves and Bisgaard, 1997] Graves, S. B., and Bisgaard, S.; "Five Ways Statistical Tolerancing Can Fail and What to Do about Them', Report 159, Center for Quality and Productivity Improvement, University of Wisconsin-Madison; www.engr.wisc.edu/centers/ cqpi

[Graves et al., 1991] Graves, S. B., Carmichael, W., Daetz, D., and Wilson, E.; "Improving the Product Development Process, *Hewlett-Packard Journal*, June, 1991, pp. 71-76

[Hounshell, 1984] Hounshell, D. A.; *From the American System to Mass Production, 1800-1932* (Baltimore, MD: Johns Hopkins U. Press)

[Juran, 1986] Juran, J. M.; "The Quality Trilogy", *Quality Progress*, August 1976, pp. 19-24

[Juran, 1993] Juran, J. M.; "Made in U.S.A.: A Renaissance in Quality", *Harvard Business Review*, July-August, pp. 42-50

[Juran and Gryna, 1993] Juran, J. M., and Gryna, F.; *Quality Planning and Analysis*, 3rd ed. (NY,: McGraw-Hill)

[Shewhart, 1939] Shewhart, W. A.; *Statistical Method from the Viewpoint of Quality Control* (Washington, DC: Graduate School, US Department of Agriculture)

[Zangwill, 1993] Zangwill, W. I.; *Lightning Strategies for Innovation* (NY: Lexington Books)

Role of Statistics in Achieving Global Consistency of Tolerances

Vijay Srinivasan
IBM Research & Columbia University
Room 2-150, IBM T. J. Watson Research Center
P. O. Box 218, Yorktown Heights
NY 10598, U.S.A.
vasan@us.ibm.com

Abstract: This paper proposes and defends a thesis that global consistency of tolerances is achievable only by consistent use of statistics across specification, production, and inspection. It starts with a global view of variation in a product of an engineering enterprise, embodied in two major axioms: imprecision in manufacturing and uncertainty in measurements. It traces the successful evolution of statistical practices in production and inspection to deal with these variations. Finally, designer specified tolerances are brought into the global picture. Challenges that still lie ahead are summarized at the end.
Keywords: tolerances, statistics, consistency.

> "The conception of statistics as the study of variation is the natural outcome of viewing the subject as the study of populations; for a population of individuals in all respects identical is completely described by a description of any one individual, together with the number in the group. The populations which are the object of statistical study always display variation in one or more aspects."
>
> R. A. Fisher, *Statistical Methods for Research Workers*

1. INTRODUCTION

The main theme of this 6th CIRP International Seminar on Computer Aided Tolerancing is *global consistency of tolerances*. Let's start with the hypothesis that global consistency warrants a uniformity of principles and practices applied across an enterprise - a set of engineering firms in our case - and that it doesn't exist today. Observation indicates that industry struggles with this problem, and has developed islands of practices that achieve local consistency by *ad hoc* means. Theory and principles to unify these practices are still evolving.

This paper advances a thesis that global consistency of tolerances is achievable only by consistent use of statistics across the engineering enterprise. To put it simply, this is because the only thing that is global is *variation*, and statistics is the appropriate discipline for studying variation as observed by Ronald Fisher [Fisher, 1950]. The rest of the paper provides supporting evidence for this thesis.

We start with a global view of variation in Section 2, and argue that inevitable variations in production and inspection influence product specification. Section 3 describes how industry uses statistical tools to deal with variations in production. Section 4 addresses the same issue for variations in inspection. Statistical metrics to characterize variations in design are discussed in Section 5. The paper concludes with a summary of challenges ahead.

2. A GLOBAL VIEW OF VARIATION

We begin with a basic question: Why do engineers tolerance industrial artifacts? The answer lies in the fact that we can never manufacture exact parts, and even if we do we can never tell by measurement that they are exactly the same This inevitability of variability is embodied in the following two axioms:

1. *Axiom of manufacturing imprecision*: All manufacturing and assembly processes are inherently imprecise and produce parts and products that vary.
2. *Axiom of measurement uncertainty*: All measurement systems and measuring instruments contain inherent sources of error, and the result of a measurement is always subjected to some uncertainty.

The axiom of manufacturing imprecision implies that no two manufactured objects are exactly the same, and they are definitely not the same as the ideal shape the designer had envisioned. The axiom of measurement uncertainty implies that measurements can never guarantee with certainty that the manufactured object has conformed to the design specification – at best, we can only assure conformance with a high degree of confidence. It may be argued that these two axioms are provable facts from basic laws of physics; we will not succumb to this temptation here and leave them as axioms without proof. There is ample industrial evidence to suggest that these axioms have never been violated.

The designer now has the challenge of dealing with the inevitable variations in the parts of the product he develops. Fortunately, the final functionality of his product is not adversely affected by the variations in its constituent parts as long as they can be kept within some bounds. So he consciously accommodates these variations in his design and documents them by specifying tolerances. In fact, a product specification is incomplete – indeed, the product is not manufacturable – without explicit information about allowable variations in the form of tolerances. The connection between product specification, production, and inspection is emphasized in Shewhart's quality control cycle [Shewhart, 1986] shown in Figure 1.

It is instructive to note that Shewhart viewed the specification-production-inspection process as a cycle, highlighting the iterative nature of the problem. Tolerance specification is not an isolated activity; it strongly affects the production and inspection phases and, in turn, is affected by them. Because of this we will first examine how variations in production and inspection are handled in Sections 3 and 4, respectively.

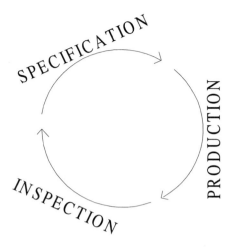

Figure 1. Shewhart's quality control cycle of specification, production, and inspection.

3. VARATION IN PRODUCTION

Controlling variation in the *output* of manufacturing processes is the chief concern of any quality conscious production operation. The output is the effect of several causes – namely the input, production system parameters, and the environment – and the goal in manufacturing process control is to reduce variations in these causes to the minimum while operating within economic constraints. The only successful practical solution known to date to this problem is provided by *statistical process control* (SPC).

3.1 Process control

A basic premise in SPC is that the causes of variation can be grouped under two categories: assignable causes and chance causes. Variation due to assignable causes can be reduced to vanishingly small level, but this is not possible for chance causes owing to economic reasons. A manufacturing process is in a "state of statistical control" if it is only under the influence of chance causes. In practice, control charts as shown in Figure 2 provide simple graphical methods for evaluating whether or not the process has attained, or continues in, a "state of statistical control" [ISO7870, 1993]. So current SPC practices, as codified by ISO, are based on an operational definition of a state of statistical control and are dominated by control chart methodology.

Is there a more precise definition of being in a "state of statistical control"? It can be argued that being in a state of statistical control is the same as a random process being stationary [Voelcker, 1996]. Intuitively a stationary random process is one whose

398

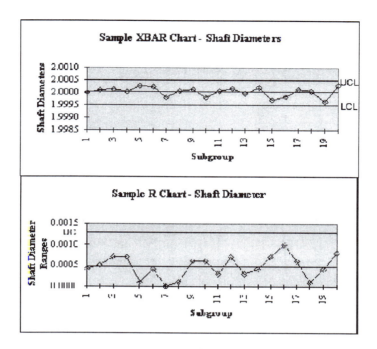

Figure 2. *Examples of X-bar and R control charts used widely in statistical process control. UCL = upper control limit, LCL = lower control limit.*

distribution remains the same as time progresses, because the underlying mechanism does not change with time [Parzen, 1962]. One can then view control charts as practical means of ensuring that a given process is indeed stationary.

An important characteristic of SPC is that the concept of statistical control is *intrinsic* to the process. The upper and lower control limits established for control charts are derived from the output of the process itself. Therefore, it is perfectly reasonable to conclude that an old milling machine is operating in a state of statistical control even though parts machined in it have large variations compared to, say, a modern CNC milling machine. This brings us to the second question on variability in production: Is the process *capable* of producing good parts? This leads us to process capability.

3.2 Process capability

The notion of process capability is relatively old. Table I is an extract from a product design handbook listing of variations that can be expected from typical manufacturing processes to produce rounded holes [Bralla, 1986]. This shows that even if a process is in control, it may or may not be capable of achieving the accuracy we want.

Table I characterizes capability by the range of variation. This is a simple statistical metric that is easy to measure and understand. It is also *intrinsic* to the process. Modern approach to process capability relies on other statistical metrics, such as the mean and the standard deviation, to characterize the variation. These are captured as non-dimensional ratios in various process capability indices (PCI) [Kotz et al., 1993], such as

Process	Usual material	Normal dimensional accuracy of 25mm hole, in mm	Surface finish, in μm	Maximum depth-to-diameter ratio
Drilling	All machinable	+0.2,-0.05	1.6-6.6	8:1
Reaming	All machinable	±0.025	0.8-3.2	8:1
Boring	All machinable	±0.04	0.4-6.3	5:1
Gun drilling	All machinable	±0.05	0.8-2.5	300:1
Sand casting	Gray iron, aluminum	±1.50	6.3-25	1.5:1
Die casting	Aluminum, zinc	±0.10	0.8-1.6	4:1
Injection molding	Thermoplastics	±0.10	0.2-3.2	2:1 (blind) 4:1 (through)
Punching	Sheet metal	±0.15	0.8-1.6	1:1
Power metal	Iron, steel, bronze	±0.06	0.2-1.6	1.35:1 (blind) 4:1 (through)

***Table I**. Capabilities of manufacturing processes that produce rounded holes.*

$$C_p = \frac{USL - LSL}{6\sigma},$$

$$C_{pk} = \min(\frac{USL - \mu}{3\sigma}, \frac{\mu - LSL}{3\sigma}), \text{ and}$$

$$C_{pm} = \frac{USL - LSL}{6\sqrt{(\mu - \tau)^2 + \sigma^2}}$$

where USL is the upper specification limit, LSL is the lower specification limit, τ is the target value, μ is the mean, and σ is the standard deviation.

Modern approach to process capability differs from the old approach in two major aspects. First is the use of statistical metrics other than just the range to characterize variation. Second, in defining PCIs as non-dimensional ratios it introduces the specification limits LSL and USL that are *extraneous* to the process itself. Thus the PCIs do not provide metrics intrinsic to the capability of a process itself; they indicate how capable a particular process is in staying within some externally specified limits.

We can now examine the relationship between process capability and process control. A popular view is that a process capability study is used to choose or qualify a particular process before mass production begins, and then process control is imposed to make sure that it stays capable. This may suggest that there is no role for process capability studies once mass production begins, but, in practice, the PCIs are computed in parallel with plotting control charts to keep a dynamic account of the capability. In fact, many researchers view PCIs computed in the absence of a state of statistical control as meaningless. Modern SPC practice includes both control charts and PCIs, thereby monitoring statistical control as well as statistical capab_lity of the manufacturing process.

4. VARIATION IN INSPECTION

Parts are inspected to ensure that they conform to design specifications or to ascertain the quality of their manufacturing processes. As noted in Section 2, measurements made during the inspection process are always subject to some uncertainty. Moreover, it is not always possible to measure and inspect every part that is produced. Shortcuts, such as inspecting only a sample, are often used which then introduce an additional level of uncertainty about the final result. Let's look at these in some detail.

4.1 Estimation by sampling

The goal of acceptance sampling is to come up with an inspection plan requiring only a small number of parts from a population to be inspected to draw conclusion about the acceptance of the whole lot [ISO8550, 1994; Duncan, 1986]. First popularized by military in its procurement operations, such sampling plans are now routine in industry. Literature on this topic is quite vast and the reader should consult the references quoted above for details. A scientific basis for acceptance sampling is provided by the statistical estimation theory.

The quantity that is estimated can differ from application to application. In some sampling plans, called inspection by attributes, the quantity that is estimated is the fraction of the population that does not conform to specification limits. Other sampling plans that deal with inspection by variables estimate the population mean, standard deviation, or some ratios of these two.

In all these plans a probabilistic statement of the degree of confidence that they are indeed good estimates of the correct values for the population accompanies the estimated values. In fact, one can design a sampling plan based on an acceptable degree of confidence (e.g., acceptable quality level (AQL)) for the consumer.

4.2 Measurement uncertainty

In spite of our earlier assertion in Section 2 about the inevitability of measurement uncertainties, only recently this problem has received much attention. One plausible reason is the increasing demand for high precision in mass produced goods. When uncertainties in measurements were an order of magnitude smaller than the variations they were expected to detect (the infamous 10:1 rule), it was possible to ignore them in practice. As we move to high precision world, however, the margin has narrowed to as low as 4:1 or worse, forcing us to pay greater attention to measurements.

The international authority on measurement uncertainty is the ISO's "Guide to the expression of uncertainty in measurement" [GUM, 1993]. It defines an expanded uncertainty U as a "quantity defining an interval about the result of a measurement that may be expected to encompass a large fraction of the distribution of values that could reasonably be attributed" to the particular quantity subject to measurement. Figure 3 illustrates how the expanded uncertainty U shrinks the specification zone for the purpose of conformance assessment. Only those parts whose measurements fall within the conformance zone can be accepted as satisfying the specification with high degree of confidence.

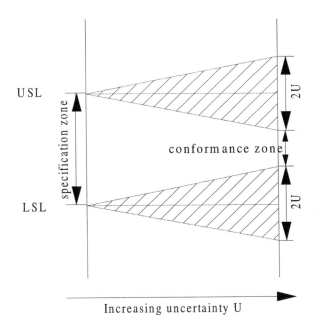

Figure 3*. Influence of uncertainty U on the assessment of measurement results. USL = upper specification limit, LSL = lower specification limit. [Weckenmann et al., 1998]*

The next natural question is how measurement uncertainty affects the assessment of a collection of parts. It clearly has an impact on sampling plans for inspection. It also affects process capability analysis [Weckenmann et al., 1998] because PCIs are calculated using measured values that are sampled from a population. These important issues are still under investigation.

5. BACK TO TOLERANCING

Based on the evidence presented in Sections 3 and 4, it is clear that statistics plays a central role in dealing with variations in production and inspection. We will now see what is the role of statistics in tolerancing. First a few words about tolerancing.

Tolerancing is a specification. Formally, it is a representation that has a syntax and associated semantics. Therefore, it is treated like a language. ISO/TC 213 is the ISO technical committee charged with standardizing the language, and it calls this the "geometrical product specification" (GPS) language. GPS replaces the older acronym

GD&T (geometrical dimensioning and tolerancing) due to the recognition that we are indeed dealing with product specification. ISO/TC 213's charter includes specification as well as verification (particularly, metrology of inspection) and is expected to harmonize these practices. Although the goal of ISO/TC 213 is to define a product specification language, it deals mainly with defining a language for specifying features. Why is this the case?

5.1 From product to features

Figure 4 shows a simple hierarchy of taxonomy that will be useful in understanding GPS. A product is recursively defined as an assembly of subassemblies and parts. Parts, in turn, are composed of features, which are then subjected to various conditions (or, synonymously, constraints). A couple of observations are worth making here.

First, as noted above, the recursive definition of a product ends in parts. It has proven to be remarkably difficult to come up with an acceptable definition for parts. A scientific approach is to define a part as that can be modeled as a solid, say, as a regular set. Industry, however, is used to calling anything that has a "part number" as a part. This has some practical appeal because it is more in line with the way an engineering organization, as opposed to an individual engineer, views a product. A second observation is that the conditions, or constraints, on features come in two types: first is intrinsic, such as size and form, and these are on the features themselves; second is relational, such as position, and these are defined for two are more features.

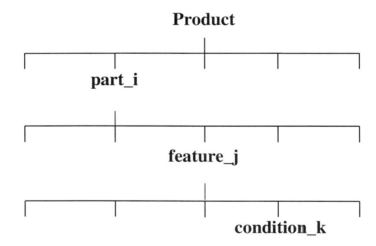

Figure 4. *A hierarchy of taxonomy in ISO's GPS language.*

It turns out that the difficulty in arriving at a good definition of parts to end the recursion on the product definition is not a showstopper. A key to overcoming this difficulty is to focus on features, for the relationships among features can be defined irrespective of whether the features belong to the same part or to different parts. This is an important observation because it allows standards to focus exclusively on features and constraints imposed on these features; these can then be combined upwards to specify parts or products as needed. This is the conclusion of a recent study conducted by an advisory group ISO/TC 213/AG 10 of international experts that looked into the needs of assembly-level tolerancing.

5.2 Risk at the product level

In tolerance analysis we start with variations at the feature-level and propagate it up to the product-level (see Figure 4). In tolerance synthesis (also called tolerance allocation) we start with variation at the product-level and flow this down the hierarchy to the feature-level. In either case, variation at the product-level is associated with some risk and this risk is minimized subject to some economic constraints. Risk, whether in tolerancing or in any other endeavor, is best characterized using the language and methods of statistics.

To be specific, let's assume a functional relationship between product-level characteristic vector y and feature-level characteristic vector x as $y = f(x)$. At this stage, these characteristics are not restricted to geometry. For example, a component of y can very well be the stress at a critical point in the product. There are several statistical metrics that can be used to characterize the variation in y at the product-level, such as:

1. *Range*. This is the simplest statistical metric. It is closely related to the following metric, which is more popular.
2. *Maximum deviation from the target*. This can be either one-sided deviation (e.g., $y \leq 0.05$, target for y is 0.00) or two-sided deviation (e.g., $9.95 \leq y \leq 10.05$, target for y is 10.00). Sometimes the two-sided variation is split asymmetrically (e.g., $9.92 \leq y \leq 10.04$, target for y is 10.00).

   This is the familiar Chebyschev metric for variation and follows the common practice of worst-case tolerancing. Its power lies in its simplicity, but it brings with it a false sense of determinism (e.g., *100%* interchangeability, zero defect). As we saw in Sections 3 and 4, we can never assure in down-stream activities in production and inspection that the characteristic in question is within specified bound(s). We can only make statistical statements about the confidence we have in making such conformance assessment.
3. *Root-Mean-Square (RMS) deviation from the target*. This is the familiar Gaussian metric for variation. It is also the metric used to control variation in Taguchi's quadratic loss function approach [Srinivasan, 1997].
4. *Fraction of the population of products falling outside a limit*. This is a better statistical metric than the maximum deviation from the target. It is used in the six-sigma approach to design for quality.

If these are some of the popular metrics for the product-level characteristics y, what are then the appropriate metrics for feature-level characteristics x? A little reflection on our discussion in Section 5.1 tells us that the issues are the same, for what is a part to a customer is a product to a supplier due to the recursive nature of the problem. So, to maintain consistency, relationship between features, whether they belong to the same part or to different parts, will be specified using the same types of statistical metrics described above. In fact, there are compelling economic pressures that favor the use of statistical metrics to specify feature-level variations. Following this reasoning, ISO/TC 213/WG 13 on Statistical Tolerancing of Mechanical Parts is working towards defining a statistical extension to the GPS language using several statistical metrics similar to those mentioned above [Srinivasan, 1997].

6. CHALLENGE AHEAD

In spite of past and recent developments in the application of statistics to specification, production, and inspection, several challenges lie ahead. As noted in the body of the paper, research and standards organizations are working on issues in each of these three areas. A sharper definition of a "state of statistical control", better understanding and definition of PCIs, quantifying measurement uncertainty and its influence on sampling plans and PCIs, and statistical tolerancing are some of the issues that are receiving attention. To a lesser extent, efforts are underway to integrate these developments to maintain consistency across them.

REFERENCES

[Bralla, 1986] J. G. Bralla, *Handbook of Product Design for Manufacturing*, McGraw-Hill Book Co., New York, 1986.

[Duncan, 1986] A. J. Duncan, *Quality Control and Industrial Statistics*, 5th Edition, Irwin, Illinois, 1986.

[Fisher, 1950] R. A. Fisher, *Statistical Methods for Research Workers*, 11th Edition, Hafner Publishing Company, New York, 1950.

[GUM, 1993] *Guide to the Expression of Uncertainty in Measurement*, International Organization for Standardization, Geneva, 1993.

[ISO7870, 1993] ISO 7870:1993 *Control charts – General guide and introduction*, International Organization for Standardization, Geneva.

[ISO8550, 1994] ISO/TR 8550:1994 *Guide for the selection of an acceptance sampling system, scheme or plan for inspection of discrete items in lots*, International Organization for Standardization, Geneva.

[Kotz et al., 1993] S. Kotz and N. L. Johnson, *Process Capability Indices*, Chapman & Hall, London, 1993.

[Parzen, 1962] E. Parzen, *Stochastic Processes*, Holden-Day, San Francisco, 1962.

[Shewhart, 1986] W. A. Shewhart, *Statistical method from the viewpoint of quality control*, Dover, New York, 1986.

[Srinivasan, 1997] V. Srinivasan, ISO Deliberates Statistical Tolerancing, In: *Proceedings of the 5th CIRP Seminar on Computer Aided Tolerancing*, Toronto, Canada, 1997.

[Voelcker, 1996] H. B. Voelcker, private communication, 1996.

[Weckenmann et al., 1998] A. Weckenmann and M. Knauer, Causes and Consequences of Measurement Uncertainty in Production Metrology, In: *Proceedings of the 6th IMEKO Symposium on Metrology for Quality Control in Production*, Vienna, Austria, Sept. 8-10, 1998.

Tolerance Analysis And Synthesis Using Virtual Joints

Luc Laperrière and Philippe Lafond
Laboratoire de Productique
Université du Québec à Trois-Rivières
C.P. 500, Trois-Rivières, QC
Canada, G9A 5H7
luc_laperriere@uqtr.uquebec.ca

Abstract: This paper presents the results of a research aiming at mathematically modeling the effects of tolerances buildups in close kinematic tolerance chains. The model consists of associating a set of six virtual joints, three for small translations and three for small rotations, to every pair of functional elements in a tolerance chain. These virtual joints can therefore simulate the effects of positional and orientational inaccuracies between two functional elements of the same part which are assumed to result from manufacturing precision limits. The mathematical model is obtained by first associating a coordinate frame to every virtual joint. Transformations matrices describing the global position and orientation of every frame with respect to a base frame are then computed. The effects of small translations and small rotations of every virtual joint on a point of interest in the chain (in particular on the functional requirement) can be computed using standard Jacobian matrices, which are easily obtained once the transformations matrices have been figured out. The model is in the form of six equations relating the new position and orientation of a point of interest in the chain (in cartesian space) to the small dispersions of the functional elements of the chain (in joint space) as simulated by possible moves about their virtual joints. A detailed example is provided that will illustrate the use of the developed model.
Keywords: functional analysis, dispersions, functional element, functional requirement

1. INTRODUCTION

Proper tolerancing of a part is crucial for function, interchangeability, "manufacturability" and cost. In the past few years, there has been a fast growing interest for every aspect of engineering, manufacturing and inspection which involves tolerancing. The current popularity of the Geometric Design and Tolerancing (GD&T) standard is a good example.

As far as research is concerned, the concept of Computer-Aided Tolerancing (CAT) is also gaining fast in popularity. Many research avenues have been developed which are all concerned with CAT: tolerance modeling [Requicha et al., 1986] [Roy et

405

al., 1988] [Gossard et al., 1988] [Clément et al., 1996]; datum specification [Tandler, 1997] [Zhang et al., 1996]; inspection methods [Nassef et al., 1996] [Mathieu et al., 1997]; statistical tolerancing [O'connor et al., 1997]. However, very few researches are concerned with one of the key aspect of tolerancing: functional analysis.

In this paper, we tackle an important problem in functional analysis: having identified which Functional Elements (FEs) influence a Functional Requirement (FR) by including them in closed kinematic chains [Laperrière, 1997], how can we mathematically model the influence of each FE in the chains?

The next section of the paper will establish the mathematical background upon which the proposed solution is based. Section three will illustrate the use of the approach through an example, providing the set of mathematical equations which models the influences. Section four will discuss some of the useful information implied by the equations obtained in section three, and section five will provide some concluding remarks.

2. MODELING DISPERSIONS

2.1 Mathematical background

The problem of modeling the influence of FEs dispersions on a FR is mapped into the problem of modeling the effects of small displacements of such FEs as a result of their moves about virtual joints associated to them. For a FE with 6 virtual joints, 3 for small translations and 3 for small rotations, one can then relate the 6 small dispersions of this FE to the 6 small displacements they can cause at another point of interest in the kinematic chain, i.e. the point where a FR must be satisfied. The functional relationship between such influence of the FEs on the FR involves the use of the familiar Jacobian matrix for small displacements:

$$\begin{bmatrix} \delta \vec{s} \\ \delta \vec{\alpha} \end{bmatrix} = \begin{bmatrix} J_1 J_2 \ldots J_n \end{bmatrix} \cdot \vec{\delta} \tag{1}$$

where:

$\delta \vec{s}$: 3-vector of point of interest's small translations
$\delta \vec{\alpha}$: 3-vector of point of interest's small rotations
J_i : i^{th} column of the Jacobian matrix
$\vec{\delta}$: vector of individual virtual joints small dispersions

Each column of the Jacobian matrix is easily computed if the coordinate frames are attached to every virtual joint such that their individual "z" axis coincide with either the rotation or translation axis of the virtual joint. If this constraint is satisfied, and if we use the notation T_0^{i-1} to represent the transformation between frames O_{i-1} and O_0 (O_0 is the basic frame or the global coordinate frame), then the ith column of the Jacobian, J_i, can be systematically computed as:

$$J_i = \begin{bmatrix} \vec{z}_0^{\,i-1} \times \left(\vec{d}_0^{\,n} - \vec{d}_0^{\,i-1} \right) \\ \vec{z}_0^{\,i-1} \end{bmatrix} \qquad (2)$$

where:

$\vec{z}_0^{\,i-1}$: 3rd column of T_0^{i-1}

$\vec{d}_0^{\,i-1}$:   last column of T_0^{i-1}

The upper term in equation (2) is the contribution of the ith rotational virtual joint to small linear displacements of the point of interest, obtained with the familiar cross product of the virtual rotation axis z_0^{i-1} with its distance from the point of interest expressed in O_0, which is (d_0^n - d_0^{i-1}). The bottom term is the contribution of the ith rotational virtual joint to small rotational displacements of the point of interest. For a translational virtual joint, there is no contribution to small rotational displacements of the point of interest and equation (2) reduces to:

$$J_i = \begin{bmatrix} \vec{z}_0^{\,i-1} \\ \vec{0} \end{bmatrix} \qquad (3)$$

2.2 Modeling virtual joints

For tolerancing purposes, we associate coordinate frames to the toleranced FE in a pair, assuming there exists a set of virtual joints that can make the toleranced FE of the pair "move" relative to the other FE of the pair, to simulate manufacturing inaccuracies.

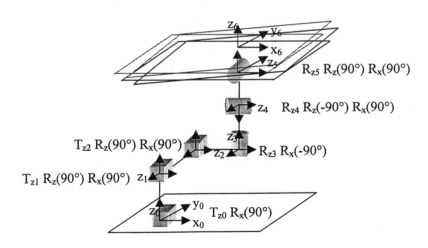

Figure 1; Adding virtual joints and coordinate frames to the toleranced FE in a pair.

Since the method relies on consecutive virtual joints each with a single DOF, the toleranced FE in a pair will be modeled with 6 virtual frames attached to the 6 virtual joints to account for the 6 possible dispersions it can undergo (3 small translations and 3 small rotations).

Figure 1 illustrates how the 6 possible dispersions are modeled for a FE pair consisting of two planes of the same part. The boxes represent translational dispersions and the cylinders represent rotational dispersions. It can be verified that the three translations and three rotations are all performed with respect to three different global axis, making the corresponding dispersions general. The actual distances and angles between FEs (linear and angular dimensions of the part) are implicitly accounted for by at least one of the transformations. For a tolerance chain involving "n" FE pairs around the FR, the mathematical model relating small dispersions of the FR (point of interest) to small dispersions of individual toleranced FEs around FR becomes:

$$
\begin{bmatrix} \vec{\delta s} \\ \vec{\delta \alpha} \end{bmatrix} = \begin{bmatrix} J_1 J_2 J_3 J_4 J_5 J_6 \end{bmatrix}_{FE_1} \cdots \begin{bmatrix} J_1 J_2 J_3 J_4 J_5 J_6 \end{bmatrix}_{FE_{n-1}} \begin{bmatrix} J_1 J_2 J_3 J_4 J_5 J_6 \end{bmatrix}_{FE_n} \cdot \begin{bmatrix} \vec{\delta}_{FE_1} \\ \cdot \\ \cdot \\ \cdot \\ \vec{\delta}_{FE_{n-1}} \\ \vec{\delta}_{FE_n} \end{bmatrix} \tag{4}
$$

where :

$\vec{\delta s}$:   3-vector of point of interest's small translations

$\vec{\delta \alpha}$:   3-vector of point of interest's small rotations

$(J_1 J_2 \ldots J_6)_{FEi}$:   6x6 Jacobian matrix associated with the toleranced FE of the ith FE pair

$\vec{\delta}_{FE_i}$:   6-vector of small dispersions associated with the toleranced FE of the ith FE pair

3. EXAMPLE

Figure 2 shows a simple 3-parts mechanism with two functional requirements, FR_1 and FR_2. Our analysis is limited to FR_1. Before going further, some terms must be defined:

D_i^j :   this term will be used whenever the translational DOF from one virtual joint "i" to another "j" also involves going through a linear dimension of the part;

δ: this term represents the small translational dispersion of one FE in a pair;

$\delta\phi$:   this term represents the small rotational dispersion of one FE in a pair;

We begin by associating 6 virtual joints to every FE pair in one possible chain around FR_1. Since the chain we are concerned with presents 5 FE pairs around FR_1, namely 3 FE pairs on the same parts and 2 kinematic pairs on different parts, we end up assigning 30 different frames. Figure 2 shows how each frame was assigned. Note that the global frame O_0 starts at one of the two FEs defining the functional requirement FR_1, namely on the upper plane of the base part, and the last frame O_{30} is attached to the other FE of FR_1, namely at the pin's axis, which is also the point of interest in this case.

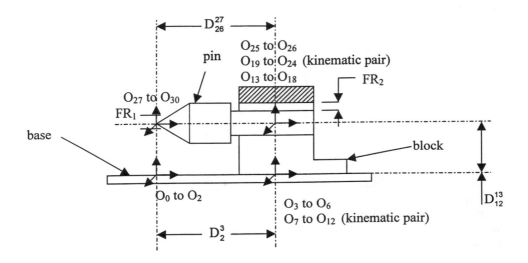

Figure 2; *Assignment of the coordinate frames associated with the virtual joints of one possible tolerance chain around FR1.*

Starting at O_0, 6 virtual joints are defined, each associated with a coordinate frame. The consecutive relative position and orientation of these frames is as defined in figure 2. These frames model the possible small relative "movement" between two regions of the upper plane of the base part, du to manufacturing inaccuracies. As is always the case for a FE pair, there is a physical dimension between these two regions. This means that the move to one of the frames will necessarily imply translating by an amount equal to such dimension. According to figure 1, this translation is in x_0 and should be performed third. It is therefore referred to as D_2^3 in the transformation matrices below.

$$T_0^1 = \begin{bmatrix} 1 & 0 & 0 & 0 \\ 0 & 0 & -1 & 0 \\ 0 & 1 & 0 & 0 \\ 0 & 0 & 0 & 1 \end{bmatrix} \quad T_0^2 = \begin{bmatrix} 0 & 0 & 1 & 0 \\ 0 & -1 & 0 & 0 \\ 1 & 0 & 0 & 0 \\ 0 & 0 & 0 & 1 \end{bmatrix} \quad T_0^3 = \begin{bmatrix} 0 & 1 & 0 & D_2^3 \\ -1 & 0 & 0 & 0 \\ 0 & 0 & 1 & 0 \\ 0 & 0 & 0 & 1 \end{bmatrix}$$

$$T_0^4 = \begin{bmatrix} 0 & 0 & 1 & D_2^3 \\ -1 & 0 & 0 & 0 \\ 0 & -1 & 0 & 0 \\ 0 & 0 & 0 & 1 \end{bmatrix} \quad T_0^5 = \begin{bmatrix} 0 & 1 & 0 & D_2^3 \\ 0 & 0 & 1 & 0 \\ 1 & 0 & 0 & 0 \\ 0 & 0 & 0 & 1 \end{bmatrix} \quad T_0^6 = \begin{bmatrix} 1 & 0 & 0 & D_2^3 \\ 0 & 1 & 0 & 0 \\ 0 & 0 & 1 & 0 \\ 0 & 0 & 0 & 1 \end{bmatrix} \tag{5}$$

At this point, we associate another set of 6 coordinate frames on the kinematic pair (as opposed to a FE pair) defined by the upper plane of the base part and the bottom plane of the block part. In this example however, we will not consider the DOFs of this kinematic pair so the transformation between the frames associated these 6 new virtual joints will not include any variable. The transformation matrices are easily obtained as:

$$T_0^7 = \begin{bmatrix} 1 & 0 & 0 & D_2^3 \\ 0 & 0 & -1 & 0 \\ 0 & 1 & 0 & 0 \\ 0 & 0 & 0 & 1 \end{bmatrix} \quad T_0^8 = \begin{bmatrix} 0 & 0 & 1 & D_2^3 \\ 0 & -1 & 0 & 0 \\ 1 & 0 & 0 & 0 \\ 0 & 0 & 0 & 1 \end{bmatrix} \quad T_0^9 = \begin{bmatrix} 0 & 1 & 0 & D_2^3 \\ -1 & 0 & 0 & 0 \\ 0 & 0 & 1 & 0 \\ 0 & 0 & 0 & 1 \end{bmatrix}$$

$$T_0^{10} = \begin{bmatrix} 0 & 0 & 1 & D_2^3 \\ -1 & 0 & 0 & 0 \\ 0 & -1 & 0 & 0 \\ 0 & 0 & 0 & 1 \end{bmatrix} \quad T_0^{11} = \begin{bmatrix} 0 & 1 & 0 & D_2^3 \\ 0 & 0 & 1 & 0 \\ 1 & 0 & 0 & 0 \\ 0 & 0 & 0 & 1 \end{bmatrix} \quad T_0^{12} = \begin{bmatrix} 1 & 0 & 0 & D_2^3 \\ 0 & 1 & 0 & 0 \\ 0 & 0 & 1 & 0 \\ 0 & 0 & 0 & 1 \end{bmatrix} \tag{6}$$

The next set of frames applies on the FE pair defined by the bottom plane of the block part and its hole's axis. The transformations associated with these frames will require a vertical translation by an amount equal to the physical distance between the two FEs concerned. According to figure 1, this translation is now performed first and is therefore noted D_{12}^{13} in the transformations below:

$$T_0^{13} = \begin{bmatrix} 1 & 0 & 0 & D_2^3 \\ 0 & 0 & -1 & 0 \\ 0 & 1 & 0 & D_{12}^{13} \\ 0 & 0 & 0 & 1 \end{bmatrix} \quad T_0^{14} = \begin{bmatrix} 0 & 0 & 1 & D_2^3 \\ 0 & -1 & 0 & 0 \\ 1 & 0 & 0 & D_{12}^{13} \\ 0 & 0 & 0 & 1 \end{bmatrix} \quad T_0^{15} = \begin{bmatrix} 0 & 1 & 0 & D_2^3 \\ -1 & 0 & 0 & 0 \\ 0 & 0 & 1 & D_{12}^{13} \\ 0 & 0 & 0 & 1 \end{bmatrix}$$

$$T_0^{16} = \begin{bmatrix} 0 & 0 & 1 & D_2^3 \\ -1 & 0 & 0 & 0 \\ 0 & -1 & 0 & D_{12}^{13} \\ 0 & 0 & 0 & 1 \end{bmatrix} \quad T_0^{17} = \begin{bmatrix} 0 & 1 & 0 & D_2^3 \\ 0 & 0 & 1 & 0 \\ 1 & 0 & 0 & D_{12}^{13} \\ 0 & 0 & 0 & 1 \end{bmatrix} \quad T_0^{18} = \begin{bmatrix} 1 & 0 & 0 & D_2^3 \\ 0 & 1 & 0 & 0 \\ 0 & 0 & 1 & D_{12}^{13} \\ 0 & 0 & 0 & 1 \end{bmatrix} \tag{7}$$

Next we model the kinematic pair consisting of the hole axis on the block part and the cylinder axis of the pin part. Again here, we are not concerned with the actual DOFs permitted by this pair so the transformations simply become:

$$T_0^{19} = \begin{bmatrix} 1 & 0 & 0 & D_2^3 \\ 0 & 0 & -1 & 0 \\ 0 & 1 & 0 & D_{12}^{13} \\ 0 & 0 & 0 & 1 \end{bmatrix} \quad T_0^{20} = \begin{bmatrix} 0 & 0 & 1 & D_2^3 \\ 0 & -1 & 0 & 0 \\ 1 & 0 & 0 & D_{12}^{13} \\ 0 & 0 & 0 & 1 \end{bmatrix} \quad T_0^{21} = \begin{bmatrix} 0 & 1 & 0 & D_2^3 \\ -1 & 0 & 0 & 0 \\ 0 & 0 & 1 & D_{12}^{13} \\ 0 & 0 & 0 & 1 \end{bmatrix}$$

$$T_0^{22} = \begin{bmatrix} 0 & 0 & 1 & D_2^3 \\ -1 & 0 & 0 & 0 \\ 0 & -1 & 0 & D_{12}^{13} \\ 0 & 0 & 0 & 1 \end{bmatrix} \quad T_0^{23} = \begin{bmatrix} 0 & 1 & 0 & D_2^3 \\ 0 & 0 & 1 & 0 \\ 1 & 0 & 0 & D_{12}^{13} \\ 0 & 0 & 0 & 1 \end{bmatrix} \quad T_0^{24} = \begin{bmatrix} 1 & 0 & 0 & D_2^3 \\ 0 & 1 & 0 & 0 \\ 0 & 0 & 1 & D_{12}^{13} \\ 0 & 0 & 0 & 1 \end{bmatrix} \quad (8)$$

Finally, a last set of frames is used to model the dispersions between the FEs of the pair consisting of the cylinder axis and cone axis of the pin part. The physical dimension is once again in x_0 so the translation is performed third to yield the following transformations:

$$T_0^{25} = \begin{bmatrix} 1 & 0 & 0 & D_2^3 \\ 0 & 0 & -1 & 0 \\ 0 & 1 & 0 & D_{12}^{13} \\ 0 & 0 & 0 & 1 \end{bmatrix} \quad T_0^{26} = \begin{bmatrix} 0 & 0 & 1 & D_2^3 \\ 0 & -1 & 0 & 0 \\ 1 & 0 & 0 & D_{12}^{13} \\ 0 & 0 & 0 & 1 \end{bmatrix} \quad T_0^{27} = \begin{bmatrix} 0 & 1 & 0 & D_2^3 + D_{26}^{27} \\ -1 & 0 & 0 & 0 \\ 0 & 0 & 1 & D_{12}^{13} \\ 0 & 0 & 0 & 1 \end{bmatrix}$$

$$T_0^{28} = \begin{bmatrix} 0 & 0 & 1 & D_2^3 + D_{26}^{27} \\ -1 & 0 & 0 & 0 \\ 0 & -1 & 0 & D_{12}^{13} \\ 0 & 0 & 0 & 1 \end{bmatrix} \quad T_0^{29} = \begin{bmatrix} 0 & 1 & 0 & D_2^3 + D_{26}^{27} \\ 0 & 0 & 1 & 0 \\ 1 & 0 & 0 & D_{12}^{13} \\ 0 & 0 & 0 & 1 \end{bmatrix} \quad T_0^{30} = \begin{bmatrix} 1 & 0 & 0 & D_2^3 + D_{26}^{27} \\ 0 & 1 & 0 & 0 \\ 0 & 0 & 1 & D_{12}^{13} \\ 0 & 0 & 0 & 1 \end{bmatrix} \quad (9)$$

Having obtained these transformation matrices, equations (2) and (3) can now be used to compute the Jacobian columns J_1 to J_{30} :

$$J_1 to J_{15} = \begin{bmatrix} 0 & 0 & 1 & 0 & 0 & D_{12}^{13} & 0 & 0 & 1 & 0 & 0 & D_{12}^{13} & 0 & 0 & 1 \\ 0 & -1 & 0 & D_{26}^{27} & -D_{12}^{13} & 0 & 0 & -1 & 0 & D_{26}^{27} & -D_{12}^{13} & 0 & 0 & -1 & 0 \\ 1 & 0 & 0 & 0 & 0 & -D_{26}^{27} & 1 & 0 & 0 & 0 & 0 & -D_{26}^{27} & 1 & 0 & 0 \\ 0 & 0 & 0 & 0 & 1 & 0 & 0 & 0 & 0 & 0 & 1 & 0 & 0 & 0 & 0 \\ 0 & 0 & 0 & 0 & 0 & 1 & 0 & 0 & 0 & 0 & 0 & 1 & 0 & 0 & 0 \\ 0 & 0 & 0 & 1 & 0 & 0 & 0 & 0 & 0 & 1 & 0 & 0 & 0 & 0 & 0 \end{bmatrix}$$

$$J_{16} to J_{30} = \begin{bmatrix} 0 & 0 & 0 & 0 & 0 & 1 & 0 & 0 & 1 & 0 & 0 & 1 & 0 & 0 & 0 \\ D_{26}^{27} & 0 & 0 & 0 & -1 & 0 & D_{26}^{27} & 0 & 0 & 0 & -1 & 0 & 0 & 0 & 0 \\ 0 & 0 & -D_{26}^{27} & 1 & 0 & 0 & 0 & 0 & -D_{26}^{27} & 1 & 0 & 0 & 0 & 0 & 0 \\ 0 & 1 & 0 & 0 & 0 & 0 & 0 & 1 & 0 & 0 & 0 & 0 & 0 & 1 & 0 \\ 0 & 0 & 1 & 0 & 0 & 0 & 0 & 0 & 1 & 0 & 0 & 0 & 0 & 0 & 1 \\ 1 & 0 & 0 & 0 & 0 & 1 & 0 & 0 & 0 & 0 & 0 & 1 & 0 & 0 & 0 \end{bmatrix} \quad (10)$$

The last step is to use equation (4) and multiply the Jacobian by the 30x1 column of possible rotational and translational dispersions "δ_s" of the 5 FEs in the chain, which gives:

$$\delta x_{30} = \delta z_2 + \delta \phi z_5 \, D_{12}^{13} + \delta z_8 + \delta \phi z_{11} \, D_{12}^{13} + \delta z_{14} + \delta z_{20} + \delta z_{26} \tag{11}$$

$$\delta y_{30} = -\delta z_1 - \delta \phi z_3 \, D_{26}^{27} - \delta \phi z_4 D_{12}^{13} - \delta z_7 - \delta \phi z_9 \, D_{26}^{27} - \delta \phi z_{10} \, D_{12}^{13} - \delta z_{13} - \delta \phi z_{15} \, D_{26}^{27} - \delta z_{19}$$
$$- \delta \phi z_{21} \, D_{26}^{27} - \delta z_{25} \tag{12}$$

$$\delta z_{30} = \delta z_0 + \delta \phi z_5 \, D_{26}^{27} + \delta z_6 + \delta \phi z_{11} \, D_{26}^{27} + \delta z_{12} + \delta \phi z_{17} \, D_{26}^{27} + \delta z_{18} + \delta \phi z_{23} \, D_{26}^{27} + \delta z_{24} \tag{13}$$

$$\delta \phi x_{30} = \delta \phi z_4 + \delta \phi z_{10} + \delta \phi z_{16} + \delta \phi z_{22} + \delta \phi z_{28} \tag{14}$$

$$\delta \phi y_{30} = \delta \phi z_5 + \delta \phi z_{11} + \delta \phi z_{17} + \delta \phi z_{23} + \delta \phi z_{29} \tag{15}$$

$$\delta \phi z_{30} = \delta \phi z_3 + \delta \phi z_9 + \delta \phi z_{15} + \delta \phi z_{21} + \delta \phi z_{27} \tag{16}$$

4. DISCUSSION OF THE RESULTS

The equations above provide a great deal of information to a designer. At this point however, it should be clear that some of these equations do not correspond to the intended FR_1. In particular, the design intent of FR_1 is clearly to bound the vertical position of the pin's endpoint, regardless of its position in the x-y plane perpendicular to the page (this at least is our assumption). This means that only equation (13) above should apply.

Looking at this equation more closely, we realize that the right side explicitly shows every dispersion that effects the pin's endpoint in the vertical position. By further realizing that groups of six consecutive indices of the dispersions (the δ_i's in equation (13)) all belong to the same FE, we end up with an equation which explicitly identifies which FE influences the FR (this was in fact the objective of the research). Consider for example the terms in the group δ_{12} to δ_{17}, namely δz_{12} and $\delta \phi z_{17}$ in equation (13). These belong to the block's hole axis. This tells us that when the dispersion $\delta \phi z_{17}$ (which rotates the hole's axis in the plane of the page) is multiplied by its lever D_{26}^{27}, it results in an influence of the vertical position of the pin's endpoint. Also, when this same axis is translated vertically in δz_{12}, there is also an influence on the point of interest. These two dispersions on the hole's axis (namely small rotation $\delta \phi z_{17}$ and small translation δz_{12}) can be assumed to result from manufacturing inaccuracies during drilling.

Another important information underlying the final equations is the effect of parts linear and angular dimensions. At the design stage, it is likely that some of these dimensions are not definitely fixed. Using the approach in this paper, we see how different choices of such dimensions affect the FR in different ways, and most importantly how these choices can make some tolerances more expensive to respect. To illustrate this using the final equations of section 3, note that some dimensions appear as multipliers of some dispersions, du to the lever effect these dimensions provide between the point at which the dispersion apply and the point at which the FR applies. Looking again at equation (13), clearly as some of these dimensions are increased in values, the final effect is an increase in the vertical position of the pin's endpoint. Assuming that such vertical displacement of the pin's endpoint is bound to some value by the designer to ensure functionality, then clearly if the levers are increased then the $\delta_{i's}$ must be decreased proportionally to respect the FR and this one becomes more expensive to obtain.

Finally, the approach provides information on how DOFs of kinematic pairs also affect the FR. Consider for example the terms δz_{18} and $\delta\phi z_{23}$ in equation (13). Both are associated to possible movements of the pin inside the block as provided by the functional fit FR_2. The former is a DOF consisting of a vertical translation, the latter a rotation in the plane of the page. Indeed, if FR_2 is loose enough, these dispersions can be assumed to be non zero and they will affect functional requirement FR_1 in the vertical direction (equation (13)). This example also shows that the models implicitly keeps track of how FRs can influence each other.

5.   CONCLUSION

The developed model is very general and finds many different applications in a design context. Assuming that a single FR has many different tolerance chains (which is often the case), the developed model can first be used to determine which among all chains has more impact, by simply counting the number of terms involved in the sets of equations for each chain (the chain modeled by equations with the most terms would be the most influent). For a set of six equations modeling a single chain, critical FEs of this chain can also be easily spotted by simply counting the number of terms associated with each FE.

By providing potential values for the small dispersions of every virtual joints (the $\delta_{i's}$), the developed model can be used for tolerance analysis to see the buildup effect of such dispersions on the point of interest. This corresponds to providing values to all the terms of the right hand side of the six equations and computing the resulting effects on the FR. On the other hand, at the design stage the designer is more likely to provide values for the FR on the left hand side of the equations to ensure functionality, and compute tolerance values of the assembly on the right hand side such that their sum complies with the FR. In this context the model is used for tolerance synthesis, i.e. to compute the tolerance values that would have to be associated to every virtual joint in order to satisfy the FR prescribed.

One important limitation of the model is that the frames associated to virtual joints necessarily model the effects of translational or rotational dispersions *at single points*. In other words, the important notion of a tolerance zone is not explicit to the model. Of course, simulations can be run which would provide a large number of random values of the $\delta_{i's}$ and which could map tolerance zones as clouds of points for every virtual joint. Another area of future work that we are currently addressing is to translate sets of δ

values that satisfy the FR into standard engineering dimensional or geometric tolerances. Preliminary work shows that this promises to be a challenging problem.

REFERENCES

[Clement et al., 1996] Clément A. Rivière A. and Serré P.: "The TTRS: a common declarative model for relative positioning, tolerancing and assembly", *MICAD proceedings*, v11, n1-2, pp.149-164

[Gossard et al., 1988] Gossard D.C. Zuffante R.P. and Sakurai H.: "Representing dimensions, tolerances and features in MCAE systems", *IEEE transactions on computer graphics and applications*, v8, n2, pp. 51-59

[Laperrière, 1997] Laperrière, L.: "Identifying and Quantifying Functional Elements Dispersions During Functional Analysis", *Proceedings of the 5th CIRP Seminar on Computer Aided Tolerancing*, Toronto, Canada, pp. 99-112

[Mathieu et al., 1997] Mathieu M. and Ballu A.: "Virtual gage with internal mobilities for the verification of functional specifications", *Proceedings of the 5th CIRP Seminar on Computer Aided Tolerancing*, Toronto, Canada, pp. 265-276

[Nassef et al. 1996] Nassef A. and ElMaraghy H.A.: "Optimization and interpolation issues in evaluating actual geometric deviations from CMM data", *Proceedings of ASME International congress and exposition*, v4, pp. 425-431

[O'Connor et al., 1997] O'Connor M.A. and Srinivasan V.: "Composing distribution function zones for statistical tolerance analysis", *Proceedings of the 5th CIRP Seminar on Computer Aided Tolerancing*, Toronto, Canada, pp. 13-24

[Requicha et al., 1986] Requicha A.A.G. and Chan S.C.: "Representation of geometric features, tolerances and attributes in solid modelers based on constructive solid geometry", *IEEE Journal of Robotics and Automation*, RA-2, pp. 156-166

[Roy et al., 1988] Roy U. and Liu C.R.: "Feature-based representational scheme of a solid modeler for providing dimensional and tolerancing information", *Robotics and computer integrated manufacturing*, v4, n3-4, pp.335-345

[Tandler, 1997] Tandler W.: "The tools and rules for computer automated datum reference frame construction", *Proceedings of the 5th CIRP Seminar on Computer Aided Tolerancing*, Toronto, Canada, pp. 49-62

[Zhang et al. 1996] Zhang X. and Roy U.: "Criteria for establishing datums in manufactured parts", *Journal of manufacturing systems*, v12, n1, pp. 36-50

Analysis of positional tolerance based on the assembly virtual state

F. Bennis - L. Pino
IRCyN -URM 6597
1, rue de la Noë
44321 Nantes FRANCE
Fouad.Bennis@ircyn.prd.fr

C. Fortin
Ecole Polytechnique de Montréal
Department of Mechanical Engineering
Montréal H3C 3A7 CANADA
clement.fortin@meca.polymtl.ca

Abstract
Functional gaging techniques can be used to check the conformance of a part with Maximum Material Condition features, but the cost of this gages can be expensive. A Coordinate Measurement Machine can simulate the gage for simple cases. Unfortunately, when the datum reference frame of the feature is not unique, the inspection of the parts defined with MMC using CMM becomes more complex, and it is currently not supported. To resolve this problem a method using a kinematic model is proposed, which simulates the gaging control. The method takes into account the possible rotation of the DRF, and it obtains all possible situations of the DRF for which the parts conforms. And the actual value of the feature can be calculated.
Keywords: Tolerance analysis, geometric tolerances, Maximum Material Condition.

1 INTRODUCTION

Where function, interchangeability and assemblability of mating part features is involved, the maximum material condition (MMC) principle may be used to great advantages. In design state MMC principles permit greater possible tolerances when part feature sizes vary from their MMC limits. The interchangeability is ensured if a functional gage can be assembled with the actual part. The manufacturing cast can be reduce by using this interdependence between size and position. MMC is applicable to tolerance and datum reference.

Functional gaging techniques are usually used in industrial applications to control the conformance of MMC position tolerances. They are easy to use and never accept bad parts. However, building, storage and maintenance of gages can be very expensive. Paper gaging techniques [Foster, 1994] can be used instead of hard gage. Jackman and al. propose a numerical method [Jackman et al., 1994] to calculate the compliance measure of the alignment of cylindrical part features. They used circle to circle intersection which can not give the actual value. The previous methods are not suitable for composite tolerances and generally, when the orientation of the Datum reference frame (DRF) affects the conformance. Lehtihet and al. [Lehtihet et al., 1991] deal with this problem using an optimization method.

In the following, a method for checking the conformance cf MMC position tolerances with MMC datum is developed. All definitions used in this paper are in agreement with the mathematical standard [ASME, 1994] and the ISO Standard.

2 GENERAL DESCRIPTION OF THE METHOD

If the DRF, as stated in the feature control frame of the drawing, is composed by an MMC datum, it can have a possible motion. If it exists, at least one situation of the DRF where the assembly of the virtual condition gage and the actual feature is possible, then the part conforms. The method proposed takes into account the variation of the orientation of a DRF with MMC feature, and it is developed in three steps:

2.1 Definition of the *actual candidate axis set*

For an internal cylinder the actual candidate axis set is the set of axes of all cylinders of MMC virtual condition size that are enclosed within the actual mating cylinder. The cylinders are constrained to be oriented by the primary datum but not located. Note that when the hole is called as a datum feature, the actual candidate axis set (ACAS) is equivalent to the candidate datum set definition. The ACAS is a cylindrical zone located at the center of the actual mating cylinder, oriented with respect to the primary datum.

To simplify the explanation of the method, only cylindrical internal features are used and the primary datum is supposed to be a plane. Actual mating surfaces, basically oriented with respect to the primary datum, of all features are known.

For each individual feature, the ACAS is defined. This set represents the possible motion of an oriented gage at virtual MMC size in the real feature. The gage is oriented but not located to the primary datum. The virtual gage is supposed to be enclosed by the actual mating cylinder at the basic orientation for all motions. This hypothesis is applied in paper gaging techniques and by the other research works as [Srinivasan et al., 1989 – Fortin et al., 1995]

For an internal cylinder, figure 1 depicts the ACAS definition and the notations used. Z represents the actual candidate axis set of the feature. Its size is equal to the difference between the oriented actual mating cylinder size and the virtual MMC cylinder size:

$$T_Z = D_{AM} - D_{CV} \tag{1}$$

where:

T_Z is the diameter of the ACAS. D_{AM} is the diameter of the actual mating envelope cylinder AM at the basic orientation. CV is a cylinder at virtual MMC size, basically oriented but not located relative to the DRF, its size is:

$$D_{CV} = D_{MMC} - t \tag{2}$$

Equations (1) and (2) lead to the following equation:

$$T_Z = D_{AM} - D_{MMC} + t \tag{3}$$

Where t is the specified tolerance value of the feature. D_{MMC} is the size of the feature at MMC state.

The ACAS is calculated for all features of the feature control frame (Toleranced features and datum features).

This step is sufficient to verify the conformance of the feature when the orientation of DRF is constant or does not affect the result. The first example of section 3 shows the application of the ACAS to verify the conformance and to calculate the actual value of a toleranced group pattern of holes.

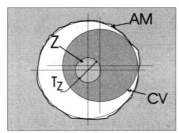

Figure 1 : ACAS definition and notations

2.2 Definition and study of the kinematic mechanism

The relative situation of the DRF with respect to the actual part is the result of the composition of all situations of each datum feature that define the DRF. If some datum features are MMC datum and they are manufactured away from MMC size, then, multiple datum feature candidate exists, and then multiple candidate DRF exist. They are called candidate datum reference frame set [ASME, 1994]. In this section, a kinematic mechanism is used to calculate this set [Bennis et al., 1998]. The mechanism is used to calculate of candidate DRF relative to an actual part.

For the internal cylindrical datum feature, the axis motion is enclosed in the ACAS of the feature. A transformation associated to each motion defines the situation axis of the virtual gage axis at virtual condition size with respect to the actual position axis of the hole (center of ACAS) relative to a candidate DRF. This must be done for each individual datum feature. On the other hand, the feature control frame defines an explicit constraint between each individual datum feature of the DRF (datum precedence). Each constraint can be defined by an homogenous transformation. Like in robotics, each homogenous transformation can be represented by an open kinematics chain composed by joints and links. The constraint transformations, may close these chains. A kinematic mechanism is obtained with serial and closed-chain. Rivest proposed a similar method to define tolerance zones for each individual feature using kinematics [Rivest et al., 1994].

The basic situation of the axis of each toleranced hole with respect to a candidate DRF can be defined by a constant transformation. To each candidate DRF corresponds a situation of the toleranced feature axis basically located to the DRF. This is a particular configuration of the kinematic mechanism. For the mechanism point of view, the set of all true position axis of the specified feature is the workspace of the point that simulates this axis. The properties of the determinant of Jacobian matrix of the mechanism for a particular point are used to calculate the boundary of this workspace. As commonly known, the singularity of the Jacobian matrix gives the boundary of the workspace of the studied point. This workspace represents all situations of the point of the mechanism. In the following, the study is done for the end point of the mechanism. To define the Jacobian matrix, the following relations are used.

The first one is the equation of the point relative to the base DRF which is written as :

$$^{base}\mathbf{T}_{end}(\boldsymbol{q}) = \mathbf{U_0} \tag{4}$$

where $\mathbf{U_0}$ describes the possible situations of the end effector. And $\boldsymbol{q}$ represents the parameter configuration of the mechanism.

The second one is the equation of closed-chain of the mechanism. This equation is given by the product of the homogenous transformations between each closed-chain element.

$$\prod {}^{a(i)}\mathbf{T}_i = \mathbf{I} \tag{5}$$

Where $a(i)$ is the previous element of the ith element of the mechanism, and $\mathbf{I}$ the identity matrix. It may have more than one closed-chain in the mechanism.

All the equations give the following system :

$$\begin{cases} \mathbf{F}(\mathbf{q}_d, \mathbf{q}_i) = 0 \\ \mathbf{\Psi}(\mathbf{q}_d, \mathbf{q}_i) = \mathbf{U}_0 \end{cases} \tag{6}$$

Where $\mathbf{F}(\mathbf{q}_d, \mathbf{q}_i)$ is the equation of close chains, $\mathbf{\Psi}(\mathbf{q}_d, \mathbf{q}_i)$ the equation of the task, $\mathbf{q}_d$ are dependent parameters, and $\mathbf{q}_i$ are independent parameters.

The derivative of the first expression gives :

$$\frac{\partial \mathbf{F}}{\partial \mathbf{q}_d} d\mathbf{q}_d + \frac{\partial \mathbf{F}}{\partial \mathbf{q}_i} d\mathbf{q}_i = 0 \tag{7}$$

$$d\mathbf{q}_d = J_d^{-1} J_i \, d\mathbf{q}_i \tag{8}$$

with $J_d = \dfrac{\partial \mathbf{F}}{\partial \mathbf{q}_d}$ and $J_i = \dfrac{\partial \mathbf{F}}{\partial \mathbf{q}_i}$

The derivative of the second expression of equation (6) leads to the following :

$$\frac{\partial \mathbf{\Psi}}{\partial \mathbf{q}_d} d\mathbf{q}_d + \frac{\partial \mathbf{\Psi}}{\partial \mathbf{q}_i} d\mathbf{q}_i = d\mathbf{U}_0 \tag{9}$$

$$d\mathbf{U}_0 = \mathbf{J} \, d\mathbf{q}_i \tag{10}$$

with $\mathbf{J} = K_d (J_d^{-1} J_i) + K_i$

$K_d = \dfrac{\partial \mathbf{\Psi}}{\partial \mathbf{q}_d}$ and $K_i = \dfrac{\partial \mathbf{\Psi}}{\partial \mathbf{q}_i}$

The analysis of the determinant of $\mathbf{J}$, gives the singularities which define the boundary of the workspace W of a particular point of the mechanism. For each toleranced feature, W is the true position set.

2.3 Conformance and actual value

A MMC internal hole conforms if all points of the surface of the hole lie outside the virtual condition zone. This zone is located at the true position axis of the feature and its size is $R_{MMC} - t_0 / 2$. The actual value of the position deviation of the hole is the smallest value of t_0 to which the feature conforms. [ASME, 1994]. For a given tolerance, the feature conforms if it exists at least one candidate DRF from the candidate DRF set for which the feature does not violate the constraints defined by the tolerance. The actual value associated with the tolerance is the minimum candidate actual value.

The ACAS defined in the previous section, represents the location of an axis of a virtual condition cylinder in the actual feature. The workspace W of the axis represents all the candidate DRF for the feature. The intersection between the workspace W and the ACAS Z, is used to check the conformance of the feature. If the intersection between W and Z is not empty, then it exists a candidate DRF for which the feature conforms. The actual value of the feature can be calculate using this intersection. If there is no intersection, then the feature does not conform. For a group pattern, each individual feature needs to conform, and each ACAS needs to have a common intersection. The actual value, calculated from the distance between the two sets, is greater than the specified tolerance value.

Traditional gage technique give only a binary result of the conformance of the part : a *go / no go* result.

Figure 2 : Positive actual value

Figure 3 : Positive actual value

Figure 4 : Negative actual value

Lets δ be the minimum distance between the workspace W and the center of the ACAS Z. The actual value of the feature can be calculated with δ . There are two cases:

1. When the center of Z is outside W (fig. 2 and 3). In this case, the actual value of the feature is :
$$\delta = D_{AM} - D_{MMC} - t_{AV} \qquad (11)$$
$$t_{AV} = e - \delta \qquad (12)$$
   with $e = D_{AM} - D_{MMC}$.

2. When the center of Z is inside W (fig. 4). In this case the feature conforms for a zero value of t_{AV} . The smallest actual value for which the feature conforms is a negative actual value. It can be calculated using equation (12). A negative value of t_0 can be interpreted as the unused portion of the bonus tolerance resulting from the departure of the feature from its MMC size.

In the following, some examples are gradually introduced to show the application of the method according to the complexity of the example. The first example studies a group pattern of holes at MMC state related to MMC datum reference. In this case, the orientation of the DRF is constant. The second example deals with an MMC hole related to a DRF with two holes as secondary and tertiary datum at MMC. In this case, the orientation of the DRF is not constant and must be taken into account. The third example studies a group pattern of holes at MMC related to a DRF with two holes as secondary and tertiary datum at MMC.

3   MMC GROUP PATTERN RELATED TO AN MMC DATUM FEATURE

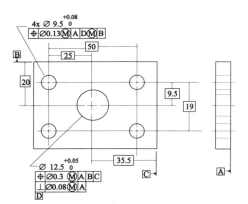

Figure 5 Example of a MMC position tolerance

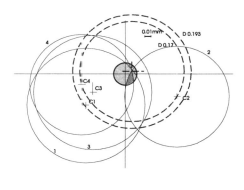

Figure 6 : Extended method of the gage paper technique

A traditional group pattern example of four holes is studied (Fig. 5). [Foster, 1994, Fortin et al., 1995]. This group is defined relative to a DRF with one MMC feature. The four holes are located by a positional tolerance with respect to datum A for orientation, datum D for location and datum B for rotation, as the four holes are related to the center of the MMC datum D. If the actual size of D is produced away from its MMC size, the datum axis D can shift, and then the relative position of the four holes with respect to DRF can shift too. The displacement of datum D must be taken into account in the analysis of the conformance of the four holes. The orientation of the four holes is defined as constant by datum B. In this case, the study of a kinematic mechanism is unnecessary.

Table I gives some data of an actual part. The sign of the coordinates indicates the relative position of each hole with respect to the DRF constructed with datum A, actual mating axis of D and datum B. This candidate DRF is use to define a global coordinate system R.

The traditional paper gaging graphic analysis technique resolves this problem by randomly moving the ACAS to try to encompass all the pattern holes center simultaneously. Figure 6 shows the four holes location relative to R. These zones are translated to the origin of R. The dotted circles represent the cylinders in which the axis of the four holes must be. Center of hole 1 and 3 must be in a $\varnothing 0,193$ zone, and center of hole 2 and 4 in a $\varnothing 0,17$ zone. The center of these cylinder is moves in the $\varnothing 0,038$ tolerance zone, which represents the ACAS of D, and all the four holes center must be in their respective cylinder.

To avoid the random search, our method calculates the intersection of all the ACAS with the ACAS of D. This intersection gives simultaneously all permissible locations of the DRF for which the group conforms. Figure 6 shows this zone in dark gray. This method can be useful for the analysis and synthesis of tolerances because it gives all situations of the DRF that gives conformed parts.

Since the orientation of DRF is constant, the possible displacement of the DRF is only translation. If only datums A and D are specified in the feature control frame, the study of the kinematic mechanism is unnecessary too because the global rotation around datum D does not affect the relation between the four holes and D. The calculation of the ACAS is sufficient to deal with this type of specification. In the next examples, the orientation of the DRF is more complex.

Hole i	x_i	y_i	T_{zi}
1	-25.064	-9.553	0.193
2	-24.916	9.462	0.17
3	24.947	9.469	0.193
4	24.929	-9.518	0.17
D	0	0	0.038

Table I: actual data from [Foster, 1994]

4 MMC FEATURE RELATED TO DRF WITH TWO MMC DATUMS

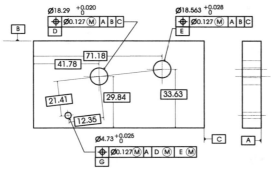

Figure 7: Example of MMC location tolerances

This example is extracted from [Rivest et al., 1994]. The MMC hole G is located with respect to datum A for orientation, to D at MMC state for location and E at MMC state for orientation. Datums D and E are MMC datums so if their actual size departs from their MMC size, datum D and E may shift. And the orientation of the DRF may shift too. So the study of the motion of the DRF is necessary to check the conformance of the hole G. To study the motion of the DRF, a kinematic mechanism is defined. This mechanism simulates the motion of the DRF (Figure 8).

The range of the parameters of the structure are calculated relative the allowable motion of the DRF. For example, in figure 7 point P_A is used to simulate the motion of the CV_D axis. P_A belongs to the candidate datum set of D (or ACAS of D). The allowable values for r_1 can vary in $[\ 0,\ T_{Z_D}/2\ ]$. To a same way, the CV_E motion is simulated by point P_B, and the allowable values of r_2 are in $[\ 0,\ T_{Z_E}/2\ ]$. T_Z is the size of the ACAS given by equation (1). In the worst case, r_1 and r_2 are maximum.

Using equation (4), the closed-chain equation can be written as the following :

$$tx = r_1\cos(\alpha_1) + r\cos(\alpha_1+\alpha_a) - r_2\cos(\alpha_2)$$
$$ty = r_1\sin(\alpha_1) + r\sin(\alpha_1+\alpha_a) - r_2\sin(\alpha_2)$$

With $\overrightarrow{O_1O_2} = \begin{bmatrix} t_x \\ t_y \end{bmatrix}$, t_x and t_y are constant values.

The derivative of these two equations gives :

$$\mathbf{J}_1 \begin{bmatrix} d\alpha_a \\ dr \end{bmatrix} = \mathbf{J}_2 \begin{bmatrix} d\alpha_1 \\ d\alpha_2 \end{bmatrix} \tag{13}$$

The task equation can be developed as :

$$x_G = r_1\cos(\alpha_1) + AH_x\cos(\alpha_a) - HG_x\sin(\alpha_a)$$
$$y_G = r_1\sin(\alpha_1) + AH_x\sin(\alpha_a) + HG_x\cos(\alpha_a)$$

The derivative of this expressions gives :

$$\begin{bmatrix} dx_G \\ dy_G \end{bmatrix} = \mathbf{J}_d\ d\alpha_a + \mathbf{J}_i\ d\alpha_1 \tag{14}$$

$$\begin{bmatrix} dx_G \\ dy_G \end{bmatrix} = \mathbf{J} \begin{bmatrix} d\alpha_1 \\ d\alpha_2 \end{bmatrix} \tag{15}$$

Where $\mathbf{J}$ is given by equation 10.

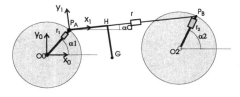

Figure 8: Kinematic mechanism   **Figure 9** : Shape of the workspace of the mechanism in the worst case

Figure 9 gives an example of the shape of the workspace of the mechanism at point G defined in figure 8. The possible displacement of the DRF is not limited to translations as in the first example. This intersection between ACAS Z_G of G and the workspace W_G gives all the possible situations of DRF for which the feature conforms. The actual value can be calculated by the smallest value of t_0 for which the ACAS is tangent to the workspace of the axis. In figure 9, the actual value can be deduced from the dotted circle. Traditional methods can not give such results.

5 MMC GROUP PATTERN RELATED TO DRF WITH TWO MMC DATUMS

In this example, a group pattern of four holes is studied. The group is located with respect of primary datum A, secondary datum D and tertiary datum E. As in the first example, each hole in the group may conform for the same candidate DRF. As in the second example, datums D and E are MMC datums, they can shift, and the orientation of the DRF is not constant. The same kinematic mechanism as in figure 8 can be used to simulate the DRF displacement, and to calculate W_i for each hole of the group pattern.

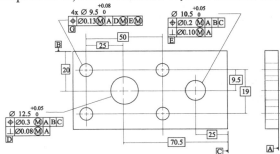

Figure 10 : Example of MMC position tolerance

Table II gives the relative variation of the coordinates of the center of each feature in the group. Figure 11 shows the configuration of the DRF for which the group pattern of holes conforms, the coordinate system used in this figure is (x_0, y_0) , as defined in figure 8. Each point in this figure represent a coordinate x_D and y_D of axis D, and α_a which represents the orientation of a candidate DRF. The dark gray zone shows the domain where the group conforms. One can see in particular that all values of α_a smaller than 0.01 radians lead to bad parts. Figure 12 shows a slide of this domain for an orientation $\alpha a = 0.03$. The dark gray zone represents the domain for x_D and y_D where the part conforms. The actual value can be obtained from the smallest actual value of all the orientations of the DRF for which the group conforms.

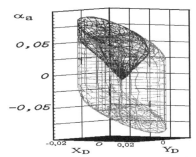

Figure 11 : Conformance domain for the actual part

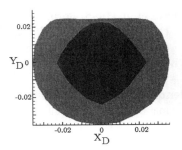

Figure 12 : Section at $\alpha_a = 0.03$

	Xi	Yi	Tzi
Cyl 1	-25.056	9.444	9.54
Cyl 2	24.944	9.556	9.54
Cyl 3	-24.994	9.056	9.54
Cyl 4	25.056	-9.44	9.54

Table II : data modification

6   CONCLUSION

In this paper a method for calculating the conformance of MMC features and of group pattern of features, relative to MMC DRF is presented. It takes into account the possible rotation of the DRF. This method can be applied to composite tolerances. It allows the calculation of the actual value of the feature. In design this method can be helpful to calculate the possible actual value for a given size of the tolerance zone, and for a given realization of the part. The algorithms may be included in a tolerance synthesis algorithms.

REFERENCES

[**Requicha, 1983**] A. A. G. Requicha. "Toward a theory of Geometric Tolerancing"; In: *The Int. _ of Robotic Research*. 2(4) pp. 45-60, 1983.

[**Srinivasan et al., 1989**] Srinivasan, R. Jayaraman. "Geometric Tolerancing: II. Conditional tolerances"; In: *IBM J. Res. Develop*.33(2) pp.105–124 33(2) 1989

[**ASME, 1994**] ASME ", Mathematical definition of dimensionnig and tolerancing principles"; In: *ASME Y14.5M-1994*, 1994.

[**Foster, 1994**] Foster L. W., "GEO METRIC IIIm the metric application of geometric dimensioning and tolerancing" Addison-Wesley Publishing company 1994.

[**Jackman et al., 1994**] Jackman J., Deng J.-J. , Ahn, H.-I., Xuo W. and Vanderman S. "A compliance measure for the alignment of cylindrical part features", In: *IIE Transactions*. 26(1) pp. 2–10, January 1994.

[**Lehtihet et al., 1991**] Lehtihet E.A. and Gunasena N.U. "On the composite position tolerance for patterns of holes", In: Annals of the *CIRP*, 40(1), pp 495-498,1991.

[**Fortin et al., 1995**] Fortin C. and Chatelain J.-F. "A soft gazing approach for complex case including datum shift analysis of geometrical tolerance"; In: *4th CIRP Seminar on computer Aided Tolerancing*. pp. 312-327, 1995

[**Rivest et al., 1994**] Rivest L.,. Fortin C, and. Morel. C "Tolerancing a solid model with a kinematic formulation", In: *C. A.D*. 26(6) pp. 465-476, 1994.

[**Bennis et al., 1998**] Bennis F., Pino L.and Fortin C., "Geometric tolerances transfer for manufacturing by an algebraic method";In : *IDMME*, 3, pp. 713-720 1998.

Operations on polytopes:

application to tolerance analysis

TEISSANDIER Denis*, DELOS Vincent**, COUETARD Yves*
*LMP – Laboratoire de Mécanique Physique – UPRES A 5469 CNRS
Université Bordeaux 1
351, Cours de la Libération
33405 Talence Cedex, France
Phone : 05 56 84 62 20, Fax : 05 56 84 69 64
Email : d.teissandier@lmp.u-bordeaux.fr, couetard@lmp.u-bordeaux.fr

**BRGM, Avenue de Concyr, Orléans-La-Source - BP 6009
45060 Orléans Cedex 2 - France
Phone : 33-(0)3-38-64-34-34 - fax : 33-(0)3-38-64-35-18

Abstract : This article presents numerical methods in order to solve problems of tolerance analysis. A geometric specification, a contact specification and a functional requirement can be respectively characterized by a finite set of geometric constraints, a finite set of contact constraints and a finite set of functional constraints. Mathematically each constraint formalises a n-face (hyperplan of dimension n) of a n-polytope ($1 \leq n \leq 6$). Thus the relative position between two any surfaces of a mechanism can be calculated with two operations on polytopes : the Minkowski sum and the Intersection. The result is a new polytope: the calculated polytope. The inclusion of the calculated polytope inside the functional polytope indicates if the functional requirement is satisfied or not satisfied. Examples illustrate these numerical methods.
Keywords: Three-dimensional Dimension-Chain - Geometric Specification - Contact Specification - Functional Requirement - Polytope - Tolerance.

1. INTRODUCTION

The variational classes by Requicha were introduced at the beginning of the 80's and propose a model of tolerances of form, orientation, position and dimension [Requicha, 1983]. A generalization of specifications by volume envelopes is based on the works of Requicha [Srinivasan and al., 1989]. Fleming presents a model for geometric tolerances and constraints from contacts [Fleming, 1988]. Among the dimension-chains models based on Small Displacements Torsor concept [Clément and al., 1988], we note the Clearance Deviation Space by Giordano [Giordano and al., 1993]. The Clearance

Deviation Space purposes an assembly method according to maximum material condition limit.

This article presents numerical methods in order to solve problems of tolerance analysis. These methods deal with Tolerance Zones constructed by offsetting as in [Requicha, 1983]. We use the Small Displacements Torsor [Bourdet and al., 1995]. Geometric specification and contact specification are characterised by the same model as in [Giordano and al., 1993].

2.   EXPRESSION OF CONSTRAINTS

A Tolerancing Tool manipulates three sources of essential information:
a. the geometric specifications (between associated surfaces of the same part).
b. the contact specifications (between associated surfaces of two distinct parts),
c. the functional requirements of an assembly (between any associated surfaces).

2.1.   Geometric constraints

An associated surface is a surface of perfect form (i.e. an ideal surface: surface described with a finite number of geometric features). A nominal surface is an ideal surface by definition.

A geometric specification is formalised by geometric constraints of position between a nominal surface S_0 and an associated surface S_1: see figure 1.

The tolerance zone (ZT) limit an area of space around S_0 within which S_1 must be situated: they are constructed by two (positive and negative) offsettings on S_0.

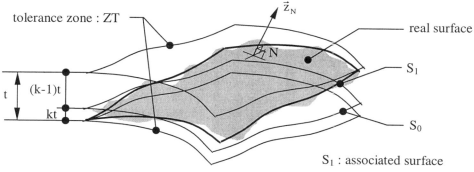

Figure 1: Geometric specification - geometric constraints.

We define geometric constraints of position between S_1 and S_0 as follows [Teissandier et al., 1997]:

$$S_1 \subset ZT \Leftrightarrow \forall N \in S_0 \quad \text{with } 0 \le k \le 1 \qquad (k-1).t \le \vec{\varepsilon}_{N,1/0}.\vec{z}_N \le k.t \tag{1}$$

A unit vector $\vec{z}_N$ is constructed such as $\vec{z}_N$ is parallel to the local normal at any point N of S_0 (see figure 1). Vector $\vec{z}_N$ is oriented in such a way that the positive direction corresponds to the side exterior of the material.

With: $\begin{cases} \vec{\varepsilon}_{N,1/0} : \text{translation vector between } S_1 \text{ and } S_0 \text{ expressed at point N.} \\ \vec{\rho}_{1/0} : \text{rotation vector between } S_1 \text{ and } S_0. \\ \forall M \in E^3 \ (E^3 : \text{Euclidean Space}) \quad \vec{\varepsilon}_{N,1/0} = \vec{\varepsilon}_{M,1/0} + \vec{\rho}_{1/0} \times \overrightarrow{MN} \end{cases}$

We use the property of linearization of displacements into small displacements [Bourdet et al., 1995], [Clément et all., 1988].

We get according to (1):

$$S_1 \subset ZT \Leftrightarrow \forall N \in S_0, \forall M \in E^3 \text{ with } 0 \le k \le 1 \quad -kt \le \left(\vec{\varepsilon}_{M,1/0} + \vec{\rho}_{1/0} \times \overrightarrow{MN} \right) \vec{z}_N \le (1-k)t \quad (2)$$

In a base $(\vec{x}, \vec{y}, \vec{z})$, the normal vector $\vec{z}_N$ of S_0 at point N can be written as follows:

$$\vec{z}_N \begin{pmatrix} a_N \\ b_N \\ c_N \end{pmatrix} \text{ and } \overrightarrow{MN} \begin{pmatrix} d_{MNx} \\ d_{MNy} \\ d_{MNz} \end{pmatrix}.$$

$(2) \Rightarrow S_1 \subset ZT \Leftrightarrow \forall N \in S_0, \quad \forall M \in E^3 \text{ with } 0 \le k \le 1 :$

$$(k-1)t \le \left(\varepsilon_{1/0x,M} + \rho_{1/0y} \cdot d_{MNz} - \rho_{1/0z} \cdot d_{MNy} \right) a_N + \left(\varepsilon_{1/0y,M} - \rho_{1/0x} \cdot d_{MNz} + \rho_{1/0z} \cdot d_{MNx} \right) b_N \quad (3)$$

$$+ \left(\varepsilon_{1/0z,M} + \rho_{1/0x} \cdot d_{MNy} - \rho_{1/0y} \cdot d_{MNx} \right) c_N \le k.t$$

We obtain an infinity of equations (3). The unknowns are the six components written at point M:

$\rho_{1/0x}, \quad \rho_{1/0y}, \quad \rho_{1/0z}, \quad \varepsilon_{M,1/0y}, \quad \varepsilon_{M,1/0z}, \quad \varepsilon_{M,1/0z}.$

Any surface can be discretised into n points N_i. So, it is possible to express a set of n equations (3). The n equations (3) characterize the n geometric constraints induced by the tolerance zone associated with S_0.

The vertices of this polytope correspond to the maximum and the minimum values of

$\rho_{1/0x}, \quad \rho_{1/0y}, \quad \rho_{1/0z}, \quad \varepsilon_{M,1/0y}, \quad \varepsilon_{M,1/0z}, \quad \varepsilon_{M,1/0z}.$

This method can be applied on any ideal surface. We consider five types of surfaces: plane, cylindrical, conic and toric surfaces.

2.2. Contact constraints

Amongst the five main types of surfaces considered in the previous paragraph : plane, cylindrical, conic and toric surfaces, figure 2 summarizes the possible cases of joint. Since complex surfaces are not used in a joint between two parts (with the exception of a few particular cases i.e. gearing) they will not be considered. Each case in the above table can be sub-classified into several other cases according to the relative position of nominal surfaces. Any joint is characterized by the types of two considered surfaces with a set of mating conditions. A mating condition is a set of constraints between geometric features of two ideal surfaces [Teissandier, 1995]. An exhaustive list of the cases has been compiled by [Clément et al., 1997]: in order to define the relative position between two any surfaces 13 constraints has been identified.

	plane	cylinder	cone	sphere	tore
plane	×	×	×	×	×
cylinder		×	×	×	×
cone			×	×	×
sphere				×	×
tore					×

Figure 2: Main cases of joints.

Following the same method as in the previous paragraph, a contact specification can be formalised by a set of n contact constraints of position between two associated surfaces of two distinct parts. Let us consider a joint made up of two planes S_1 and S_2. S_1 and S_2 are nominally parallel and separated by a distance « d » (see figure 3). Vectors $\vec{z}_1$ and $\vec{z}_2$ are respectively constructed such as $\vec{z}_1$ and $\vec{z}_2$ are normal vectors of S_1 and S_2 (see figure 3) oriented in such a way that the positive direction corresponds to the side exterior to the material.

Let us define surface S such as : $S = S_1 \cap S_2$ with $d = 0$. The set of mating conditions is:

$$\{\vec{z}_1 \times \vec{z}_2 = \vec{0}, \quad \vec{z}_1 \cdot \vec{z}_2 < 0, \quad S \text{ is a plane surface}\} \tag{4}$$

If the set of mating conditions is satisfied, we can express a constraint of positioning : $0 \le d \le D$ (5)

A permanent contact between S_1 and S_2 is such as: $D = 0 \Rightarrow 0 \le d \le 0 \Rightarrow d = 0$.

If S is not a plane (line, point or empty hole), the previous constraint can not be defined: the set of mating condition is not verified.

Let us consider the boundary (C) of S.

At any point M of the Euclidean space E^3, it is therefore:

$$\forall N \in S, \forall M \in E^3 (4, 5) \Rightarrow \left(\vec{\varepsilon}_{M,1/2} + \vec{\rho}_{1/2} \times \overrightarrow{MN}\right)\vec{z}_N \le d \Rightarrow 0 \le \left(\vec{\varepsilon}_{M,1/2} + \vec{\rho}_{1/2} \times \overrightarrow{MN}\right)\vec{z}_N \le D \tag{6}$$

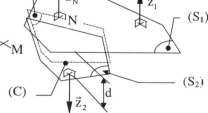

Figure 3: Contact specification between two nominal parallel planes.

Following the same method as geometric constraints, we can write:

$$\forall N \in S, \quad \forall M \in E^3 : (4, 5, 6) \Rightarrow 0 \leq \left(\varepsilon_{1/0x,M} + \rho_{1/0y} \cdot d_{MNz} - \rho_{1/0z} \cdot d_{MNy}\right) a_N +$$

$$\left(\varepsilon_{1/0y,M} - \rho_{1/0x} \cdot d_{MNz} + \rho_{1/0z} \cdot d_{MNx}\right) b_N + \left(\varepsilon_{1/0z,M} + \rho_{1/0x} \cdot d_{MNy} - \rho_{1/0y} \cdot d_{MNx}\right) c_N \leq D \tag{7}$$

If $\vec{z}_N = \vec{z}$, this corresponds to the particular case where: $a_N = b_N = 0$ and $c_N = 1$.

Thus (7) can be written as follows:

$$\forall N \in S, \quad \forall M \in E^3 : (4, 5, 6) \Rightarrow 0 \leq \varepsilon_{1/0z,M} + \rho_{1/0x} \cdot d_{MNy} - \rho_{1/0y} \cdot d_{MNx} \leq D \tag{8}$$

Contact constraints (8) traduce the three degrees of freedom of the studied joint: 1 displacement in rotation and 2 displacements in translation at point M. As in the previous paragraph, we can obtain a finite set of n equations (7).

2.3. Functional constraints

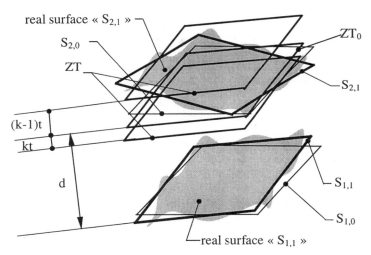

Figure 4: Functional requirement.

———— geometric specification ▬▬▬▬ combination of sets of
════ functional requirement geometric and contact specifications

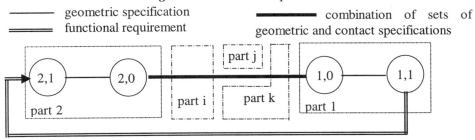

Figure 5: Surface graph according to figure 4.

The figure 4 illustrates an example of two surfaces $S_{2,1}$ and $S_{1,1}$ of any mechanism. $S_{2,1}$ is an associated surface of nominal surface $S_{2,0}$ and $S_{1,1}$ is an associated surface of nominal surface $S_{1,0}$: see figure 4. The relative position between $S_{2,1}$ and $S_{1,1}$ is the result of a combination of sets of contact specifications and geometric specifications on different parts: see figures 4 and 5. In the figure 4 $S_{2,0}$ and $S_{1,0}$ are two nominal parallel planes. A functional requirement between $S_{2,1}$ with regards to $S_{1,1}$ is defined by a tolerance zone ZT which limits an area of space around ZT_0 within which $S_{2,1}$ must be situated: see figure 4. ZT is composed of two parallel planes separated by the dimension t of ZT. These two surfaces are parallel to ZT_0, a parallel plane to $S_{1,1}$. The relative position between $S_{1,1}$ and ZT_0 is specified by the dimension d. A functional requirement is formalised by a finite set of n functional constraints between any associated surface with regards to any surface of the same mechanism. The method is the same as geometric specification.

$$S_{2,1} \subset ZT \Leftrightarrow \forall N \in S_{2,0}, \quad \forall M \in E^3 \text{ avec } 0 \le k \le 1 : (k-1)t \le \vec{\varepsilon}_{2,1/1,1x,M}.\vec{z}_N \le k.t \Rightarrow$$

$$(k-1)t \le \left(\varepsilon_{2,1/1,1x,M} + \rho_{2,1/1,1y}.d_{MNz} - \rho_{2,1/1,1z}.d_{MNy}\right)a_N + \tag{9}$$

$$\left(\varepsilon_{2,1/1,1y,M} - \rho_{2,1/1,1x}.d_{MNz} + \rho_{2,1/1,1z}.d_{MNx}\right)b_N + \left(\varepsilon_{2,1/1,1z,M} + \rho_{2,1/1,1x}.d_{MNy} - \rho_{2,1/1,1y}.d_{MNx}\right)c_N \le k.t$$

We obtain a finite set of n equations (8). The surfaces $S_{1,1}$ and $S_{2,1}$ can be (or can not be) specified from the same part. That means that a functional requirement can be reduced to a combination of two geometric specifications.

3.   OPERATIONS ON POLYTOPES

3.1.   Definition of a polytope

Geometric specifications, contact specifications and functional requirements can be respectively characterized by a finite set of geometric constraints (3), a finite set of contact constraints (8) and a finite set of functional constraints (9).

Each constraint of (3), (8) and (9) corresponds to a n-face (hyperplan of dimension n: $0 \le n \le 6$) in the real affine space R^d. Mathematically, (3) defines a geometric n-polytope (polytope of dimension n). By analogy, (8) defines a contact n-polytope and (9) defines a functional n-polytope.

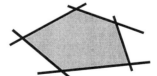

Convex hull of a finite set points in R^2 Intersection of finitely many closed halspaces in R^2

Figure 6: *Definition of a 2-polytope in R^2.*

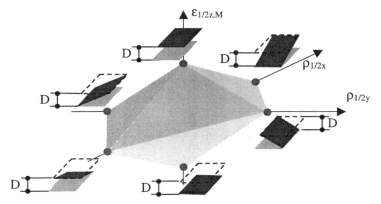

Figure 7: *3-polytope in R^3 from contact specification between two rectangular planes.*

A polytope is a point set $P \subseteq R^d$ which can be presented either as [Ziegler, 1995]:
- a bounded intersection of finitely many closed halfspaces in some R^d,
- a convex hull of a finite set of points in some R^d.

The figure 6 illustrates the two definitions of a polytope. The dimension of a polytope is the dimension of its affine hull. A n-polytope is a polytope of dimension n in some R^d. For example, we have defined a finite set of contact constraints (7) specified on two nominal parallel planes. (7) is a finite set of 3-faces in R^6 and (8) is a finite set of 3-faces in R^3. It is possible to give a graphic representation of a n-polytope if $1 \leq n \leq 3$. A 0-polytope is a point, a 1-polytope is a segment line and a 2-polytope is a polygon. For example, the finite set of contact constraints (7) (i.e. finite set of 3-faces) between two nominal parallel planes such as S is rectangular plane can be illustrated in R^3 by figure 7.

3.2. Minkowski sums of polytopes

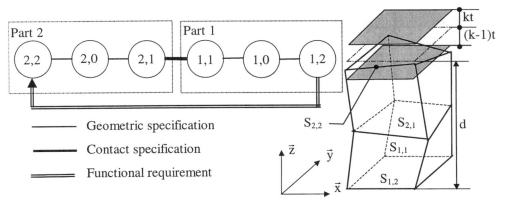

Figure 8: *Association of specifications in series.*

Let us consider the mechanism of figure 8. The contact specification between $S_{2,1}$ and $S_{1,1}$ is a permanent contact. The corresponding contact polytope is a 0-polytope (i.e. a point). We can express the relative position of $S_{2,2}$ with regards to $S_{1,2}$. We have according to the properties of small displacements:

$$\forall N \in S_{2,2} \quad S_{2,2} \subset ZT \Leftrightarrow (k-1)t \leq \vec{\varepsilon}_{N,S2,2/S1,2}.\vec{z}_N \leq k.t \tag{10}$$

$$\left.\begin{array}{l} \vec{\varepsilon}_{M,2,2/1,2} = \vec{\varepsilon}_{M,2,2/2,0} + \vec{\varepsilon}_{M,2,0/2,1} + \vec{\varepsilon}_{M,2,1/1,1} + \vec{\varepsilon}_{M,1,1/1,0} + \vec{\varepsilon}_{M,1,0/1,2} \\[2mm] \forall N \in S_{2,2} \text{ with } 0 \leq k \leq 1: (k-1)t_{2,2} \leq \vec{\varepsilon}_{N,2,2/2,0}.\vec{z} \leq k.t_{2,2} \\[2mm] \forall N \in S_{2,1} \text{ with } 0 \leq k \leq 1: (k-1)t_{2,1} \leq \vec{\varepsilon}_{N,2,1/2,0}.(-\vec{z}) \leq k.t_{2,1} \\[2mm] \forall N \in S_{2,1/1,1} \quad \vec{\varepsilon}_{N,1,2/2,2}.\vec{z} = 0 \\[2mm] \forall N \in S_{1,2} \text{ with } 0 \leq k \leq 1: (k-1)t_{1,2} \leq \vec{\varepsilon}_{N1,2/1,0}.(-\vec{z}) \leq k.t_{1,2} \\[2mm] \forall N \in S_{1,1} \text{ with } 0 \leq k \leq 1: (k-1)t_{1,1} \leq \vec{\varepsilon}_{N1,1/1,0}.\vec{z} \leq k.t_{1,1} \end{array}\right\} \tag{11}$$

Figure 9: Minkowski sum of 3-polytopes in R^3.

(11) characterizes the Minkowski sum of five polytopes. These five polytopes are the geometric specifications of $S_{2,2}$, $S_{2,1}$, $S_{1,1}$, $S_{1,2}$ and the contact polytope between $S_{1,2}$ and $S_{2,2}$ (in this case this polytope is a 0-polytope). The Minkowski sum of a two polytopes P_1 and P_2 is a polytope $P_1 + P_2$ [Gritzmann et al., 1993]:

$$P_1 + P_2 = \left\{ x \in R^d / \exists x_1 \in P_1, \exists x_2 \in P_2 : x = x_1 + x_2 \right\} \tag{12}$$

The association of geometric specifications and contact specifications in series are mathematically formalised by Minkowski sums of d-polytopes [Srinivasan, 1993]. In the example of figure 8, we can illustrate the Minkowski sum of five 3-polytopes in R^3: see figure 9. The Minkowski sum is an commutative and associative operation.

3.3. Intersection of polytopes

In the example presented in figures 10 and 11, the relative position between $S_{1,0}$ and $S_{2,0}$ must satisfy the two following relations:

$$\forall M \in E^3 \begin{cases} \vec{\varepsilon}_{M,2,0/1,0} = \vec{\varepsilon}_{M,2,0/2,1} + \vec{\varepsilon}_{M,2,1/1,1} + \vec{\varepsilon}_{N,1,1/1,0} \\[2mm] \vec{\varepsilon}_{M,2,0/1,0} = \vec{\varepsilon}_{M,2,0/2,2} + \vec{\varepsilon}_{M,2,2/1,2} + \vec{\varepsilon}_{N,1,2/1,0} \end{cases} \tag{13}$$

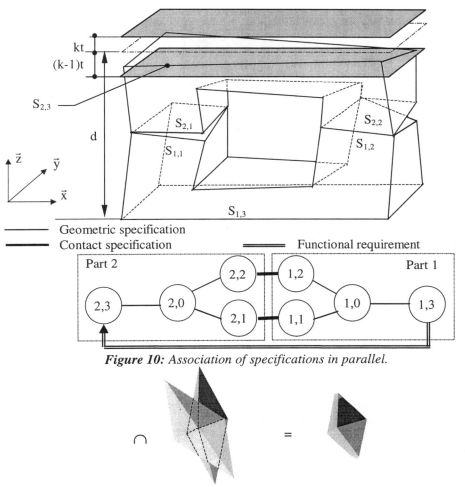

Figure 10: *Association of specifications in parallel.*

Figure 11: *Intersection of 3-polytopes.*

Each relation of (12) can be formalised by a Minkowski sum of three 3-polytopes. The result of (12) is an Intersection of two Minkowski sums of 3-polytopes: see figure 11.

4. CONCLUSION: TOLERANCING ANALYSIS WITH POLYTOPES

With Minkowski sums and Intersections of geometric d-polytopes and contact d-polytopes, the relative position between any surfaces of a mechanism can be calculated. The result is a new polytope: the calculated polytope. If the calculated polytope is included inside the functional polytope, the functional requirement is satisfied: see figure 12.

434

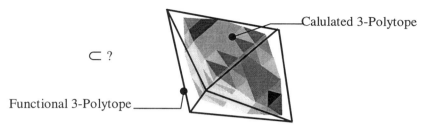

$\subset$?

Functional 3-Polytope

Calulated 3-Polytope

Figure 12: *Verification of inclusion of 3-polytope.*

REFERENCES

[Bourdet et al., 1995] Bourdet, P.; Mathieu, L.; Lartigue, C.; Ballu, A.; « The concept of the small displacement torsor in metrology », *Inter. Euroconf., Advanced Mathematical Tools in Metrology*, Oxford, 1995.

[Clément et al., 1988] Clément, A.; Bourdet, P.; « A Study of Optimal-Criteria Identification Based on the Small-Displacement Screw Model », *Annals of the CIRP* Vol. 37/1/1988.

[Clément et al., 1997] Clément, A.; Rivière, A., Serré P., Valade C; « The TTRSs : 13 Constraints for Dimensioning and Tolerancing », *5th CIRP Seminar on CAT*, pp. 73-82, Toronto (Canada), April 27-29, 1997.

[Giordano et al., 1993] Giordano, M.; Duret, D.; « Clearance Space and Deviation Space, Application to three-dimensional chains of dimensions and positions », *Proceedings of 3rd CIRP Seminar on CAT*, Eyrolles, pp. 179-196, 1993.

[Gritzmann et al., 1993] Gritzmann, P.; Sturmfels, B.; « Minkowski addition of polytopes: computational complexity and applications to Gröner bases », *Siam Journal of Discrete Mathematics*, Vol. 6, n°2, pp. 246-269, 1993.

[Fleming, 1987] Fleming, A.D.; « Analysis of geometric tolerances and uncertainties in assemblies of parts, *PhD Thesis,* Dep. of Artif. Intel., Univ. of Edinburgh, 1987.

[Requicha, 1983] Requicha, A. A. G.; « Toward a theory of geometric tolerancing », *The International Journal of Robotics Research*, Vol. 2, n°4, pp. 45-60, 1983.

[Srinivasan et al., 1989] Srinivasan, V.; Jayaraman, R.; « Conditional tolerances », *IBM Journal of Research and Development*, Vol. 33, n°2, pp 105-124, 1989.

[Srinivasan, 1993] Srinivasan, V.; « Role of Sweeps in Tolerancing Semantics », *Inter. For. Dimensional Tolerancing and Metrology*, CRTD-Vol. 27, pp. 69-78, 1993.

[Teissandier, 1995] Teissandier, D.; « L'Union Pondérée d'espaces de Liberté : un nouvel outil pour la cotation fonctionnelle tridimersionnelle », *PhD Thesis*, Laboratoire de Mécanique Physique - Université Bordeaux I, 1995.

[Teissandier et al., 1997] Teissandier, D.; Couétard, Y.; Gérard, A.; « Three-dimensional Functional Tolerancing with Proportioned Assemblies Clearance Volume: application to setup planning », *5th CIRP Seminar on CAT*, pp. 113-124, Toronto (Canada), April 27-29, 1997.

[Ziegler, 1995] Ziegler, G. M.; « Lectures on polytopes », *Springer Verlag*, 1995.

Author index

Keyword index